Literary Mapping in the Digital Age

Drawing on the expertise of leading researchers from around the globe, this pioneering collection of essays explores how geospatial technologies are revolutionizing the discipline of literary studies. The book offers the first intensive examination of digital literary cartography, a field whose recent and rapid development has yet to be coherently analysed. This collection not only provides an authoritative account of the current state of the field, but also informs a new generation of digital humanities scholars about the critical and creative potentials of digital literary mapping. The book showcases the work of exemplary literary mapping projects and provides the reader with an overview of the tools, techniques and methods those projects employ.

David Cooper is Senior Lecturer in English Literature at Manchester Metropolitan University, UK.

Christopher Donaldson is Lecturer in Romanticism at the University of Birmingham, UK.

Patricia Murrieta-Flores is Director of the Digital Humanities Research Centre at the University of Chester, UK.

Digital Research in the Arts and Humanities
Series Editors: Marilyn Deegan, Lorna Hughes, Andrew Prescott and Harold Short

Digital technologies are becoming increasingly important to arts and humanities research, expanding the horizons of research methods in all aspects of data capture, investigation, analysis, modelling, presentation and dissemination. This important series will cover a wide range of disciplines with each volume focusing on a particular area, identifying the ways in which technology impacts on specific subjects. The aim is to provide an authoritative reflection of the 'state of the art' in the application of computing and technology. The series will be critical reading for experts in digital humanities and technology issues, and it will also be of wide interest to all scholars working in humanities and arts research.

Also in the series:

Critical Gaming: Interactive History and Virtual Heritage
Erik Champion

Performing Digital: Multiple Perspectives on a Living Archive
David Carlin and Laurene Vaughan

Crowdsourcing our Cultural Heritage
Mia Ridge

Digital Archetypes: Adaptations of Early Temple Architecture in South and Southeast Asia
Sambit Datta and David Beynon

Paradata and Transparency in Virtual Heritage
Anna Bentkowska-Kafel, Hugh Denard and Drew Baker

Art Practice in a Digital Culture
Hazel Gardiner and Charlie Gere

Literary Mapping in the Digital Age
Edited by David Cooper, Christopher Donaldson and Patricia Murrieta-Flores

Forthcoming:

Digital Palaeography
Stewart Brookes, Malte Rehbein and Peter A. Stokes

Shakespeare's Language in Digital Media
Jennifer Roberts-Smith and Janelle Jenstad with Mark Kaethler

Historic Newspapers in the Digital Age
Paul Gooding

Literary Mapping in the Digital Age

Edited by
David Cooper, Christopher Donaldson
and Patricia Murrieta-Flores

Routledge
Taylor & Francis Group
LONDON AND NEW YORK

First published 2016
by Routledge

2 Park Square, Milton Park, Abingdon, Oxfordshire OX14 4RN
711 Third Avenue, New York, NY 10017

Routledge is an imprint of the Taylor & Francis Group, an informa business

First issued in paperback 2017

British Library Cataloguing in Publication Data
A catalogue record for this book is available from the British Library

Library of Congress Cataloging-in-Publication Data
A catalogue record for this book is available from the Library of Congress

ISBN: 978-1-4724-4130-0 (hbk)
ISBN: 978-0-8153-6678-2 (pbk)

Typeset in Baskerville
by Apex CoVantage, LLC

Contents

Figures

Plates

Tables

Contributors

Mark Algee-Hewitt is Assistant Professor of Digital Humanities in the department of English at Stanford University, USA, and Co-Associate Director for Research of the Stanford Literary Lab.

Jon Anderson is a Reader in Human Geography at Cardiff University, UK.

Miranda Anderson was the initiator of and a Research Fellow on the Palimpsest Project. She also initiated and is now a Research Fellow on the AHRC-funded project, *A History of Distributed Cognition* at the University of Edinburgh.

David J. Bodenhamer is Executive Director of the Polis Center, Professor of History, and Adjunct Professor of Informatics, and Co-Director of the Virtual Center for Spatial Humanities at Indiana University–Purdue University, Indianapolis, USA.

Dan Bonenberger is Associate Professor of Geography at Eastern Michigan University, USA.

Sally Bushell is Professor of Romantic and Victorian Literature at Lancaster University, UK.

David Cooper is Senior Lecturer in English Literature at Manchester Metropolitan University, UK, and an associate of the European Research Council-funded *Spatial Humanities: Texts, GIS, Places* project.

Christopher Donaldson is Lecturer in Romanticism at the University of Birmingham, UK; he is also Co-Investigator on the Leverhulme Trust-funded *Geospatial Innovation in the Digital Humanities* project (2015–2018) and an associate of the European Research Council-funded *Spatial Humanities: Texts, GIS, Places* project (2012–2016).

Ian Gregory is Professor of Digital Humanities in the Department of History at Lancaster University, UK, and Principal Investigator on both the European Research Council-funded *Spatial Humanities: Texts, GIS, Places* project (2012–2016) and the Leverhulme Trust-funded *Geospatial Innovation in the Digital Humanities* project (2015–2018).

Trevor M. Harris is Eberly Distinguished Professor of Geography at West Virginia University, USA.

Ryan Heuser is a Ph.D Candidate in English at Stanford University, USA, and Co-Associate Director for Research of the Stanford Literary Lab.

H. Frank LaFone is a Geographic Information Science specialist, who has been working professionally in the field of GIS for 15 years.

Annalise Lockhart worked on *Mapping the Emotions of London*, whilst completing her BA (awarded by Stanford University, USA, June 2014).

James Loxley is Professor of Early Modern Literature at the University of Edinburgh, UK, and Principal Investigator on the AHRC-funded *Palimpsest* project.

Patricia Murrieta-Flores is Director of the Digital Humanities Research Centre at University of Chester and Senior Researcher in Archaeology at the University of Chester, UK, and an associate of the European Research Council-funded *Spatial Humanities: Texts, GIS, Places* project.

Barbara Piatti is Project Leader and Researcher at the Institute for Cartography and Geoinformation, Zurich, ETH Zürich, Switzerland, and the co-leader of the *Literary Atlas of Europe* project.

Gary Priestnall is an Associate Professor within the School of Geography, the University of Nottingham, UK.

Les Roberts is a Lecturer in the School of the Arts at the University of Liverpool, UK.

Tania Rossetto is Senior Researcher and Assistant Professor of Cultural Geography at the University of Padua, Italy.

Angharad Saunders is Senior Lecturer in Human Geography at the University of South Wales, UK.

Erik Steiner is the Creative Director of the *Spatial History Project* in the Center for Spatial and Textual Analysis (CESTA) at Stanford University, USA.

Van Tran is a BA Candidate in English at Stanford University, USA.

Charles Travis is Post-Doctoral Research Fellow with the Trinity Long Room Hub, Trinity College, Dublin, Ireland, and the designer of the *Digital Literary Atlas of Ireland, 1929–1949.*

Series Preface

This series explores the various ways by which engagement with digital technologies is transforming research in the arts and humanities. Digital tools and resources enable humanities scholars to explore research themes and questions which cannot be addressed using conventional methods, while digital artists are reshaping such concepts as audience, form and genre. Digital humanities is a convenient umbrella term for these activities, and this series exemplifies and presents the most exciting and challenging research in the digital humanities. Digital humanities encompass the full spectrum of arts and humanities work, and scholars working in the digital humanities are strongly committed to interdisciplinary and collaborative methods. Consequently the digital humanities are inextricably bound to a changing view of the importance of the arts and humanities in society and provide a space for restating and debating the place of arts and humanities disciplines within the academy and society more widely. As digital technologies fundamentally reshape the sociology of knowledge, they challenge humanities scholars and artists to address afresh the fundamental cognitive problem of how we know what we know. Computing is the modelling of method, and this series reflects the belief that digital humanities proceeds by examining from many different perspectives the methods used in the arts and humanities, in some cases modifying and extending them, and in others drawing on relevant fields to develop new ones. The volumes in this series describe the application of formal computationally-based methods in discrete but often interlinked areas of arts and humanities research. The distinctive issues posed by modelling and exploring the archives, books, manuscripts, material artefacts and other primary materials used by humanities scholars, together with the critical and theoretical perspectives brought to bear on digital methods by the arts and humanities, form the intellectual core of the digital humanities, and these fundamental intellectual concerns link the volumes of this series. Although generally concerned with particular subject domains, tools or methods, each title in this series is accessible to the arts and humanities community as a whole. Individual volumes not only stand alone as guides but collectively provide a survey of 'the state of the art' in research on the digital arts and humanities. Each publication is an authoritative statement of current research at the time of publication and illustrate the ways in which engagement with digital technologies are changing the methods, subjects and audiences of digital arts and humanities. While reflecting the historic emphasis of the digital humanities on methods, the series also reflects the increasing consensus that digital humanities should have a strong theoretical grounding and offers wider critical perspectives in the humanities. The claim that digital humanities is an academic discipline is frequently controversial, but the range and originality of the scholarship described in these volumes is in our view compelling testimony that digital humanities should be recognised as a major field of intellectual and scholarly endeavour. These publications originally derived from the work of the AHRC ICT Methods Network, a multi-disciplinary partnership which ran from 1 April 2005 to 31 March 2008 providing a national forum for the exchange and dissemination of expertise, with funding from the UK's Arts and Humanities Research Council. The success of this network in generating strong synergies across a wide community of researchers encouraged the continuation of this series, which bears witness to the way in which digital methods, tools and approaches are increasingly featuring in every aspect of academic work in the arts and humanities.

Acknowledgments

Thanks are due to the Humanities Research Centre at Manchester Metropolitan University for the award of research leave to support the completion of this project, as well as to the School of English, Drama and American & Canadian Studies at the University of Birmingham, the Department of History and Archaeology at the University of Chester and the Department of History at Lancaster University, each of which has provided additional assistance. We would also like to thank Lianne Sherlock and Dymphna Evans for their enthusiastic support in the early stages of the project, as well as Denise File and her colleagues at Taylor & Francis who have expertly and patiently guided the collection through to publication. Thanks are due, too, to the anonymous reviewers for their insightful and helpful feedback. In addition, we would like to extend our gratitude to each of the participants at the Lancaster University symposium 'Digital Texts and Geographical Technologies in the Digital Humanities', from which this collection originates. The research leading to this collection has received funding from the European Research Council (ERC) under the European Union's Seventh Framework Programme (FP7/2007–2013) / ERC grant 'Spatial Humanities: Texts, GIS, Places' (agreement no. 283850).

Introduction

Rethinking Literary Mapping

David Cooper, Christopher Donaldson and Patricia Murrieta-Flores

> We organize information on maps in order to see our knowledge in a new way. As a result, maps suggest explanations; and while explanations reassure us, they inspire us to ask more questions, consider other possibilities.
>
> (Peter Turchi, *Maps of the Imagination* 11)

> As a subversive force, the Digital Humanities should not be considered a panacea for whatever ails the humanities, for they bring their own challenges and limitations. The point, to my mind, is not that it is better (or worse) but rather that it is *different*, and the differences can leverage traditional assumptions so they become visible and hence available for rethinking and reconceptualizing.
>
> (N. Katherine Hayles, *How We Think* 23–4)

This book is about the relationship between the practice of mapping, the application of geospatial technologies and the interpretation of literary texts. The contributors have been selected from a range of disciplines and they approach this relationship from different perspectives. Yet, notwithstanding these differences, their contributions are collectively defined by a shared preoccupation with the use of digital mapping tools and techniques in literary studies and cultural-geographical research. Each of the following chapters, that is to say, explores the dynamic ways that the creation of literary maps can confirm meaning and challenge critical assumptions. Each, moreover, reflects a common interest both in how digital technologies create new ways of conceptualising and practising literary map-making, and, furthermore, in how such map-making in turn changes the way we use and think about digital technologies. Accordingly, although they showcase the work of different projects and stake out their own critical positions, the chapters comprising this volume all engage with digital mapping technologies as a means of rethinking the spatial interconnections – and tensions – that link literary texts with writers, readers and the material world. In this way, the chapters that follow aim to initiate conversations and to contribute to ongoing discussions about the practice of literary mapping in the digital age.

Digital Maps in Everyday Life

The various projects, perspectives and positions presented in this collection are located at the confluence of human geography, literary studies and the digital humanities. The purpose of this Introduction is to contextualise this collocation, firstly by considering the presence of digital maps and mapping tools within society at large, and secondly by considering their influence on the development of literary cartography as an area of research. In attending to the first of these two tasks, it is useful to recall Michel de Certeau's notion of 'the practice of everyday life' (*l'invention du quotidien*), because – as we shall see – mapping technologies constitute an innovative means through which individuals can orientate themselves and, in the process, renegotiate their relationship with their surroundings. Digital maps are, in any case, ubiquitous. They are on our desktops and in our cars. They are in our newspapers and on our smartphones. They are on our aeroplanes and in our museums and galleries. In each of these manifestations the principal function of the digital map is to convey and to facilitate the interpretation of spatial information. For instance, it is commonplace for news agencies to use Geographic Information Systems (GIS) to visualise the spatiality of statistics ranging from crime rates to election results. The Global Positioning System (GPS), on the other hand, is commonly employed by personal navigation devices that allow individual users to monitor their location and movement. Digital mapping tools, however, serve more than purely utilitarian functions in our daily lives. In public museums and galleries, mapping software such as Google Maps and its cognate virtual globe platform Google Earth are used to enhance the visitor experience by presenting interactive interfaces that spatially curate individual objects and artefacts or even entire collections. In the intimate spaces of our homes, moreover, many of us use these online resources to undertake virtual flights to far off locations and to revisit the half-remembered landscapes of our childhoods. Certainly, such digital mapping experiences are by no means universal, and (as we might expect in light of de Certeau's ideas) the individuals who engage with such technologies devise their own unique mapping procedures and practices. Yet, to adapt a term that Rachel Hewitt has applied to the development of analogue cartography in the eighteenth century, we demonstrably live in an age of digital 'map-mindedness' (*Map of a Nation* 203).

The ubiquity of geospatial technologies is, however, far from universally celebrated. Indeed, they are as likely to be praised for generating new interest in mapping and map-making as they are to be blamed for triggering a general decline in geographical literacy. In *On Roads: A Hidden History*, for example, the cultural historian Joe Moran considers the suspicion – and, in some cases, scorn – aroused by the increasingly widespread use of satellite navigation systems (Sat Navs) in cars. As he suggests, the wariness expressed by many commentators about such 'cyborg' technology feeds into a 'fear that cold-blooded modernity is defeating vernacular knowledge, that roads no longer lead to real places but around and through them' (Moran 82, 86). This 'fear', as Moran reminds us, is not new – 'even before the arrival of the car, people . . . worried that maps sever us from real places' – and its anxiogenic influence has spurred the counter-cultural mapping practices of 'radical thinkers from Guy Debord to Rebecca Solnit' (Moran 86). Crucially, however, this fear has become even more prevalent with the emergence of geospatial technologies: a widespread response which has found expression in the remarks of Rita Gardner, who during her term as Director of the Royal Geographical Society, condemned Sat Nav equipment for 'destroying our ability to read maps and undermining our sense of self' (Axon, Speake and Crawford 170). For Gardner, as this statement implies, Sat Navs worryingly – and unequivocally – diminish both our knowledge of the topological relationships between places and our awareness of the textural complexity of the material landscape.

This anxiety about the diminution of our spatial awareness and attunement is similarly threaded through the landscape writings of Robert Macfarlane. In *The Wild Places*, Macfarlane voices his concern for the way contemporary mapping practices – and, by extension, the popular spatial imagination – are shaped by the Cartesian 'grid map', which 'places an abstract geometric meshwork upon a space, within which any item or individual can be co-ordinated' (141). Although, as Macfarlane acknowledges, he is neither advocating 'the abolition of the grid map' nor denying that his own journeys are aided by trigonometric plans, he is nonetheless outspoken about the way such rigid cartographic instruments inhibit 'our sense of the worth of map-as-story: of cartography that is self-made, felt, sensuous' (*The Wild Places* 143). Notwithstanding the grid map's 'rigorous' geometrical precision, Macfarlane concludes, it deadens our intuitive senses by 'suppress[ing] touch, feel and provisionality' (143). Such anxieties are reasserted in *Landmarks*, Macfarlane's more recent exploration of the 'power of language . . . to shape our sense of place' (1). Here, perhaps unsurprisingly, we find Macfarlane invoking the late writings of Martin Heidegger in order to bemoan how technology 'has bequeathed to us an inadequate and unsatisfying relationship with the natural world, and with ourselves too' (*Landmarks* 25). The plane surface of the digital screen, for Macfarlane, removes us from our environment and depletes our phenomenological experience.

Macfarlane's techno-sceptical narratives of loss and disconnectedness raise a series of issues for the literary cartographer interested in the complex relationship between text, geography and digital technologies, especially when we consider them in the light of the reservations expressed about such technologies by scholars such as Gardner. Specifically, they emphasise the need to evaluate how the process of digital mapping and map-making alters the way we perceive and engage with the geographies that surround us. What is it that such maps are capable of revealing? What new kinds of spatial awareness might they help us to acquire? What is it that they displace or conceal? What might they be unable to convey? More generally, considering Macfarlane's praise for vernacular cartographic practices, in tandem with Hewitt's reflections on the beauty and value of old Ordnance Survey maps, prompts us to wonder if digital mapping tools are really as deleterious as they have been made out to be. Like the technologies of place-naming and paper maps, cannot digital maps also open up new spatial practices and spatial imaginings? Can they not, equally, contribute to and facilitate a re-enchantment of place? These are all issues that the contributions to this collection explore.

Conceptualising Digital Mapping Practices

How, then, might we begin to understand and to conceptualise the ways digital mapping tools have transformed the practice of everyday life and the production of geographical knowledge? The most prominent and popular manifestations of 'maps in cyberspace' (Zook and Dodge) are Google Maps and Google Earth. According to Trevor M. Harris et al., the 'release of Google Maps in 2005 fundamentally changed the landscape of Web mapping', effectively setting a new standard for the digital projection of 'base vector data and imagery' (Harris et al. 132). Google Maps was then supplemented, later on in the same year, by Google Earth: a light form of GIS which built upon an earlier platform, EarthViewer 3D, in compiling 'satellite imagery and aerial photographs into a 3D virtual globe' (Farman 872). Since the introduction of Google Street View in April 2008, moreover, the platform has included an ever-increasing number of 'panoramic images at street level' (Farman 872). Jerry Brotton asserts that, through these affordances, Google has developed 'the world's most popular geospatial application', and he supports this claim by indicating that, by November 2011, Google's 'market share in the United States was over 65 per cent' and that globally, Google Earth had been downloaded 'more than half a billion' times (Brotton 406). Google Earth offers the user 'an extraordinary ten petabytes of potential geographical information distributed across the globe's surface' with a simple click of a mouse (Brotton 406). The availability and accessibility of Google's applications distinguishes them from the highly specialised and expensive geographical software that was once the preserve of digital cartography. As a result, for many people maps now mean Google: a cartographic 'sea change' which is at least partly responsible 'for the waning popularity of the paper map' lamented by writers such as Macfarlane and Hewitt (Hewitt, 'Turn Around When Possible' 2).

Crucially, however, users of Google's applications do not simply and passively absorb the geographic information provided to them, but actively participate in the process of spatial discovery. As the interdisciplinary collection *Rethinking Maps* affirms, such cartographic interactions are underpinned by a 'processual' framing through which both 'map making and map use' are understood to be 'embodied and dynamic . . . cultural practices involving action and affects' (Kitchin et al. 17). For the co-editors of

Rethinking Maps, maps are no longer to be conceived as materially stable providers of objectively surveyed geographical data. Rather, they are to be understood (à la J.B. Harley) as cultural texts which are performed, and brought-into-being, with each individual mapping practice (Kitchin et al. 17). This theorisation of a 'post-representational cartography' is demonstrably applicable to all forms of geospatial technology (Kitchin et al. 10). That said, it seems especially pertinent for thinking about the nature of digital maps, which encourage and empower the user to make an active contribution to the unfolding – and even the creation – of new cartographies.

Cartographic dynamism is cinematically foregrounded in the way the user first accesses the virtual globe of Google Earth. In *A History of the World in Twelve Maps*, Jerry Brotton describes the spectacle of launching Google's virtual globe with narrative flare: 'From 11,000 kilometres above its surface, the planet earth spins into view out of the black void of deep space. The sun's rays illuminate its surface, which appears free of clouds and water, although its ocean floors still sparkle ultramarine blue, the continents a beguiling patchwork of greens, browns and pinks' (405). From this extra-terrestrial viewpoint, the user can then navigate his or her way through virtual space, zooming in on self-selected locations and manipulating the geo-visualisation by rotating the map, adjusting the cartographic scale and toggling between cartographic representations and satellite photographs. According to Harris et al., the emphasis Google Earth places on facilitating this semi-immersive experience corresponds to a broader trend in the development of online tools. 'The technology used to build Web content has', he explains, 'shifted toward a dynamic user experience that uses asynchronous technologies to reduce the load times for content to create a seamless user experience' (Harris et al. 131). Saliently, though, the sense of user interactivity extends far beyond the capacity to control the way the digital map manifests on screen. 'In addition', continue Harris et al., 'Google Earth and Google Maps support embedded multimedia such as photographs, text, oral narrative, sketches, video, and audio within the map or globe representation thereby allowing users and communities to upload and share spatialized qualitative information'. Thus, for Harris et al., the slippy interface of the desktop map constitutes 'a collaboratory space' which has the potential to provide unique, user-generated 'insights into aspects of place' (Harris et al. 131–2).

The new spatialities created by the processual, interactive and even participatory nature of contemporary digital mapping practices are likewise evident if we move beyond desktop applications. In an article published in 2004, Nigel Thrift formulated a post-humanist analysis, contending that motorists' use of Sat Navs in urban environments brings about 'new forms of embodiment-cum-spatial practice which are sufficiently subtle and extensive to have every chance of becoming a new background to everyday life' (Thrift 52). Here, then, Thrift – the influential non-representational theorist – self-consciously opposes the reactionary critiques typified by Gardner to demonstrate that new conceptual framings are required for the new phenomenological experiences created by geospatial technologies. Moran, moreover, goes further than this by suggesting that, 'despite the Orwellian echoes of "ground-truthing"' – the system by which the accuracy of the digital map is continuously monitored through fieldwork – the 'very complexity [of Sat Nav technology] gives it something of the collective consciousness of an internet wiki'. As Moran makes clear, the digital maps in our cars are always provisional and are open to corrections and updates from 'public-spirited' drivers (Moran 87). Even the digital maps to be found in Sat Navs, therefore, can be understood as collaborative cartographies. Such social practices extend to mobile phone technologies, as a smartphone application such as Waze – which provides free traffic information – is exclusively reliant upon user input. This app thereby fosters a sense of a (virtual) community based upon the sharing of spatial knowledge.

At the same time, the development of smartphone technology has, in the words of Chris Speed, taken digital map users 'out of the car and back on foot' (Speed 161). Most advanced mobile operating systems, as Speed explains, 'feature GPS technologies and a mapping application of one form or another that can stream a street or satellite map onto the screen and pinpoint the user within it' (161). The result is that 'the locative properties' of mobile devices are changing 'the way that we navigate physical and social spaces' (Speed 160). Clancy Wilmott expands on this analysis in an article on 'mobile mapping as a contemporary urban practice', which focuses on the Australian city of Sydney. According to Wilmott:

> The Sydney that I find on a mobile phone is a cartographic city. It is so completely dominated by and reliant on geo-coding systems, that it is impossible to avoid maps or to express the city without them. Maps form the architecture of the mobile city: they direct flows, produce spaces and position places (Wilmott).

Whereas Thrift updated de Certeau's famous account of 'Walking in the City' for the Sat Nav generation, Wilmott, writing eight years later, emphasises how the imbrications of digital mobile mapping and pedestrian practice are 'disrupting traditional spatial paradigms' (Wilmott). In addition, she stresses the embodied performativity that is integral to the use of digital maps accessed on mobile devices: 'depending on the device . . . a specific kind of tactility is required – tapping, clicking, pressing, swiping – all of which eventually form a subconscious part of the process of calling on and adjusting the map'. As Wilmott points out, 'the assumption that "the map" is a static representational object is no longer accurate', since the smartphone map unfolds as the user moves his or her finger across the screen and/or moves through physical space (Wilmott). Fundamentally, then, digital maps are to be understood 'as always in a state of becoming; as always mapping; as simultaneously being produced *and* consumed, authored and read, designed *and* used, serving as a representation *and* practice' (Kitchin et al. 17).

Literary Cartography

Digital technologies, therefore, have transformed our daily spatial practices. At the same time, they have had a demonstrable influence on the ways scholars are thinking spatially, and understanding spatial patterns and networks, in a range of intersecting disciplines. Before turning to consider how such technologies have shaped the processes and procedures of researchers working on the literature of landscape, space and place, however, it is first necessary to reflect on literary cartography as an interpretive practice. It is, moreover, necessary to acknowledge that, in spite of its increased prominence as both a concept and a procedure, a standard definition of literary cartography has yet to take hold. In a special issue of *The Cartographic Journal* devoted to 'The Geographies of Fictional Worlds', Barbara Piatti and Lorenz Hurni classify literary cartography as an 'ancillary science' of literary geography, explaining that whereas the latter constitutes a 'topic', the former designates an 'approach' (Piatti and Hurni 218). This distinction (which follows the traditional, hierarchical classification of cartography as a sub-discipline of geography) similarly underpins Andrew Thacker's influential conceptualisation of a 'critical literary geography'. In wondering whether cartography might be one way of understanding both the textual representation of geographical space and the spatial contexts in which literary works are produced and read, Thacker identifies two extant forms of literary mapping. He rightly acknowledges that many 'contemporary critics seem to have taken a . . . metaphorical route for maps, perhaps taking a lead from [Frederic] Jameson's use of "cognitive mapping" as a master-trope for interpreting contemporary culture' (Thacker 61). This metaphorical understanding of mapping continues to shape one form of contemporary literary cartographic practice and is a defining characteristic, we would argue, of the essays brought together by Robert T. Tally Jr. in a recent addition to his *Geocriticism and Spatial Literary Studies* series (Tally Jr. 2014). Citing the pioneering example of Franco Moretti's *Atlas of the European Novel 1800–1900,* however, Thacker identifies an alternative form of literary cartography in which actual maps are created out of the spatial analysis of a literary text or set of texts. In the pages that follow, we define literary cartography along these lines. That is to say, we – and the contributors to this collection – are interested in the potential and problems associated with the making of literary maps and other closely related forms of literary geovisualisation.

Broadly conceived in this way, the practice of literary cartography can be seen to take its orientation from a critical conviction in mapping as a practice that enriches the reader's appreciation of the literary work of art. Such a conviction, though possibly apparent as far back as the geographical treatises of Polybius and Strabo (Kerrigan 3), is, by and large, indicative of a distinctly modern literary sensibility. As Thomas Cachey Jr. has clarified, early modern literary maps, such as Alessandro Velutello's extraordinary plan of the landscape of Petrarch's *Canzoniere* (published in 1525), reflect not only contemporary advances in cartographical knowledge, but also the rise of a new critical preoccupation with spatially locating the literary work (Cachey Jr. 456). Within the Anglophone world, in particular, this readerly preoccupation with seeking out the world behind the word, though evident in early literary-cartographic projects such as Matthew Parris's *Chronica Majora*, gained special prominence during the nineteenth century and attained its apotheosis in the Victorian 'love-affair with biography' and literary celebrity, and the concomitant 'impulse to naturalise' authors and their works 'to particular places' (Watson 106, 201). It is for this reason that the earliest works to be labelled literary geographies, in English, were typically tourist texts: literary itineraries, which, whether intended for the armchair or the field, were often embellished with maps and plans documenting sights of either literary-biographical or purely imaginative significance. William Sharp's now oft-cited 1904 volume *Literary Geography*, a collection of articles first published in *The Pall Mall Magazine*, a handful of which contain maps, is indicative of the quality of such early literary-cartographic endeavours.

Over the past century the field of literary cartography has come to be defined by a series of influential studies, which range from H.C. Darby's exploratory 1948 analysis of Thomas Hardy's *Wessex* to Malcolm Bradbury's popular *Atlas of Literature*, which was published in 1996. Saliently, however, the recent preoccupation with the making of literary maps has been informed by an increased theoretical awareness of the complexity of cartography and the inherent 'slipperiness' of literary maps (Bushell). This complements broader developments in the interdisciplinary field of literary geography. As the diverse – and, in some instances, divergent – work of geographers such as Jon Anderson and Angharad Saunders and literary critics such as Sally Bushell and Damian Walford Davies suggests, contemporary literary cartography is, by no means, unified by a common methodology or even a common perspective. Yet, in spite of these differences, the work of these scholars takes its orientation as a field of inquiry from a collective interest, on the one hand, in 'the place-bound nature of literary forms' (Moretti 5), and a widely shared responsiveness to 'the spatial ontology of literary production' (Saunders 444) on the other. This new critical literary cartography is, moreover, uniformly informed by a willingness to reflect on what Thacker identifies to be the potential 'problems' of using 'cartography as a guide to literary interpretation' (Thacker 60).

It would overextend the scope of this Introduction to summarise the various forms and instantiations of contemporary literary cartographic thought and practice. Instead, therefore, we want to turn to the ways the processes of literary map-making have been manifestly enriched by the consideration and implementation of geospatial technologies, including many of the digital tools described above. A crucial development here has been the emergence of an array of digital mapping projects including large-scale atlases, such as ETH Zurich's *A Literary Atlas of Europe*, Trinity College Dublin's *Digital Literary Atlas of Ireland* and the University of Queensland's *Cultural Atlas of Australia*, as well as projects focused on the literary cartographies of specific landscapes and cityscapes, such as the University College London-funded *Mapping St Petersburg: Experiments in Literary Cartography*, Lancaster University's *Mapping the Lakes* and the University of Edinburgh's *Palimpsest: Literary Edinburgh*. Other project developers have focused on the mapping of individual texts: so, for example, Barbara Hui's *Litmap* employs Google Maps to visualise the difficult-to-define spatial networks articulated in W. G. Sebald's *Rings of Saturn*; whereas Boston College's *Walking Ulysses* uses both historic and contemporary maps of Dublin to chart the locational geographies of Joyce's great Modernist novel. Elsewhere, projects have been informed by ambitious theoretical considerations and have self-consciously addressed the question of how geospatial technologies might radically recalibrate the fundamental practices of reading and analysing literary texts. The work of Anouk Lang, for example, has drawn upon methodologies traditionally used in quantitative spatial analysis to facilitate new understandings of the global networks of literary Modernism. Elsewhere, a playful preoccupation with geovisualisation tools and technologies has shaped the experiments in computational criticism developed in the *Stanford Literary Lab* under the directorship of Moretti.

The Challenges of Digital Literary Mapping

In what follows we propose briefly to consider four interpenetrating issues, or challenges, which need to be confronted by all digital literary map-makers: Why map? How to map? What to map? And, finally, is there potential – and perhaps even a need – to go beyond the map? According to David M. Berry, the 'digital humanities . . . have had a rather interesting history': 'Originally called "computing in the humanities", or "humanities computing", in the early days they were often seen as a technical support to the work of the "real" humanities scholars' (Berry 2). The first wave of digital humanities scholarship, therefore, was invariably characterised by the uncritical 'application of the computer' (Berry 3). As Todd Presner, David Shepard and Yoh Kawano explain, though, 'over the past decade, the methods, media, and materiality of humanities research have undergone [a] dramatic change . . . and the humanities have developed new research methods through their encounter with the computational sciences (Presner et al. 20). The second wave of the digital humanities, therefore, 'is an emerging field that explores the deeply productive tension and precarious linkage between computational practices and humanities scholarship' (Presner et al. 20). By extension, then, digital literary map-makers can no longer claim that the value of their work resides exclusively in the novel application of computational tools. Instead, the practice of digital literary map-making ought to be predicated on that most fundamental of questions: Why? That is to say, why might it be helpful to geovisualise literary texts? Moreover, what is the purpose of the digital literary map? Who is it for?

Many extant projects have sought to exploit the popular appeal of digital maps, highlighted earlier on in the Introduction, to use geovisualisation as a communication tool. Some of these ambitions have been overtly pedagogical, with a particular emphasis on the use of geospatial technologies to enhance the literary-geographical knowledge of students. The *Mapping Dubliners* project at Oklahoma State University, for instance, employs Google Maps and Google Earth to assist students in visualising the complex geographies of mobility and paralysis evident in James Joyce's canonical short story collection. Similarly, *Google Lit Trips* – devised by Jerome Burg – aims to assist course instructors in 'creating engaging and relevant literary experiences for [their] students' by furnishing them with 'free downloadable files that mark the journeys of characters from famous [works of] literature on the surface of Google Earth' (*Google Lit Trips*). Alongside the development of these online resources, there has been a growing tendency to place digital maps in literary heritage centres and museums in an attempt to use geovisualisations as a means of curating literary texts and historical artefacts. Intriguingly, the practice of digital map-making is also increasingly embedded within major literary research projects whose scholarly aims extend beyond the literary-geographical. A notable example here is the University of Glasgow's *Editing Robert Burns for the 21st Century* project, which incorporates three digital mappings of the poet's tours in the Highlands and Borders of Scotland. Crucially, these maps – which allow the user to trace Burns's movements, as plotted by Nigel Leask, across historic maps digitised from the National Library of Scotland's Map Department – form an integral part in the project's 'major innovative website', which aims to offer a point of virtual connection with the 'wider global community' (*Editing Robert Burns for the 21st Century*).

Clearly, then, desktop digital literary maps carry significant potential both for enriching pedagogical practices and for engaging new audiences. Yet, further communication opportunities are afforded by the use of geospatial technologies in the development of literary applications for mobile devices. Unsurprisingly, the commercial potential of such applications has been recognised by publishing houses as exemplified by Penguin's 'amplified' edition of Jack Kerouac's *On the Road* which charts Sal Paradise's motorised movements across the United States. At the same time, the rich pedagogic potential of apps has been noted by the Joyceans at Boston College whose students, over several years, have collaboratively contributed to *JoyceWays*: a crowd-sourced mapping of the city of Dublin that allows users to access texts, maps and photographs via mobile devices.

It is important to stress, however, that, for others, digital literary map-making is an unapologetically scholarly practice which is underpinned by the ambition to advance knowledge. Damian Walford Davies, for example, ends his recent monograph *Cartographies of Culture* by speculatively gesturing towards the possibility of creating a 'Digital Literary Atlas of Wales' (Walford Davies 206). Walford Davies's vision for this projected atlas corresponds with the concept of 'deep mapping', articulated by Bodenhamer et al., as he envisages a GIS which would 'prompt a critical and affective inhabitation of the cultural *dimensionality* of a literary work' (Walford Davies 206). More specifically, he imagines a digital enterprise which would collate a range of visualisations including a 'palimpsest of charts . . . graphic images, and 2- and 3-dimensional cartographic modelling and animation' (Walford Davies 206). There is little doubt that the 'Digital Literary Atlas of Wales' would appeal to a wide constituency beyond the academy. Central to Walford Davies's vision, though, is the way 'maps and other graphic mappings invoked by the atlas would play critical methodologies off against each other' so that, for example, 'cartography as a trigger for historicist "embedding" would contend with cartography as a formalist provocation' (Walford Davies 206–7). In the end, therefore, the creation of Walford Davies's ambitious digital atlas would be informed by the belief that the process of thinking cartographically about literary texts is, in itself, a form of self-reflexive critical practice. Moreover, as he stresses, the critical potential of encountering the unmappable reminds the digital literary map-maker of the need for 'a literary theory and criticism attentive to the complex modalities of the "folding landscapes" of literary cartography': a theory and criticism which 'must remain, . . . in the words of Tim Robinson, "faithful to more than the measurable" ' (Walford Davies 209). As Derek Schilling puts it, in a robust critique of what he terms 'the new literary geography', there is a need to 'examine the effects of cartographic reason[ing] on the interpretive process, asking ourselves what thinking in two dimensions, regardless of the granularity achieved, pushes us to ignore' (Schilling 226). The process of literary mapping, then, has a dual scholarly purpose: it has the potential to facilitate a literary-geographical understanding of literary texts; and, at the same time, it has the potential to facilitate further conceptual thinking about what it means to map.

Having considered the core question 'why map?', the digital map-maker must decide what geospatial technologies to use in the creation of the new literary cartography. As the various disciplinary and interdisciplinary perspectives represented in this collection testify, at present digital literary cartographers work with a wide variety of tools and techniques. Yet, as the chapters that follow also illustrate, there has been a collective pull towards a few key technologies. Researchers across the arts and humanities have become preoccupied with the geovisualisation and interpretive possibilities afforded by GIS, a development which David J. Bodenhamer, John Corrigan and Trevor M. Harris have labelled 'the spatial humanities'

(*The Spatial Humanities* vii). Early digital literary mapping projects fed off and back into this wider interest as many of the maps produced by pioneering projects such as the *Digital Literary Atlas of Ireland, 1922–1949, A Literary Atlas of Europe* and *Mapping the Lakes* were created using proprietary software such as ArcGIS: a specialist platform which is able 'to integrate, analyze, and visualize large amounts of both spatial and temporal data' (Richardson 210). The ability of GIS to visualise a layered landscape has been especially conducive to digital literary map-makers interested in palimpsestic – or 'stratigraphic' (Mitchell) – topographies. That is to say, the technology has been perceived as an appropriate medium for geovisualising terrains which have been subjected to years, decades and, in some cases, even centuries of cultural over-determination. Saliently, given that the scholarly use of GIS has been historically rooted in quantitative spatial analysis (Gregory), early literary GIS projects were committed, at least in part, to testing the methodological possibilities of quantitative approaches to the literature of space, place and landscape. This commitment to the quantitative can likewise be traced in Matthew L. Jockers' concept of 'macroanalysis' and Franco Moretti's ideas about 'distant reading'. Collectively, such critical paradigms contribute to what Anouk Lang has described as the 'disciplinary making-strange, or "analytical ostranenie"', which 'goes to the heart of why digital humanities is such an exhilarating field' (Lang 221). At the same time, early literary GIS contributed to a wider interdisciplinary move towards the possibility of a post-positivist qualitative GIS (Elwood and Cope).

Less specialised forms of geographic technology, such as Google Maps and Google Earth, have inevitably become integral to digital literary mapping. This turn to web-based spatial technology was framed in the *Mapping the Lakes* project as a progressive development as it moved 'literary GIS away from the two-dimensionality of conventional plane cartographies and into the two-and-a-half dimensional contoured landscapes represented in Google Earth' (Cooper and Gregory 104). It is unsurprising, then, that Google Maps and Google Earth have been used in a plethora of digital projects which range from the mapping of unprecedentedly large literary corpora (*Spatial Humanities: Texts, GIS, Places*) to individual canonical texts (*LITMAP*), and from the geospatialisation of the literature of a nation (*Cultural Atlas of Australia*) to quotidian reading practices (*Placing Literature*). Moreover, the use of web-based technology has been instrumental in the movement of digital literary mapping from the desktop to the smartphone and, by extension, from the study to the street (*Cardiff Plotlines*). Still, even though Google technologies dominate, they are not the exclusive medium for digital literary mapping practices. For example, the researchers behind the *Mapping St Petersburg* project praised Google for sparking 'a veritable revolution in digital cartography'. At the same time, however, they expressed anxiety about the perceived geovisual rigidity of Google Maps and gestured towards the possibilities that might be afforded by the use of Open Street Map: 'the "crowdsourced" internet map, that being free and open source allows one to change the look and content of the map tiles' (*Mapping St Petersburg*). As many of the contributors to this collection demonstrate, it is possible – even desirable – for the digital humanist to acknowledge the limitations of geospatial technologies, such as those offered by Google, and to consider alternative forms of map-making.

The third challenge facing the digital map-maker is to determine what is to be mapped in order for the literary map to be brought-into-being. Many literary mapping projects have constructed toponymic maps by identifying and geo-referencing the place-names found either in specific literary works or collections of literary texts. This practice has been fundamental to the development of literary GIS, and major advances have been made in facilitating the automatic extraction of place-names within large corpora, a process which is often applied in the macro-mapping of spatial patterns and trends. Significantly, the geo-referencing of place-names has likewise been integral to the way some digital literary mapping practices have emerged outside academia. The popular *Placing Literature* website, for example, frames the development of 'an online database of places from scenes in literature' thereby calling attention to the widespread appeal of rooting the geographies of fiction in recognisably real-world environments:

> Setting is a vital component in [many works] of literature – from Charles Dickens's London to Mark Twain's Hannibal to Jack Kerouac's Lowell. Putting a story in the context of a real physical location gives stories context, placing readers in a familiar locale as the story unfolds around them[.] (*Placing Literature*).

In such instances, then, the ambition to identify place-names is demonstrably shaping the reading of the literature of space, place and landscape and, in turn, is implicitly privileging the textual representation of actual – that is to say, empirical – geographies. At the same time, this preoccupation with toponyms is also inflecting the writing processes of some creative practitioners. Clarissa Draper – a Canadian novelist, based in Mexico, whose debut mystery novel, *The Sholes Key* (2012), is set in contemporary London – uses her blog to codify a seven-step guide for using Google Maps to 'help plan your book'. Draper's authorial advice involves both the making of customised maps which allow creative writers to locate 'where your story takes place' and the insertion of annotations (a brief description, for example, of the imagined event which is to occur at a particular geographical site) which serve the function of geo-located compositional notes' (Draper). As a result, Draper's evangelical enthusiasm for the way that Web-based maps can enable creative writers to authenticate the geographical fidelity of their represented worlds corresponds with what Annika Richterich identifies to be the 'new literary realism' which has emerged out of a widespread fascination with geospatial technologies (Richterich).

Notably, however, most works of literature are not predicated on the aggregation of unambiguously mappable data. What happens, then, when the digital map-maker encounters toponyms which do not correlate to any real-world geography? How are purely fictional place-names geovisualised within the cartographic parameters of GIS software? Clearly, such challenges are particularly profound when dealing with what Anne-Kathrin Reuschel and Lorenz Hurni refer to as 'spatial uncertainty in fiction' (Reuschel and Hurni). Yet, saliently, such issues are similarly encountered when mapping, for example, a work of Victorian travel writing. That is to say, how does the map-maker handle the misspelling of place-names or the misidentification of geographical features? How does he or she differentiate between geographical sites through which the embodied author actually experiences and those locations which the author only visits in imaginative space? The questions, and the cartographic challenges, proliferate. In contrast to quantitative data (such as, say, historical mortality rates), therefore, these ambiguous forms of spatial information provide the reader with 'fuzzy geographies' (Evans and Waters), which, in turn, lead to the creation of 'slippery' maps (Bushell).

A fourth key issue is the consideration of whether there is scope – and perhaps even a need – for the makers of digital literary maps to move beyond online tools or even GIS. In an interdisciplinary collection first published in 2009, Sarah Elwood and Meghan Cope proposed that the development of qualitative GIS depends upon 'a mixed methods approach' that 'emphasizes the infinitely creative and political possibilities of bringing together multiple ways of knowing and making knowledge' (Elwood and Cope 6). In this experimental vision, then, GIS is integrated within a programme of research based on the interweaving of methodologies from the humanities and geographic information science. As a result, GIS maps appear alongside alternative presentational forms such as grid images of urban space produced by computer-aided qualitative analysis software (CAQDAS) and audio recordings of the residents of a particular community. The ambition 'to create new forms of knowledge based on the intentional, reflexive mining of [such heterogeneous] approaches' (Knigge and Cope 96) likewise underpins the initiative outlined by Bodenhamer, Corrigan and Harris to advance from the spatial humanities to 'deep maps and spatial narratives' (*Deep Maps*). An exploratory form of geospatial practice, as Bodenhamer et al. explain, deep maps offer 'a finely detailed, multimedia depiction of a place and the people, animals, and objects that exist within it' (*Deep Maps* 3). Deep mapping, in so many words, does not simply allow for multiple modes of presentation. Instead, it constitutes 'a platform, a process, and a product' that remains subject to addition and amendment by users (*Deep Maps* 3). In contrast to the fixity of early literary GIS visualisations, therefore, deep mapping offers 'a new creative space that is visual, structurally open, genuinely multimedia and multilayered', and which, moreover, is capable of fostering creative collaboration (*Deep Maps* 4).

'Mapping' the Collection

The chapters that have been brought together in this collection offer different (and, at times, divergent) responses to the four questions – or challenges – set out above. These differences can largely be attributed to the fact that the contributors come from a range of disciplines, including human geography, literary studies, and geographic information science. The collection as a whole presents multidisciplinary perspectives on the core questions of what literary mapping might be and do in the digital age. Saliently, though, several contributions are either multi-authored or based on the work of research projects which bridge disciplinary boundaries, and thereby bring together researchers from different scholarly communities in one, genuinely interdisciplinary, textual space. As a result, there is an emphasis on conversation running throughout the volume and, over the following pages, it will become clear to the reader that digital literary mapping is invariably predicated upon collaboration, exchange and even productive tensions. As editors, we strongly believe that the chapters brought together here encapsulate this pluralism and polyphony. Yet, even though the collection is characterised by heterogeneity, it is simultaneously possible to identify a series of cardinal preoccupations and practices at work within it and within the field of digital literary mapping in general. The fourteen chapters in this book, then, are clustered into three thematic sections: an arrangement that highlights both conceptual and methodological intersections as well as common points of departure.

The first of these three sections, 'Systems, Approaches and Innovations', provides a survey of current trends focussing on the work of a few foundational literary mapping projects. The section features five chapters which, though distinct, all explore how the mapping of digital corpora can facilitate the analysis and interpretation of complex literary geographies and networks. The first chapter, 'Mapping the Emotions of London in Fiction', comes from a team of researchers at *Stanford Literary Lab,* led by Ryan Heuser. Combining advanced computing methods such as crowdsourcing, Natural Language Processing and spatial analysis, Heuser et al. mine a corpus of eighteenth- and nineteenth-century fiction for information about the affective associations of places in London. Using contemporary demographic data, including Charles Booth's 1889 *Descriptive Map of London Poverty,* moreover, Heuser et al. move beyond measuring the 'affective investment' of individual texts in specific places, towards a revealing consideration of how fictional and historical accounts of London coalesce. This critical geospatial engagement with a literary metropolis complements the focus and methodology of the second chapter, 'The Digital Poetics of Place-names in Literary Edinburgh'. Here Miranda Anderson and James Loxley offer an insight into the practical and theoretical underpinnings of *Palimpsest: Literary Edinburgh,* a digital literary mapping resource that, in geolocating Edinburgh-based literary works to their settings, aims to sharpen our sensitivity not only towards the tales and traditions that populate the Edinburgh cityscape, but also towards those memories and associations which inhere in all places and in all acts of place-naming. Ian Gregory and Christopher Donaldson's chapter, 'Geographical Text Analysis: Digital Cartographies of Lake District Literature', picks up on this interest in local associations, but approaches the topic from a different methodological perspective. Drawing on a customised corpus of historical accounts of the English Lake District (developed by the *Spatial Humanities: Texts, GIS, Places* project), Gregory and Donaldson model the integration of geographical and corpus analysis in order to map historical responses to the Lakeland landscape and to evaluate how those responses correlate with one another and with the material geography of the Lakes region itself. Collectively, these three chapters set the stage for two further explorations of the geographies of distinct corpora. The first of these, 'Mapping Fiction: Theories, Tools, Limits and Potentials of Literary Cartography', comes from Barbara Piatti, who offers a survey of case studies completed by the *Literary Atlas of Europe* project in order to outline the possibilities and problems of creating cartographic representations of fictional works and fictional worlds. The second, Charles Travis's 'Bloomsday's Big Data: GIS, Social Media and James Joyce's *Ulysses*', develops a Web 2.0 model of literary cartographic practice to facilitate an exploration of the global literary celebration of Bloomsday.

Building on the foregoing examinations of the interplay between text and place, the second section of this collection – 'Places, Writers and Readers' – includes four chapters that address the spatiality of specific literary works and the geographies immanent within and central to their creation. The first of these is Sally Bushell's 'Mapping Fiction: Spatialising the Literary Work', which draws on Thomas Hardy's *The Return of the Native* in order to meditate both on the difficulties of applying concepts of mapping to literature and on the affordance of current mapping technologies. Similarly, Angharad Saunders's 'The Spatial Practices of Writing: Arnold Bennett and the Possibilities of Literary GIS' draws out the latent spatiality of the works of Hardy's contemporary, Arnold Bennett, in order to consider those elements of a literary work's geographical investment which GIS technology struggles to convey. A comparable interest in mapping writing practice distinguishes Jon Anderson's chapter, 'Between "Distant" and "Deep" Digital Mapping: Walking the Plotlines of Cardiff's Literary Geographies', which employs GIS and mobile technology to trace the creative practices of three writers, Peter Finch, Gillian Clarke and Lloyd Robson, investigating how their journeys through the city of Cardiff inform the characters, plots and themes of their literary work. Finally, Les Roberts's chapter, '*The Cestrian Book of the Dead*: A Necrogeographic Survey of the Dee Estuary', which concludes this section, traces the narrative psychogeography of the Dee Estuary, using an experimental blend of GIS and storytelling to bring the literary and cultural heritage of this shifting landscape to the surface.

The third and final section – 'Collecting, Curating and Creating' – builds on these discussions of future research paths and possibilities. A specific focus here is on how the application of digital mapping tools, in both critical and creative literary mapping, is informing the development of new geospatial technologies, practices, and approaches. In 'Making the Invisible Visible: Place, Spatial Stories and Deep Maps', the first chapter in this section, David J. Bodenhamer expands the idea of exploring the inherent complexities of space and literature through an extended consideration of the concept of deep mapping. This experimental method of geographical representation, as Bodenhamer explains, approaches space – or better said, place – in a multidimensional way, offering an exploratory path to geographies, stories, narratives and experiences. In a similar fashion, Trevor Harris, Frank Lafone and Dan Bonenberger's chapter, 'From Mapping Text in Space to Experiencing Text in Place: Exploring Literary Virtual Geographies', extends this discussion of exploratory multidimensional geovisualisation by showcasing an immersive virtual environment called the CAVE to bring to life the historical world of the nineteenth-century American author Rebecca Harding Davies. The third chapter in this section, Gary Priestnall's 'Spatial Frames of Reference for Literature using Geospatial Technologies', moves from the lab into the field with analyses of the creative and critical application of projected augmented relief models and geovisualisations as approaches that blend digital and non-digital forms to create tangible models of literary worlds. Following on from this elaboration of experimental geovisualisation techniques, Tania Rossetto's 'Geovisuality: Literary Implications' offers a theoretically informed assessment of the affordances of geovisuality for literary studies by considering the implications of geospatial technologies on the practices of readers and writers alike.

Extending the discussion of geovisualisation and literary creativity initiated by Rossetto's chapter, this collection ends with a chapter that takes a deliberately contrasting approach to the overarching theme of literary mapping in the digital age. Instead of concentrating on the use of geospatial technologies for the visualisation and analysis of literary texts, David Cooper's critical gaze focuses on some ways contemporary British writers have represented and used digital mapping technologies in their own creative practices. This final chapter, therefore, represents a departure from critical literary map-making by offering a movement towards the critical exegesis of works of literature created in the digital age. In making this shift, the chapter shows how a range of contemporary writers, working in different literary genres, has engaged with many of the key ideas, issues and anxieties raised in this Introduction. Alongside this, the chapter suggests that the focused exploration of the experimental strategies of creative practitioners might help to shape the future approaches, techniques and innovations of critical literary map-makers. Fundamentally, then, Cooper's chapter is undergirded by the belief, voiced by Walford Davies, that the digital mapping of extant literary texts ought to be predicated on a willingness 'to analyse critically the cartographic "gene" of the literary imagination' (Walford Davies 209).

Literary Mapping in the Digital Age

Before making way for the contributors, we want to conclude this introduction by briefly considering the position this collection occupies within the broader intellectual contexts of both cartography and the digital humanities. These twin interests have been recently brought together under the name of Doreen Massey: a geographer whose theoretical reflections on place have figured prominently in the interdisciplinary practices of many critical literary geographers. In March 2015, the seventh Doreen Massey Annual Event, held at the Open University in Milton Keynes, focused on 'Digital Geographies' and was framed by key research questions that chime with the interests of this collection. 'How is the digital reformulating geographies' objects and methods?' And, moreover, 'how have [our] objects of concern altered and how is geographical practice – its tools and techniques – altering to reflect the ubiquity of digital technologies?' Saliently, the online advertisement for the interdisciplinary event began with the assertion that: 'Digital technologies are now so diverse that, as David [M.] Berry has suggested, the category of "the digital" is becoming almost meaningless' (*OpenSpace Research Centre*). Berry's argument, that the label 'digital' is on the verge of redundancy, returns us to ideas explored at the beginning of this Introduction. That is to say, we are moving towards an age in which maps – of all kinds – will invariably be digital and will continue to evolve with and through the development of digital cultures. This collection, therefore, is an attempt to document a specific moment in the history of literary mapping; a transitional moment at which map-makers – from a range of disciplines – are constructing new theoretical frameworks, and fresh critical paradigms, in an attempt to understand the potential and limitations of geospatial technologies.

This collection emerges at a time when digital literary mapping is still coming-into-being. Accordingly, it is organised not around a single, unified thesis, but around an emphasis on the processual nature of current (literary) mapping practices. We are, therefore, conscious that the contents of this Introduction, and the collection as a whole, are inescapably partial, selective and incomplete. So, for instance, we are sensitive to the fact that both the Introduction and the contents concentrate, almost exclusively, on Anglophone literature. In addition, we are alert to the fact that the book might benefit from more sustained interrogations of both the politics of mapping and the mapping of different literary genres. Ultimately, however, a processual understanding of maps and mapping practices corresponds positively with the conceptualisation of digital literary mapping which is in evidence throughout this collection. In spite of the differing methodological approaches on display, there is a collective sense in which the practice of map-making is framed as an interpretive act and the literary map is understood as being repeatedly remade with each individual engagement. To return to the epigraphic quotations at the start of this Introduction, digital literary maps prompt the asking of further questions through both their geovisualisations and their lacunae. They may not always offer literary-geographical explanations then; but invariably, and crucially, the act of map-making opens up space for the processes of rethinking and reconceptualising which, for Hayles, define the work of the digital humanist.

Works Cited

Axon, Stephen, Janet Speake and Kevin Crawford. '"At the Next Junction Turn Left": Attitudes Towards Sat Nav Use'. *Area,* 44.2 (2012): 170–77.

Berry, David M., ed. *Understanding the Digital Humanities.* Basingstoke: Palgrave Macmillan, 2012.

Bodenhamer, David J., John Corrigan and Trevor M. Harris. 'Introduction: Deep Maps and the Spatial Humanities'. *Deep Maps and Spatial Narratives.* Eds, David J. Bodenhamer, John Corrigan and Trevor M. Harris. Bloomington: Indiana University Press, 2015. 1–6.

———. 'Introduction'. *The Spatial Humanities: GIS and the Future of Humanities Scholarship.* Eds, David J. Bodenhamer, John Corrigan and Trevor M. Harris. Bloomington: Indiana University Press, 2010. vii–xv.

Bradbury, Malcolm, ed. *The Atlas of Literature.* London: De Agostini Editions, 1996.

Brotton, Jerry. *A History of the World in Twelve Maps.* London: Allen Lane, 2012.

Bushell, Sally. 'The Slipperiness of Literary Maps: Critical Cartography and Literary Cartography'. *Cartographica.* 47.3 (2012): 149–60.

Cachey Jr., Theodore J. 'Maps and Literature in Renaissance Italy'. *History of Cartography.* Ed. David Wooward. Vol. 3. Chicago: University of Chicago Press, 2007, 450–60.

Cardiff Plotlines. 2015. Web. 20 March 2015. <http://www.cardiffplotlines.com/>

Cooper, David and Ian N. Gregory. 'Mapping the English Lake District: A Literary GIS'. *Transactions of the Institute of British Geographers.* 36.1 (2011): 89–108.

Cultural Atlas of Australia. Web. 20 March 2015. <http://australian-cultural-atlas.info/CAA/index.php>

Darby, H.C. 'The Regional Geography of Thomas Hardy's Wessex'. *Geographical Review* 38.3 (1948): 426–43.

A Digital Literary Atlas of Ireland. Web. 20 March 2015. <http://www.tcd.ie/trinitylongroomhub/digital-atlas/>

Draper, Clarissa. 'Planning Your Novel with Google Maps'. *Clarissa Draper: Listening to the Voices.* Web. 20 March 2015. <http://clarissadraper.blogspot.co.uk/2014/05/planning-your-novel-with-google-maps.html>

Editing Robert Burns for the 21st Century. N.d. Web. 20 March 2015. <http://burnsc21.glasgow.ac.uk/>

Elwood, Sarah and Meghan Cope. 'Introduction: Qualitative GIS: Forging Mixed Methods Through Representations, Analytical Innovations, and Conceptual Engagements'. *Qualitative GIS: A Mixed Methods Approach.* Eds, Meghan Cope and Sarah Elwood. Los Angeles: Sage, 2009, 1–12.

Evans, Andy and Tim Waters. 'Mapping Vernacular Geography: Web-Based GIS Tools for Capturing "Fuzzy" or "Vague" Entities'. *International Journal of Technology, Policy and Management,* 7.2 (2002): 134–50.
Farman, Jason. 'Mapping the Digital Empire: Google Earth and the Process of Postmodern Cartography'. *New Media & Society,* 12.6 (2010): 869–88.
Google Lit Trips. 2006. Web. 20 March 2015. <http://www.googlelittrips.org/>
Gregory, Ian N. and Paul S. Ell. *Historical GIS: Technologies, Methodologies and Scholarship.* Cambridge: Cambridge University Press, 2007.
Gregory, Ian, David Cooper, Andrew Hardie and Paul Rayson. 'Spatializing and Analysing Digital Texts: Corpora, GIS and Places'. *Spatial Narratives and Deep Maps.* Eds. David J. Bodenhamer, John Corrigan and Trevor M. Harris. Bloomington: Indiana University Press, 2015, 150–78.
Harris, Trevor M., L. Jesse Rouse and Susan Bergeron. 'The Geospatial Semantic Web, Pareto GIS, and the Humanities'. *The Spatial Humanities: GIS and the Future of Humanities Scholarship.* Eds, David J. Bodenhamer, John Corrigan and Trevor M. Harris. Bloomington: Indiana University Press, 2010, 124–42.
Hayles, N. Katherine. *How We Think: Digital Media and Contemporary Technogenesis.* Chicago: University of Chicago Press, 2012.
Hewitt, Rachel. *Map of the Nation: A Biography of the Ordnance Survey.* London: Granta, 2010.
———. 'Turn Around When Possible'. *The Guardian: Review.* 19 April 2014: 2–3.
JoyceWays. iTunes App. 2012. Web. 20 March 2015. <http://joyceways.com/>
Kerouac, Jack. *On the Road.* Penguin Books Amplified Edition. iTunes App. 2011. Web. 20 March 2015. <http://www.penguin.com/static/pages/features/amplified_editions/on_the_road.php>
Kerrigan, John. 'The Country of the Mind'. *Times Literary Supplement.* 11 September 1998: 3–4.
Kitchin, Rob, Chris Perkins and Martin Dodge. 'Thinking about Maps'. *Rethinking Maps.* Eds, Martin Dodge, Rob Kitchin and Chris Perkins. London: Routledge, 2009. 1–25.
Knigge, LaDona and Megan Cope. 'Grounded Visualization and Scale: A Recursive Analysis of Community Spaces'. *Qualitative GIS: A Mixed Methods Approach.* Eds, Meghan Cope and Sarah Elwood. Los Angeles: Sage, 2009, 95–114.
Lang, Anouk. 'Canadian Magazines and their Spatial Contexts: Digital Possibilities and Practical Realities'. *International Journal of Canadian Studies,* 48 (2014): 213–32.
A Literary Atlas of Europe. 2015. Web. 20 March 2015. <http://www.literaturatlas.eu/en/>
LITMAP: Mapping Literature. N.d. Web. 20 March 2015. <http://barbarahui.net/litmap/>
Macfarlane, Robert. *Landmarks.* London: Hamish Hamilton, 2015.
———. *The Wild Places.* London: Granta, 2007.
Mapping the Lakes: A Literary GIS. N.d. Web. 20 March 2015. <http://www.lancaster.ac.uk/mappingthelakes/>
Mapping St Petersburg: Experiments in Literary Cartography. 2011. Web. 20 March 2015. <http://www.mappingpetersburg.org/site/>
Mitchell, Peta. '"The Stratified Record Upon Which We Set Our Feet": The Spatial Turn and the Multilayering of History, Geography, and Geology'. *GeoHumanities: Art, History, Text at the Edge of Place.* Eds, Michael Dear, Jim Ketchum, Sarah Luria and Douglas Richardson. Abingdon: Routledge, 2011. 71–83.
Moran, Joe. *On Roads: A Hidden History.* London: Profile, 2009.

Moretti, Franco. *Atlas of the European Novel, 1800–1900.* London: Verso, 1998.
OpenSpace Research Centre. N.d. Web. 20 March 2015. <http://www.open.ac.uk/research centres/osrc/events/the-7th-doreen-massey-annual-event-digital-geographies>
Palimpsest: Literary Edinburgh. N.d. Web. 20 March 2015. <http://www.ed.ac.uk/literatures-languages-cultures/english-literature/research-activities/palimpsest>
Piatti, Barbara and Lorenz Hurni, 'Cartographies of Fictional Worlds'. *The Cartographic Journal.* 48.4 (2011): 218–23.
Placing Literature. Web. 20 March 2015. <http://www.placingliterature.com/>
Presner, Todd, David Shepard and Yoh Kawano, (eds) *HyperCities: Thick Mapping in the Digital Humanities.* Cambridge MA: Harvard University Press, 2014.
Reuschel, Anne-Kathrin and Lorenz Hurni. 'Mapping Literature: Visualisation of Spatial Uncertainty in Fiction'. *The Cartographic Journal.* 48.4 (2011): 293–308.
Richardson, Douglas. 'Geohistories'. *GeoHumanities: Art, History, Text at the Edge of Place.* Eds, Michael Dear, Jim Ketchum, Sarah Luria and Douglas Richardson. Abingdon: Routledge, 2011. 209–14.
Richterich, Anna. 'Cartographies of Digital Fiction: Amateurs Mapping a New Literary Realism'. *The Cartographic Journal.* 48.4 (2011): 237–49.
Saunders, Angharad. 'Literary Geography: Reforging the Connections'. *Progress in Human Geography.* 34.4 (2010): 436–52.
Schilling, Derek. 'On and Off the Map: Literary Narrative as Critique of Cartographic Reason'. *Literary Cartographies: Spatiality, Representation, and Narrative.* Ed. Robert T. Tally Jr. New York: Palgrave Macmillan, 2014. 215–28.
Spatial Humanities: Texts, GIS, Places. Web. 20 March 2015. <http://www.lancaster.ac.uk/fass/projects/spatialhum.wordpress/>
Speed, Chris. 'Walking Through Time: Use of Locative Media to Explore Historical Maps'. *Mapping Cultures: Place, Practice, Performance.* Ed. Les Roberts. Basingstoke: Palgrave Macmillan, 2012. 160–80.
Tally, Robert T. Tally Jr. (ed.) *Literary Cartographies: Spatiality, Representation, and Narrative.* New York: Palgrave Macmillan, 2014.
Thacker, Andrew. 'The Idea of a Critical Literary Geography'. *New Formations.* 57 (2005/6): 56–73.
Thrift, Nigel. 'Driving Through the City'. *Theory, Culture & Society.* 21, 4/5 (2004): 41–59.
Turchi, Peter. *Maps of the Imagination: The Writer as Cartographer.* San Antonio, Texas: Trinity University Press, 2004.
Walford Davies, Damian. *Cartographies of Culture: New Geographies of Welsh Writing in English.* Cardiff: University of Wales Press, 2012.
Sharp, William. *Literary Geography.* London: Pall Mall Publications, 1904.
Watson, Nicola. *The Literary Tourist: Readers and Places in Romantic and Victorian Britain.* Basingstoke: Palgrave Macmillan, 2006.
Wilmott, Clancy. 'Cartographic City: Mobile Mapping as a Contemporary Urban Practice'. *Refractory: A Journal of Entertainment Media* (2012). Web. 20 March 2015. <http://refractory.unimelb.edu.au/2012/12/28/wilmott/>
Zook, Matthew and Martin Dodge. 'Mapping Cyberspace'. *The International Encyclopedia of Human Geography.* Eds, Rob Kitchin and Nigel Thrift. New York: Elsevier, 2009, 356–67.

Part I

Mapping Methods

Systems, Approaches and Innovations

1 Mapping the Emotions of London in Fiction, 1700–1900

A Crowdsourcing Experiment

Ryan Heuser, Mark Algee-Hewitt, Annalise Lockhart, Erik Steiner and Van Tran

Introduction

How does literary geography change as it becomes digital? In the new spatial humanities, the turn towards digital methods offers both greater scale and precision, creating opportunities to study new kinds of critical objects. A concept such as Matthew Wilkens's 'geographic investment' (the number of words in a text naming particular places) is made possible by our ability to algorithmically identify and count place-names within a large literary corpus (Wilkens 804). Such a concept is radically new to literary studies and, we argue, important for understanding the complex relationship between fiction and its represented geography, as it allows macroscopic literary-geographic patterns to emerge. However, although the identification of text's geographic investment requires an algorithmic approach, such an approach is unable to parse the meaning of the patterns it identifies. Without knowledge of the ways places were invoked in fiction, maps of general geographic investment are, to a certain degree, intractable to interpretation.

A computer cannot add meaning to place; it can only count place-names. In the foundational works of literary geography, such as Franco Moretti's *Atlas of the European Novel* and Barbara Piatti's project *A Literary Atlas of Europe*, the advantage of the spatial turn lies in the critic's ability to uncover the nuanced relationship between spatial pattern and textual meaning. What is needed, we argue, is a synthesis of these methodological innovations: the ability to uncover geographic information on a significantly new scale whilst at the same time preserving our ability to understand how each place functions within its unique textual environment. Our goal in this project is to put the computational logic of Wilken's geographic investment into contact with the detailed geographical hermeneutics practised by such critics as Moretti and Piatti, so that we can interpret the meanings of places in fiction across thousands of texts.

This chapter is a record of one such attempt at contact, mapping eighteenth- and nineteenth-century novels' affective investments with places in

London.[1] During this period, London's unprecedented expansion fundamentally transformed its social organisation, a transformation deeply imbricated with the contemporaneous rises of the bourgeoisie, literacy, the publishing industry and the novel (Moretti, *The Bourgeoisie*; Barker 52–4; St Clair; Watt). In mapping the ways in which the novel affectively imagined the city at the heart of its own British publication, we hope to construct a geography of London that makes visible these reflexive, literary-sociological 'structures of feeling' (Williams 132; Porter 35). Drawing on the affect-theoretical work of Paul Fisher and Sianne Ngai, we attempt to operationalise the idea that emotions can act 'as a mediation between the aesthetic and the political in a nontrivial way' by mapping fiction's affective engagements with real places in London onto the city's own evolving urban landscape (Ngai 3). In this way, we hope to bring literature into spatial contact with aspects of its own social organisation, superimposing fiction's qualitatively distinct affective representations onto the quantitative dimensions of geographic space and historical sociological data. We believe that digital literary geography can thus provide an important and productive deconstruction of what James English has called 'the false but pervasive perception of a great divide between literature and sociology, with the former all irrational devotion and interpretative finesse and the latter all scientific rigor and verifiable "results"' (xiv). Without conflating them, or subordinating either to the other, such a geography quite literally creates a space in which concepts of literary form and social forces can meaningfully cohabit.

In the following two sections, we present two distinct phases of our project's methodology. In the first, we use predominantly computational methods to measure novels' geographic investment with London through their explicit mentions of known place-names. In the second, we leverage emerging techniques in crowdsourcing to distribute the reading and annotation of passages mentioning London places, in order to gather a readerly consensus on two ways in which each place functions within each passage: (a) whether the passage is set in the place; and (b) whether a particular emotion is associated with the place.

Geographic Investment in Fictional London

Methodology

Mapping a text's geographic investment consists of mapping the known place-names it mentions; however, we recognise that place-names do not exhaust a novel's relationship to place. Novelists of both the eighteenth

1 This chapter and project were made possible by a grant from the Andrew W. Mellon foundation to Stanford's Center for Spatial and Textual Analysis (CESTA). The authors would like to thank Maria Santos for her assistance with statistics and remote sensing techniques; as well as Gabriel Wolfenstein and Zephyr Frank, for their guidance and support. The chapter is part of an ongoing research project in the Stanford Literary Lab, 'Emotions of London'. Here, we give a general presentation of the project's methodology and some of its overall findings.

and nineteenth centuries employed a wide range of obfuscatory techniques when mentioning a setting's location, such as Austen's shortening of English shires to '——shire'. Moreover, settings were often not explicitly related to geographic space at all (Piatti 184). These alternate spatial practices, however, only highlight the specificity and intentionality of an explicit place-naming practice in fiction. Place-names, we believe, do important fictional and cultural work, suturing narrative and geographic space whilst also calling upon and contributing to connotations that have accrued through wider cultural circulation.

The corpus of fiction in which we investigated London place-names derives from the Literary Lab's fictional corpus, which extends through the eighteenth and nineteenth centuries and derives from a variety of sources.[2] Filtering to include only works of English-language fiction first published between 1700 and 1900, and which were digitised accurately enough for 90 per cent or more of its words to be recognised in a dictionary (representing its OCR accuracy), we were left with exactly 4,862 texts.[3] To find passages in this corpus mentioning London places, we developed a hybrid methodology that combines current techniques in computational toponym discovery with a traditional research-based approach. Although Named Entity Recognition (NER) software is a staple of current digital literary-geographic research, most NER packages are trained on contemporary journalistic sources and lack the precision required to recognise many specific locations in historical London (Manning et al.; Finkel and Manning). We therefore decided to supplement the results from an NER analysis of our corpus with historical gazetteer research into a variety of digital, print and map-based resources (Jackson; Paterson; Matthew). From a compiled list of 382 locations, we chose 161 to investigate for this chapter, in an attempt to represent the most frequently mentioned places as well as to retain a spatial representativeness over the wide extent of London.[4] Our list is certainly not exhaustive, but we believe that, given the multiplicity of its sources and

2 Texts in the eighteenth century are those texts in ECCO marked as fiction by the ESTC, or by researchers in the Lab through its Eighteenth-Century Fictional Marketplace project. Texts in the nineteenth century derive from the Chadwyck-Healey Nineteenth-Century Fiction corpus, as well as from a collection released on the Internet Archive by the University of Illinois.

3 The median OCR accuracy rate is 98.3 per cent. The number of included works by half-century is: 304 novels from 1700–49; 1,079 from 1750–99; 1,290 from 1800–49; 2,189 from 1850–99.

4 A place-name was included due to its frequency if it constituted 0.2 per cent or more of the total occurrences of London places in any half-century of fiction from 1700 to 1900. For example, in all our fiction from 1850–99, London places were mentioned a total of 64,000 times; if a place were to appear 128 times or more in this period, it was included in our list of locations to be searched for in any period. To ensure geographic representativeness, we included at least one place-name for each of the districts and neighbourhoods identified on Booth's 1889 map. This sampling was influenced primarily by the pragmatic needs of mitigating costs in the crowdsourcing experiment in fictional London described below.

sampling methods, it is broadly representative of the most salient locations of the period.

However, due to the ambiguity of place-names – 'the Tower', for instance, does not always refer to the Tower of London, nor 'Richmond' to the London suburb – it was not enough to simply rely on algorithmic counts of place-name mentions. We decided to generate a random sample of passages in which each place-name occurred. Dividing the two centuries into four half-century 'periods', we read through the generated passages per place, annotating whether it (e.g. 'the Tower') actually referred to the place in question. We continued this process until we identified at least ten such legitimate passages per place, per period, to be used in the crowdsourcing experiment of section 3. From these annotations, we were also able to estimate how often each place-name legitimately refers to its place in a given half-century of fiction. For instance, we estimate that 'Richmond' refers to Richmond the suburb about 60 per cent of the time in fiction published 1850–99.[5] These likelihoods were then used as 'scaling factors', and were multiplied by the number of mentions of a given place-name in all the novels of a given period. For example, 'Richmond' occurs about 2,630 times in fiction published 1850–99, so we estimate that it actually refers to the suburb about 1,578 times, or 60 per cent of the total. These scaled frequencies were then normalised, divided by the number of all words in the novels of the respective half-century, in order to express the likelihood, per million words of fiction, of encountering a legitimate mention of each place in each period. These frequencies were then represented using GIS, through which we geographically encoded each place as a polygon extending over its visible boundaries on Charles Booth's 1889 map of poverty in London. In certain cases, we draw a distinction between 'buildings', object-like places including buildings, streets and squares; and 'districts', area-like places including neighbourhoods and districts. However, in all but Plate 2, we represent the respective data as circles pinpointed to the centroid of the place's polygon in order to provide comparability between places as spatially diverse as towers and slums. In Figure 1.1, both the size and shading of a circle symbolise the place's frequency per million words in fiction of the period. Here, Booth's map is made transparent, but the boundaries of our 'districts' are outlined for geographic reference.

Interpretation: Investment and 'Stuckness'

In reviewing William Sharp's 1904 map of the chief localities of Walter Scott's novels, Piatti notes that these early maps of literary geography 'succeeded [in visualising] a couple of important aspects' of the emerging field,

5 The remaining 40 per cent of mentions refer primarily to titles (e.g., 'Duchess of Richmond'), and secondarily to personal names and other locations outside of London.

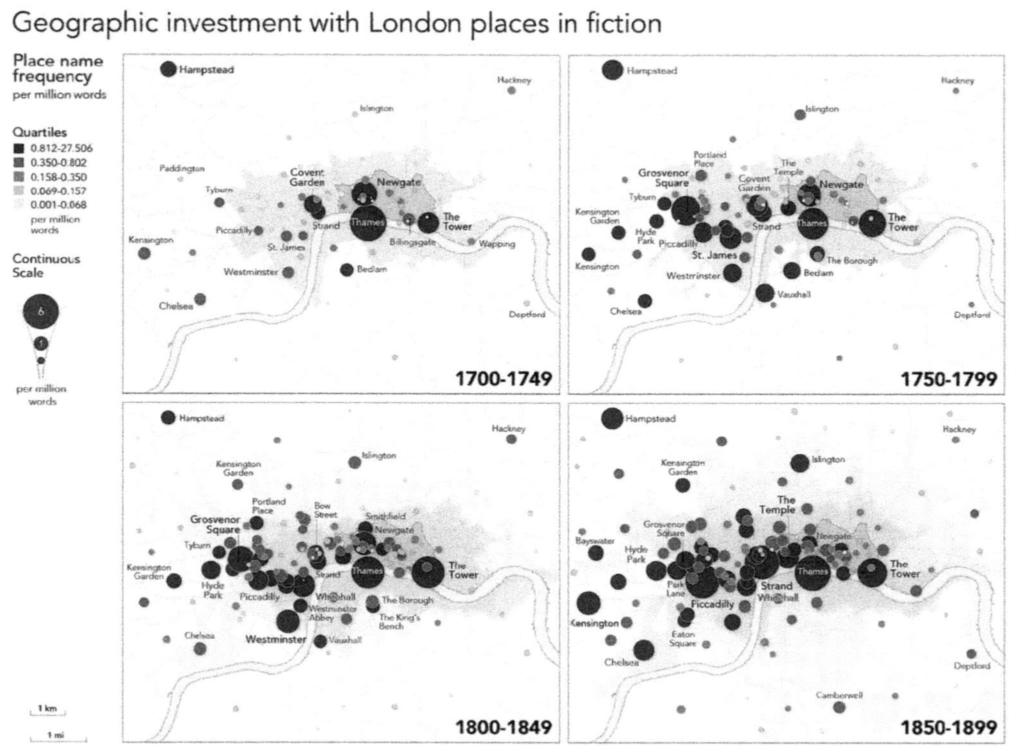

Figure 1.1 Geographic investment with London places in fiction.

Source: © The Authors.

namely 'the distribution of fictional settings ("gravity centers" vs. "unwritten regions")' (181). This tension between novels' actual geographic investments and their unexplored possibilities seems especially evident in Figure 1.1. These maps pose an important question: were novels really invested in representing London as a whole, or just the City and West End? London grew in population from 600,000 to 6.5 million over the two centuries represented here, expanding far beyond the City's historic walls and gates ('Historical Census Population'; Thirsk 6). Fictional London, however, is remarkably concentrated. Broadly representing the most frequently occurring London place-names in our fictional corpus, this map reveals a deep-rooted 'gravity centre' in fictional investment centred on the point at which the City and the West End meet. In the process of remediating the city, London fictions distort its geography by compressing and centring it in the West. This concentration of attention is apparent even amongst novels published in the last half of the nineteenth century, at which point London's urban development had already absorbed most of its outlying communities. Fictional attention, then, offers an alternate map through which we can understand the space of London, a map whose contours are only loosely defined by the contemporary urban geography of London.

But to what extent should we expect this distortion, given London's historical evolution? The City is, after all, the original extent of London, containing many of its most ancient and iconic locales: St Paul's Cathedral, the Guildhall, Smithfield and Newgate, just to name a few. And it was only during the plague and fire of London during the seventeenth century that many aristocratic Londoners began to relocate from the City to newly built residences in the West End. Consequently, the West End still contains many of the most fashionable locales in London, as sites such as St. James' Square organised a new, aristocratic and orderly social space in contradistinction to the piecemeal and mercantile space of the City (Black 118). In a way, then, the concentration of fictional attention reflects London's actual social geography – but in its seventeenth century configuration, suggesting a kind of literary-historical 'stuckness'. This stuckness manifests as a discontinuity between the place as it might have functioned in the wider geographic system of eighteenth- and nineteenth-century London, and the place as it would seem to function in the more concentrated geographic system of London's fictional representation. It is as if this earlier configuration strengthens the capacity of these places to act as cultural symbols for readers of the text, functioning as a shorthand, replete with associations that have accrued through history.

In reality, London in the eighteenth and nineteenth centuries expanded rapidly, dramatically transforming its two oldest districts – the City and the West End – from the city's residential focal point to the commercial centre around which more populous suburban districts grew. Over the course of the nineteenth century, the residential population of the City fell from 129,000 to 27,000. Although the West End doubled its population from

231,000 to 460,000, this doubling is not proportional to the increase in London's population overall ('Historical Census'). But in spite of their dwindling demographic centrality, the City and the West End remained the centre of literary London across two centuries of fiction. Comparing the history of literary attention to rates of population change, we can see that the boroughs most often mentioned in fiction – Westminster, the City and Camden – are those that least correlate with population changes. For each decade of the nineteenth century, starting from the first census for Great Britain in 1801, we compared the residential population of a borough in London to the percentage of words, from novels published in that decade, which are names for places within that borough. All three boroughs experience a decline in relative population over the course of the nineteenth century, but consistently dominate the relative share of place mentions in fiction. This inverse relationship between relative population and place mentions demonstrates that the tendency in novels to reference a place explicitly is not correlated with population size. Instead, these novels seem to emphasise sites of historical, public and even commercial significance at the expense of a more thorough representation of London's changing residential and domestic geographies.

From these maps, we can observe fiction's geographic investment with London as such; but what is the nature of this investment? How were these places incorporated into fiction? Did they serve as the setting of action, or were they simply mentioned in passing? What kinds of emotions were associated with these places? We have witnessed an historical stuckness in the spatial distribution of fictional attention, but how does the kind of attention given to these places change over time?

Crowdsourcing Emotion in Fictional London

Methodology

Here we discuss the second of our methodological phases: annotating our passages for their emotional association with place, if any. Although relatively underdeveloped in humanities research, crowdsourcing offers the advantage of gathering annotations across a large group of participants. To leverage these emerging techniques, we used the Amazon platform, Mechanical Turk: an online marketplace that allows institutions such as businesses or research groups to publish human intelligence tasks (HITs), which are then completed by anonymous online participants for a small fee paid by the requesting institution ('Amazon Mechanical Turk FAQ').

After numerous pilot experiments using batches of thirty randomly-selected passages – performed to ascertain the responsiveness, abilities and potential biases of the participants – we developed our final, most successful, experimental model. For each place, in each period for which at least ten legitimate passages were found (see section headed 'Methodology' above), we asked twenty participants to read and annotate each passage. Per passage, participants were asked to annotate whether the boldened place-name at the centre of the 200-word passage was the setting of the passage: in other words, does the passage take place here, or is it simply being mentioned? Half of the participants per passage were asked to annotate whether an emotion of fear was associated with, or experienced in, the location; the other half annotated for happiness.[6]

We decided to begin our study with these two emotions for a number of reasons. Experimental and cost-based constraints restricted to two the number of emotions we could initially study. Consequently, we hoped to capture both a positive and negative valence to emotionality. For Paul Fisher, fear is the most primordial of the negative passions, defining 'for us the very opposite of all that we will or choose or desire, and for that reason it is the negation of our own self-understanding' (15). Its long history in political theory, too, as in Hobbes' theory of society as arising out of 'mutual fear', gives to fear's self-negations a social dimension that is especially interesting for urban geography. Conversely, happiness was newly theorised in the eighteenth century into a pleasurable mental, rather than ethical, state – forming a more modern understanding of the concept that was paradoxically both individualistic and necessarily social (Norton 48). In their polarity, then, between positive and negative sociality, engagement and withdrawal, happiness and fear seemed appropriate as two initial cardinal points of emotion along which to map a range of affective representations of urban experience.

Due to the interpretive nature of our questions, assessing the results provided by the crowd was more complex than simply checking for a predetermined set of 'correct' answers. In a separate experiment, we discovered that the consensus of the crowd significantly correlated with the consensus of a group of Stanford English PhD students, suggesting that although consensus rates ranged around an average of 70 per cent, the same passages that were difficult for the crowd were also difficult for the students and vice-versa. In addition, we determined the reliability of individual annotations by closely examining the agreement amongst readers who had tagged the same passages. Using a variation of a weighted Cohen's Kappa score, we compared the answers provided by each reader on each passage to the

6 In experiments, we discovered that each question-type's average consensus – measured as the highest percentage of readers in agreement on a given passage annotation with two options – did not significantly improve beyond five readers, whilst the standard deviation did not significantly improve beyond ten. We chose ten readers (per emotion, so twenty readers per passage) to capture this advantage whilst also remaining within our budget. The annotation of each passage would earn a reader $0.07, a rate we found was sufficient to produce hourly wages that hovered around the U.S. federal minimum wage of $7.25. Each passage then cost $1.40 to annotate; and each place cost upwards of $56 if it were present in all four periods with ten passages.

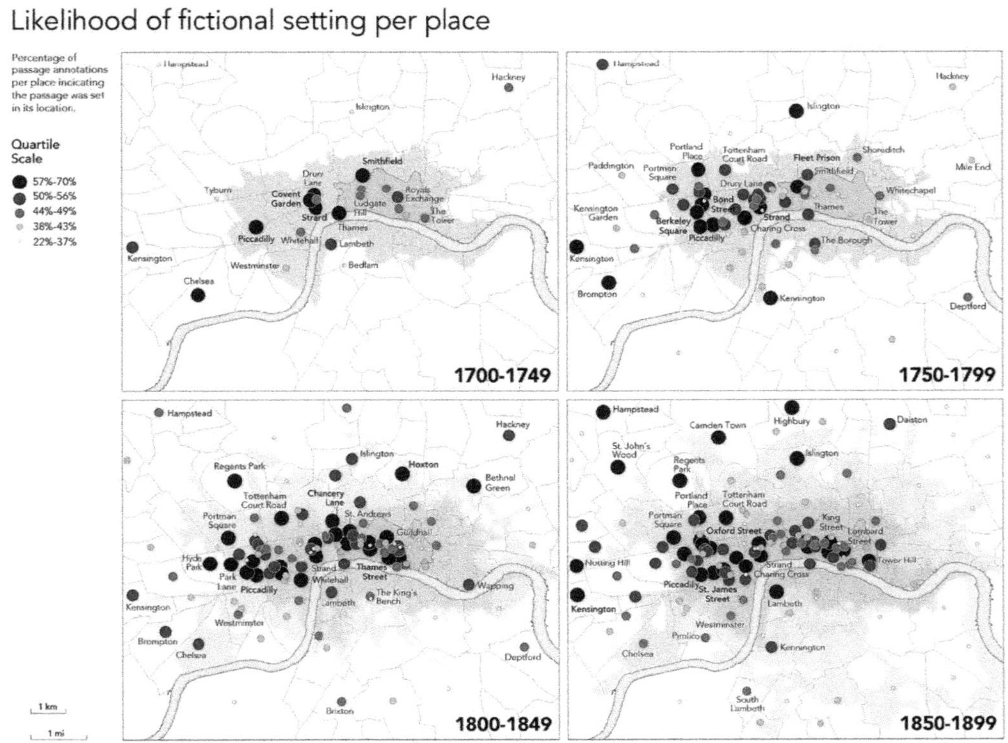

Figure 1.2 Likelihood of fictional setting per place, expressed as the percentage of readers' annotations indicating a place was the setting of its passage.

Source: © The Authors.

consensus of all of its other readers, weighting the importance of each passage by the strength of its consensus (Artstein and Massimo; Brants; Reidsma and Carletta). This process allowed us to identify readers who agreed with their peers no more often than chance; these readers' annotations were then manually inspected and possibly excluded from our analysis.

From the resulting data, we created two maps: one for setting (Figure 1.2) and one for emotion (Plate 1). For both Figures 1.2 and Plate 1, we represent all places in the manner of Figure 1.1, as circles pinpointed to their underlying polygons' centroid. To map setting, we quantified the likelihood that, in passages from a particular period mentioning a particular place, the place would act as the setting of the passage, based on the percentage of readers' positive responses to our setting question for those passages. The size and darkness of the circles in Figure 1.2 symbolise this likelihood: larger, darker circles indicate that a passage mentioning that place is more likely to be set there in that period. Similarly, for the emotion data, we quantified the fearfulness and happiness of each place in each period by calculating the percentage of positive answers to both questions on the emotion of the place. Fearfulness and happiness are therefore represented on a single spectrum, with the percentage of fearful responses subtracted from the percentage of happy ones. This spectrum, ranging mostly within –15 per cent to +15 per cent, is represented on the map as a red to green colour gradient. The colour of the circles indicates the emotional valence of the place along this fear-happiness spectrum, with darker reds representing a stronger 'net' likelihood of fear, and darker greens representing a stronger net likelihood of happiness. Conversely, by summing rather than subtracting these percentages, the size of the circles indicates an overall strength of emotionality. Small circles in the middle of the colour spectrum (i.e. grey) represent 'neutral' places rarely found to be emotional; large grey circles indicate emotionally ambivalent places that attracted fear and happiness in equal measure.

Interpretation

The resulting map demonstrates a spatial pattern of emotion drawn from fiction across the eighteenth and nineteenth centuries. Rather than visualising the relative presence of London places in fiction, this map reveals the structures of feeling associated with them. London places are represented as embedded within the intersecting cultural networks of associations, attitudes and social relations through which they operated as places in fiction.

Across the four periods, we see a bifurcation between the happy emotionality of the West End and the fearful tendencies of the City and East End. In truth, the bifurcation is more complex. Not only the West End, but many places then on the periphery of London, such as the suburbanising villages of Barnsbury or Hampstead, are also associated with happiness; whereas some locations in the West, such as the notorious Seven Dials or

Drury Lane, are associated with fear. But these complications only nuance, rather than challenge, the general geographic pattern of emotional polarity. In the previous section, we noted a stuckness in the spatial distribution of fictional attention. Here, the affective bifurcation is only strengthened by a similar kind of historical stuckness, this time playing out within our two centuries. Although particular places may change in emotional association, the overall spatial pattern remains intact across two hundred years of fiction.

Also worth noting is the sheer number of places designated as 'neutral': that is, passages in which a majority of readers found neither happiness nor fear. In total, 84 per cent of the passages sent to the crowd were considered neutral. The sheer number of neutral passages is a significant finding of our project. Uncovering the many possible meanings of the absence of affect in these places, whether related to the narratological effects of place invocation, to a dampening of emotional register, or, potentially, a bias within our participants' non-expert reading, lies outside of the scope of this chapter. Instead, we will focus here on interpreting the geography of emotional presence, in fear and happiness, more broadly.

To investigate the possible causes that led the crowd to reach a majority opinion on whether a passage was fearful or happy, we read through the passages for which a majority was formed, annotating each with one or several possible reasons for the crowd's decision. The predominant causes of fear in our passages seem to relate to representations of intimated violence, death, imprisonment, illness and social disgrace. It is as if the novels, in representing fear, turned to the most visceral and powerful representations imaginable. For instance, the most fearful passage in our corpus, by unanimous agreement from the crowd, comes from Mary Braddon's 1887 novel *Like and Unlike*:

> She spends only one day of every week in this house, but she works for us out of doors, going about the streets at night, and talking to wretched women whom few girls of her age would have the courage to approach. That fragile looking girl has penetrated the darkest alleys about Clare Market, the most dangerous streets in Ratcliff Highway, where even the police go at the risk of their lives. (Braddon 210)

Although the 'fragile looking girl' is herself likely unafraid, Braddon's narrator associates Clare Market with the fear of danger and perhaps death. This deep-rooted emotional association with place is exactly what our project sought to uncover and map. Moreover, it seems important to note that the group who does fear Clare Market, unlike the girl, is the police: the group that ideally projects middle- and upper-class juridical power. The location, and therefore the passage, is fearful not to the lower classes, but to the wealthier residents of the city. Fear, here, is both invested in a particular place and, in turn, through the geography of this investment, organises London into zones of safe travel for implicitly middle- and upper-class characters and readers.

Although no passage by Dickens happened to appear in our random sample of Clare Market passages, Dickens' *Sketches by Boz* includes Clare Market as a place where one might find one of the infamous gin-shops it sketches. However, Dickens mentions Clare Market amongst a handful of other locations only to orient the reader geographically. His sketch is really of 'a' gin-shop, which might be in any one of these locations. This fictional space of the indefinite article seems disproportionately used to represent sites of poverty and modern horror, suggesting a possible imbalance in the type of representation given to fearful and happy places. For instance, we looked at a handful of passages in Dickens' work that we knew to describe slums and urban poverty. Relatively few actually situate the reader in geographic space. Introducing 'Tom-all-Alone's', a fictional slum in *Bleak House*, Dickens writes: 'Jo lives . . . in a ruinous place known to the like of him by the name of Tom-all-Alone's. It is a black, dilapidated street, avoided by all decent people' (256). Like the gin shop, Tom-All-Alone's is 'a' slum, a fictional example that simulates 'the' slums of London. These virtual spaces define a horizon of possibility for our project, which depends upon explicit place names (see section on 'Methodology' above). From their presence within certain texts, however, we can infer that the explicitness of place-names in fiction may vary with the social conditions of the place, for example with whether it is possible for middle and upper-class citizens to travel there.

In contrast to the life-threatening danger of fictional representations of fear in London, happiness seems to relate predominantly to representations of positive sociality. Scenes of friendship, romantic experiences, high society, family and marriage are the most frequent amongst our passages annotated as happy. The only noticeable exception to this trend can be found in idealised natural descriptions – 'the luxuriant banks of the Thames' or the 'the rose-coloured [flowers] which we saw at Kew' – which are typically set in the former villages lying on the periphery of London, such as Hampstead or Richmond. But these moments usually only set the scene for the social. It seems significant that we find relatively few passages in which happiness is represented as a joyful response of an individual to a place; instead, positive responses to places are deeply mediated through the social encounters and relations that they enable. In the following passage from Luisa Keir Grant's 1854 novel *Charles Stanley*, natural description sets the scene for a party:

> Her spirits rose as the weather cleared up. Long before they got to Richmond, the sun was shining brightly, and she was talking and laughing with Charles, as if there had been no Sir James in the world. Most of the party had already arrived. Lady Ramsay was hailed by all with pleasure. Some went on the river, others wandered in the park. They met

> at dinner, apparently well satisfied with their morning's amusement. (Grant 195)

In this passage, Richmond is not associated with happiness through its agreeable environs, so much as with the social relations which its environs make possible.

Modelling the Literature and Sociology of London

Above, we have noted how London has mediated the emotional polarity of fear and happiness into a geographic polarity of east and west. But what is the logic of this divide? London's east/west polarity is often invoked as a shorthand for class differences; however, is it only class associations that have separated Newgate, Fleet Prison and the Tower of London from Grosvenor Square, Eaton Square and Hyde Park? Their difference in social function, from disciplinary sites to open spaces, seems as relevant as their location in space. It is also notable that Newgate, Fleet Prison and the Tower are amongst the most ancient of London's buildings, originating from the Norman period, whereas development for Grosvenor Square did not begin until 1725 (Sheppard).

Given our data, we considered four sociological factors possibly explaining why certain places have been imbued with a particular emotionality: location in space (the City, or north, east, west or south of it); the class characteristics of its surroundings; the social function of the place (e.g. prison, church, park); and when it was first developed. We also considered a narratological factor possibly influencing the emotion of a place: whether the passage was set there. Ascertaining the contribution of each of these factors to a place's emotionality is a complex problem. Rather than investigate each separately, we elected to measure and compare their combined explanatory effect. Although data on the location, type and age of each place is readily available, measuring its class-constituency was less straightforward. Our solution was to extract data from nineteenth-century sociological research, namely Booth's 1889 London map of poverty and income, using contemporary GIS technology. We had already placed each of our locations onto Booth's map, representing each place as a polygon covering its identifiable boundaries. Using remote sensing techniques (often used with satellite imagery) available in the software ArcGIS, we simplified the map's range of colours, which together represent Booth's seven class distinctions. For example, we compressed the range of reds that signifies Booth's second of seven classes, the 'Middle-class' and 'Well-to-do', into a single hue. These compressed colours were then quantified for each place, as the percentage of the total pixel count that each colour occupies. For districts or other areas of London, we quantified the colours within the area boundaries. For buildings, streets and other types of places, we quantified the colours within a hundred meter walking distance of their perimeter, hoping to capture the micro class neighbourhoods made visible by Booth's map. Overlaying our colour-coded literary-emotional data onto Booth's map produced Plate 2, a geographic representation of the relationship between class and affect in fictional London.

Even at a glance, relationships between emotion and class are apparent. For Booth, gold indicated the wealthiest blocks of London, and the patterns that they form around our green polygons indicate a relationship between places of fictional happiness and urban concentration of wealth. Cavendish Square, Portland Place, Grosvenor Square, Park Lane and Berkeley Square were all upper-class locations by Booth's time, and were all associated with happiness in fiction by the crowdsourcing participants. In contrast, Seven Dials and King's Bench Prison are often represented as fearful in fiction, and are embedded in areas bordering on blue and black, representing Booth's struggling classes. Unfortunately, the distribution of our locations, combined with the design of Booth's map, makes it difficult to explore more deeply the relationship between fearful and lower-class locations: Booth did not include the City in his map, which houses many of our places associated with fear, on account of its having already transformed from a residential to a commercial space ('Poverty Maps of London'). For this reason, we excluded locations in the City from our class analysis, but retained them when considering non-class factors.

How, then, do all of these apparent correlations – between emotion on the one hand, and on the other hand location, class, era and type of location – compare quantitatively? By using odds ratios, we are able to compare the explanatory weight of our factors. For example, to assess how well class explains fearfulness, we measure: (a) how likely a passage will associate happiness with its location when centred on an upper-class place; and (b) how likely it will associate happiness with its location when *not* centred on an upper-class place. Dividing these two likelihoods (a/b) produces an odds ratio.[7] Because odds ratios require categorical data, it was necessary to translate our continuous data into discrete categories. First, we aggregated Booth's seven class distinctions into three.[8] Then, we divided the age of locations into pre-1485 (Roman and medieval) and post-1485 (Tudor, Stuart and later London). We also transformed our emotional and setting

7 For example, passages centred on upper-class locations are: (a) 17.6 per cent likely to associate happiness with their place; whereas passages that are not centred on upper-class are only (b) 10.7 per cent likely to associate happiness. Dividing (a) by (b) produces an odds ratio of 1.65, indicating that passages about upper-class locations are 1.65 times more likely to associate happiness with their places than are passages not about upper-class locations.

8 The first category (gold) was considered upper-class. The next two (red and pink) were considered middle-class. The next four, the wealthiest of which is the first to mention 'poor', represented the lower classes. To discretely categorise this quantitative data, of the percentage of pixels in each of these three classes, we used the following cut-offs: Upper-class = 10 per cent or above for upper-class pixels. Middle-class = 35 per cent or above for middle-class pixels. Otherwise, the place was considered lower-class.

Table 1.1 Social and narratological factors most strongly associated with fear.

Factor	*Likelihood of* fear *when factor is absent*	*Likelihood of* fear *when factor is present*	*Odds Ratio*	*Examples of Places*
Type: Hill	4.0%	17.5%	4.43	Tower Hill, Ludgate Hill
Type: Prison	3.7%	14.6%	3.98	Fleet Prison, Newgate, The Savoy Prison
Type: Government	4.1%	9.5%	2.33	Guildhall, Old Bailey
Space: City	3.3%	7.6%	2.31	Leadenhall Street, Thames Street, The Monument
Set: Setting	3.1%	5.6%	1.79	(Almost all)
Era: Pre-1485	3.3%	4.9%	1.49	Bedlam, Newgate, Smithfield, Tyburn, The Tower
Type: Market	4.1%	5.8%	1.42	Billingsgate, Smithfield

data. A passage was considered to contain fear or happiness if five or more of the 10 respondents considered it to have the respective emotion. Similarly, a passage was considered to be set in its location if 10 or more of the 20 respondents considered it to be so. The factors that most significantly increase the likelihood of a passage to express fear are:

This table reveals a strong correlation between a place's social function and the degree of fearfulness attributed to it. A passage centred on a prison is about four times more likely to be fearful than a passage not centred on a prison. Of the two hills found to be even more strongly associated with fear, both have prisons associated with them. Ludgate Hill shares its name with a prison, whereas Tower Hill, sitting outside the Tower of London, served as a site of execution until well into the eighteenth century. Similarly, one of the two places classified under 'government', the Old Bailey, is a criminal justice court. The final association with social function, between fearfulness and the marketplace, primarily owes its place on the chart to Smithfield and Billingsgate. Smithfield was an execution site before it became a cattle market, whereas Billingsgate was infamous for its vulgarity and served as a household word for crassness.

An identifiably Foucauldian theme runs through these powerful factors for fictional fear in London: discipline and punishment. Despite the apparent obviousness of this association, it seems worthwhile to reconsider Foucault's description of the transfer of juridical power, from the penal institution to the entire social body at large, through the affective geography of literary representation (Foucault 298). If literary representation does indeed reinforce social power, then the nexus of fearful passages set near or within legislative institutions, such as prisons, courts and government offices, fulfils the social function of the institution whilst, in exchange, the affective experience of the text gains legitimacy through association with real-world sites of social discipline.

Beyond disciplinary sites, fear takes place in the City significantly more often than not. Interestingly, this spatial form of the past – London's original expanse with its oldest buildings – is more strongly associated with fear than medieval and Roman locations by themselves. The City, in other words, seems more productive of fear than would be expected from its association with the past. Perhaps this combined spatio-historical identity of the City helps explain why, in the *Atlas of the European Novel*, the only Gothic novel set in London is set in Renaissance London, before any large-scale development of the West End had occurred (Moretti 16).

In contrast to the disciplinary functions of sites of fear, the table of odds ratios for happiness (Table 1.2) reveals that it is primarily associated with churches, theatres, squares and parks. Although the role of weddings in the happiness of churches is strong, these places also seem to share a common role: they all produce a reciprocal social space in which one sees and is seen. The sociality noticed in the 'happy' passages articulates itself primarily in these places. Moreover, it is again social function that is more strongly associated with happiness than is the geography of the West End. At the same time, this spatial division of the City and the West actually articulates our observed emotional polarity of fear and happiness better than the more nuanced geography of class embedded in Booth's map. Because of the aristocratic impetus for the development of many of London's post-medieval iconic locales, modern and upper-class locations largely overlap, so it is not unexpected for them to have an almost equal odds ratio.

Still, the fact that modern, upper-class locations are significantly more likely to be represented as happy clarifies the relationship between readers and the fictional geography of the texts they consumed. Above all, the upper-class locales that are correlated with happiness would be *familiar* to the eighteenth- or nineteenth-century English reader. Whether because these social spaces were the locations in which middle- and upper-class residents of London themselves would socialise, or because they would be recognisable as famous London locales to readers outside the city, these locations carry positive associations that predominantly middle- and upper-class consumers would recognise (St Clair 114–5). This familiarity is intimately connected with the tendency of these places to carry names that reflect the actual geography of London. In contrast, many lower-class locations such as slums were likely unfamiliar by name to even those readers who resided in London. Perhaps, then, the displacement of slums into purely invented proxies (for example, Dickens' Tom-All-Alone's) provided a locus for the fearful affect without requiring any previous knowledge of

Table 1.2 Social and narratological factors most strongly associated with happiness.

Factor	*Likelihood of* happiness *when factor is absent*	*Likelihood of* happiness *when factor is present*	*Odds Ratio*	*Places*
Set: Setting	8.2%	17.5%	2.13	(Almost all)
Type: Church	11.6%	24.6%	2.11	St. George's Church, St. Paul's Cathedral
Type: Theatre	11.8%	23.5%	2.00	Covent Garden Theatre, Drury Lane Theatre
Space: West	8.3%	15.2%	1.82	Bloomsbury, Brompton, Chelsea, Hampstead, Hyde Park, Notting Hill, Strand
Type: Square	11.5%	19.2%	1.67	Berkeley Square, Grosvenor Square, Portman Square
Era: Post-1485	9.2%	15.3%	1.66	Bond Street, Marylebone, Mayfair, Piccadilly, Portland Place
Booth: Upper	10.7%	17.6%	1.65	Belgravia, Harley Street, Knightsbridge, Kensington Gardens, Regent's Park
Type: Park	11.8%	17.3%	1.46	Hyde Park, Kensington Gardens, Regent's Park

the specifics of any given slum. Positive affect, on the other hand, is associated with specific places whose nominal articulation could already perform a large part of the affective work.

However, urban factors are not as strongly associated with happiness as is a factor of narrative form: setting. We have seen how passages set in their named locations ('set passages') are significantly more likely to be either fearful or happy, rather than neutral. Evaluating fear and happiness together, set passages are more than twice as likely to associate emotion with a location. Narrative setting, in its phenomenological immediacy, seems deeply connected to emotionality. This result lends evidence to our project's premise: fictional representations of emotion can not only be mapped, but this mapping is also intimately connected to narrative. If the set-ness of a passage, in its spatio-temporal localisability, actually calls forth emotion, mapping these spatio-temporal moments seems uniquely important for understanding the relationship between place and emotion within narrative.

Setting is also the only factor associated with both fear and happiness. No selection of places – by function, class, era or location – is similarly bimodal: setting situates itself in the middle of this emotional polarity. We can visualise this mediating position diagrammatically by translating the data from Tables 1.1 and 1.2 into the network shown in Figure 1.3. Two factors are linked if they are associated with each other, and the thickness of the line indicates the strength of association.

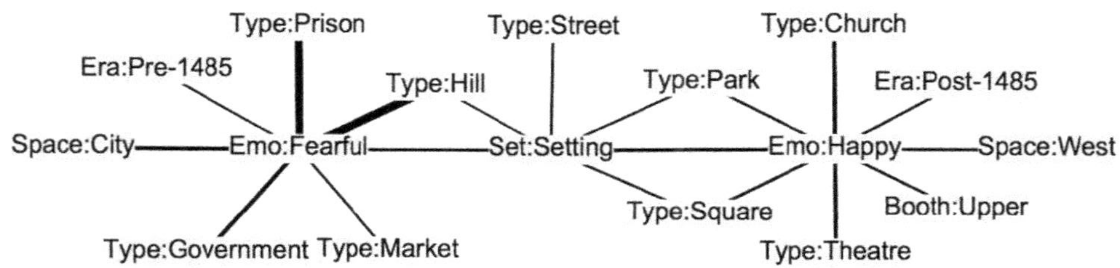

Figure 1.3 Odds ratio network built from Table 1.1 and Table 1.2.

Source: © The Authors.

From this network, we can see how starkly polarised the affective-geographic system of fictional London really is. Sites of fearfulness are never also sites of happiness. Fear and happiness – through their associations with the functions, spaces and eras of London places – essentially articulate two different Londons. Fearfulness can be located in a constellation of prisons, markets, courts of law and the ancient buildings of the City. Happiness is found in the constellation of parks, squares, theatres, churches and the modern buildings of the West. Setting is the only link mediating these two 'half-Londons' (Moretti 107). Interestingly, what makes possible this mediating role is setting's relationship to emotionality as such. Set passages are more likely to associate emotion with their location; but unlike every other factor, they are not strongly associated with a *particular* emotional valence. If setting can be said to give rise to emotional expression, it can also be said to make possible and mediate the emotional polarity that so powerfully structures the fictional geography of London.

Conclusion

What do we gain from mapping the spatial and affective geographies of fictional London? Spatially, we've seen some evidence for a chronological conservatism to London's fictional representation, a conservatism we've called stuckness. This stuckness in history implies that population growth did not substantially alter the imaginative contours of the city. Moreover, we have seen that this 'stuck' spatial system harbours a hidden social-affective logic. The modern, upper-class and public-space locations in the West End cluster together representations of the social pleasures of modernity, whereas the pre-modern and disciplinary sites of the City cluster together representations of vehement fear.

The very simplicity and precision of this rendering of London ultimately belies the possibility of a more emotionally ambiguous relationship to urban modernity. Although slums such as Clare Market and Old Nichols are occasionally represented fearfully, urban fear seems preoccupied with the ancient and juridical institutions of London, in a manner not unlike the Gothic novel. Representations of the horrors of modernity, when they occur, are likely relegated to the Tom-All-Alone's of London: authorial inventions without a precise real-world referent. In contrast, representations of the pleasures of modernity rely heavily on the existing urban geography of the West End, with its nominative parks, squares and theatres. Not coincidentally, these are also upper-class spaces. The correlation we have observed between class and affect may thus also help explain the readiness, or lack thereof, with which fiction engages with existing urban spaces. Finally, we have discovered that narrative setting plays a central role in London's fictional geography – by mediating both fearful and happy emotions, setting makes possible the emotional polarity that structures so many literary representations of London.

Stepping back from London to look at the technologies that have enabled our research, what have we learned about the capabilities and limitations of a digital literary geography? Through a combination of algorithmic and reading-based methods, we were able to uncover a literary geography of London as it evolved over two hundred years of fiction. Algorithmic methods have allowed us to track hundreds of place-names across thousands of texts. Equally, if not more importantly, reading-based methods have allowed us to do more with these places than simply locate them on a map. By annotating for setting and emotion, we designed maps that deepen and qualify our understanding of the ways in which London is represented in historical works of fiction. These reading-based annotations make possible nuanced questions about the meaning of places in fiction, expanding digital literary geography from one rooted in geographic investment to one able to qualify the nature of that investment. They also produce new layers of geographic information that interact both with each other and with socio-historical facts of London's geography, in ways that can be quantified and compared. But our methods also have important limitations, as is made evident by the differential visibility of fear within an explicitly named geographic space. In addition, our turn to crowdsourcing methods and Amazon's Mechanical Turk revealed the potential advantages of a large group of readers distributed across a corpus of texts; however, the for-profit aspect of the platform also introduced potential complications, as our results were subject to potential manipulation by participants seeking to maximise their individual profit.

Literary interpretation on this scale and to this level of complexity would not have been possible without the combination of technology and close critical attention that we describe. By informing our quantified crowdsourced data with our specific disciplinary and historical understanding of London, we are able to reveal a complex web of socio-cultural, historical, economic and narratological factors that relate emotion to place within fictional representations of the city. Digital literary geography has opened up for us not just Frances Burney's or Dickens' London, but rather a whole metropolis as mediated by the structures of feeling embedded in two centuries of fiction.

Works Cited

'Amazon Mechanical Turk FAQ'. *Amazon Web Services, Inc.* Web. 20 March 2015. <https://requester.mturk.com/help/faq>.

Artstein, Ron, and Massimo Poesio. 'Inter-Coder Agreement for Computational Linguistics'. *Computational Linguistics* 34.4 (2008): 555–96.

Black, Jeremy. *London: A History.* Lancaster: Carnegie Publishing, 2009.

Booth, Charles and David A. Reeder. *Charles Booth's Descriptive Map of London Poverty 1889.* London: London Topographical Society, 1984.

Braddon, Mary Elizabeth. *Like and Unlike, by the Author of 'Lady Audley's Secret'.* Vol. 3. London: Spencer Blackett, 1887.

Brants, Thorsten. 'Inter-Annotator Agreement for a German Newspaper Corpus'. *Proceedings of the Second International Conference on Language Resources and Evaluation, ELRA* 31 May–2 June 2000.

Cohen, J. 'A Coefficient of Agreement for Nominal Scales'. *Educational and Psychological Measurement* 20.1 (1960): 37–46.

Dickens, Charles. *Bleak House.* Ed. Nicola Bradbury. New York: Penguin, 1996.

English, James F. 'Everywhere and Nowhere: The Sociology of Literature after "the Sociology of Literature"'. *New Literary History* 41.2 (2010): v–xxiii.

Finkel, Jenny Rose and Christopher D. Manning. 'Joint Parsing and Named Entity Recognition'. *Proceedings of the North American Association of Computational Linguistics* 2009.

Fisher, Philip. *The Vehement Passions.* Princeton, NJ: Princeton University Press, 2002.

Foucault, Michel. *Discipline and Punish: The Birth of the Prison.* Trans. Alan Sheridan. New York: Vintage, 1995.

Grant, Louisa Keir. *Charles Stanley: A Novel.* Vol. 2. 1854. Reprint. London: Forgotten Books, 2013.

Greenwood, Christopher, and John Greenwood. 'Map of London, from an Actual Survey Made in the Years 1824, 1825 & 1826'. 1827. *British Library Online Gallery.* Web. 20 March 2015. <http://www.bl.uk/onlinegallery/onlineex/crace/m/0070 00000000006u00225000.html>.

'Historical Census Population'. *Greater London Authority.* Web. 20 March 2015. <http://data.london.gov.uk/dataset/historic-census-population>.

Horwood, R. 'PLAN of the Cities of LONDON and WESTMINSTER the Borough of SOUTHWARK and PARTS Adjoining Shewing Every HOUSE. By R. Horwood'. 1792. *British Library Online Gallery.* Web. 20 March 2015. <http://www.bl.uk/onlinegallery/onlineex/crace/p/007zzz000000005u00173000.html >.

Jackson, Lee. *Dictionary of Victorian London.* Lee Jackson, 2012. Web. 20 March 2015. <http://www.victorianlondon.org/>.

Manning, Christopher, Mihai Surdeanu, John Bauer, et al. 'The Stanford CoreNLP Natural Language Processing Toolkit'. *Proceedings of 52nd Annual Meeting of the Association for Computational Linguistics: System Demonstrations* 2014: 55–60.

Matthew, Adam. 'Interactive Maps'. *London Low Life.* Indiana University. 2011. Web. 20 March 2015.

Moretti, Franco. *Atlas of the European Novel, 1800–1900.* London: Verso, 1998.

———. *The Bourgeois: Between Literature and History.* London: Verso, 2013.

Ngai, Sianne. *Ugly Feelings.* Cambridge MA: Harvard University Press, 2004.

Norton, Brian Michael. *Fiction and the Philosophy of Happiness: Ethical Inquiries in the Age of Enlightenment.* Lewisburg, PA: Bucknell University Press.

'Ordnance Survey Maps, London, Five Feet to the Mile, 1893–1896'. *National Library of Scotland.* Web. 20 March 2015. <http://maps.nls.uk/os/london-1890s/>.

Paterson, Michael. *Inside Dickens' London.* Cincinnati OH: David & Charles, 2011.

Piatti, Barbara, Hans Rudolf Bär, Anne-Kathrin Reuschel, Lorenz Hurni and William Cartwright. 'Mapping Literature: Towards a Geography of Fiction'. *Cartography and Art.* Eds, William Cartwright, Georg F. Gartner and Antje Lehn. Berlin: Springer, 2009. 179–94.

Porter, Roy. *The Creation of the Modern World: The Untold Story of the British Enlightenment.* New York: W.W. Norton & Co., 2000.

'Poverty Maps of London'. *Charles Booth Online Archive.* Web. 20 March 2015. <http://booth.lse.ac.uk/static/a/4.html>.

Reidsma, Dennis, and Jean Carletta. 'Reliability Measurement without Limits'. *Computational Linguistics* 34.3 (2008): 319–26.

Rocque, John. 'An Exact Survey of the Citys of London, Westminster, Ye Borough of Southwark, and the Country near Ten Miles round'. 1746. *British Library Online Gallery.* Web. 20 March 2015. < http://www.bl.uk/onlinegallery/onlineex/crace/a/007zzz000000019u00018000.html>.

Sheppard, F.H.W., (ed.) 'Grosvenor Square: Introduction'. *Survey of London, Volume 40: The Grosvenor Estate in Mayfair, Part 2 (The Buildings)*: 112–17. British History Online. Web. 20 March 2015. <http://www.british-history.ac.uk/survey-london/vol40/pt2>.

St Clair, William. *The Reading Nation in the Romantic Period.* Cambridge: Cambridge University Press, 2004.

Thirsk, Joan, (ed.) *Chapters from The Agrarian History of England and Wales: 1550–1750.* Cambridge: Cambridge University Press, 1990.

Watt, Ian. *The Rise of the Novel: Studies in Defoe, Richardson and Fielding.* Berkeley: University of California, 1957.

Wilkens, Matthew. 'The Geographic Imagination of Civil War-Era American Fiction'. *American Literary History* 25.4 (2013): 803–40.

Williams, Raymond. *Marxism and Literature.* Oxford: Oxford University Press, 1977.

2 The Digital Poetics of Place-Names in Literary Edinburgh

Miranda Anderson and James Loxley

SHEPHERD. . . . Just suppose yoursel speakin to some stranger or ither frae England, come to see Embro' – and astonish the weak native.

NORTH. Stranger! wilt thou take us for thy guide . . . thou shalt have the history of many an ancient edifice – tradition after tradition, delightful or disastrous – unforgotten tales of tears and blood, wept and shed of old by lungs and princes and nobles of the land? . . . Or threading our way through the gloom of lanes and alleys shall we touch your soul with trivial fond records of humbler life, its lowliest joys and obscurest griefs [sic]?

(Wilson 246)

Introduction

In this invitation, Christopher North – the literary persona of the nineteenth-century writer and critic, John Wilson – conjures up some of the possible encounters with the Edinburgh into which visitors could find themselves drawn. North's invitation is echoed, to some extent, by the ambition of *Palimpsest* – a digital literary mapping project.[1] *Palimpsest* seeks to model Edinburgh's literary cityscape on a much larger scale than has hitherto been accomplished, including both its 'unforgotten tales' and its 'lowliest joys and obscurest griefs'. In doing so, the project aims to make an innovative contribution to the geocritical exploration of the mutual implication of space, place and literature.

The relationship between these terms has a long and various intellectual history. As Sten Pultz Moslund has put it, 'There are numerous approaches to the study of place in literature – place as mapped by discourses and power; place as a transplatial contact zone; place as a dynamic process or event; place as emotional, imagined, remembered, or experienced by

1 The *Palimpsest* project, including the research undertaken for this paper, has been generously funded by the Arts and Humanities Research Council, UK.

the senses' (30). The now canonical, and still productive, work of scholars including Edward S. Casey, Yi-Fu Tuan and Franco Moretti has been highly influential; equal stimulus has been taken from the conceptualisation of space developed in different ways by Michel de Certeau, Gaston Bachelard and Henri Lefebvre.[2] Such thinking has recently been undergoing redefinition and elaboration at what can sometimes seem like an accelerating rate. The last three years alone have seen important contributions from Robert T. Tally Jr., Eric Prieto and the translation into English of two monographs by Bertrand Westphal. As this collection amply demonstrates, the exploration of literary place is at the same time taking explicitly cartographic, and increasingly digital, form: Ian Gregory and David Cooper's *Mapping the Lakes*, Barbara Piatti's *A Literary Atlas of Europe*, Shelley Fisher Fishkin's collaborative *Digital Palimpsest Mapping Project* and the 'z-axis' work of the *Modernist Versions Project*, to name only four recent examples, have all generated distinctive forms of geocritical engagement.[3]

One of the most interesting features of some of the theoretical work on space and place has been the invocation of narrative as a key element in the conceptual matrix, suggesting that there is a structural kinship between place and fiction which geocritical work might seek to capture. As de Certeau, for example, famously argued:

> Stories . . . carry out a labor that constantly transforms places into spaces or spaces into places. They also organize the play of changing relationships between places and spaces. The forms of this play are numberless, fanning out in a spectrum reaching from the putting in of an immobile and stone-like order (in it, nothing moves except discourse itself, which, like a camera panning over a scene, moves over the whole panorama), to the accelerated succession of actions that multiply spaces. (118)

The tracing out of narrative, that is, of the tour or the itinerary, allows us to grasp the practice of spatial inhabitation that other accounts of place would occlude. For Nicholas Entrikin, too, it is 'the vast realm of narrative forms' that allows the thinker of place to bridge its 'existential and naturalistic conceptions' – to give us, in other words, 'a sense both of being "in a place" and "at a location", of being at the center and being at a point in a centerless

2 For a succinct overview see Cresswell.

3 Fishkin's use of the term 'palimpsest' is an interesting, if coincidental, point of connection with our own project, though we are interested in rather different implications of the term. In particular, we are less keen to describe the palimpsestic nature of our mapping solely in terms of depth, but as with the Modernist Versions Project, also see our activities in terms of the exploration of surfaces.

world' (134). Such understandings of place, emphasising the implication in each other of story and locus, might usefully be described as diegetic: the tellable world, or the world as told.

Palimpsest's Development

The prototype *Palimpsest*,[4] undertaken in 2012, created a small, mobile-friendly website featuring extracts from literary texts set in Edinburgh, geolocated on a map of the city according to the setting evoked in each extract. This offered a very different experience of literary Edinburgh from traditional literary-historical surveys, existing walking tours, or indeed other online guides to the city's cultural heritage such as *Our Town Stories*. For example, surveys and tours customarily offer only a single topographical organisation of, or a single linear route through, the city's literature (Campbell, Foster, Lownie); whereas *Our Town Stories* includes the geolocation of narratives, it offers only very limited literary content. The *Palimpsest* prototype showed that website and app users could instead be furnished with the means to find their own way through a more densely populated and richly layered literary city. In this diegetic cityscape they could encounter works from different periods clustered at one site, or elements of one literary work spread across sites within the city. The coincidence of location could allow for surprising intertextual encounters. Filters allowed users to expand or narrow the range of available works according to external properties such as genre, period, author or internal properties including grammatical and narratological features.

Although the prototype offered proof of concept, it was nonetheless fairly limited in its scope and functionality. Its database was populated and curated manually, and this constraint meant that it contained only 200 excerpts from, for the most part, the better known literary works in the Edinburgh canon. These extracts, manually gathered from poetry, prose and fiction, ranged in date from the early modern era to the present. The extracts were primarily focused on locations around Edinburgh's Royal Mile and generally ranged between 50 to 450 words in length. Lists of short sample geolocated extracts were offered with the option for the reader to choose a longer version and also to choose to read a range of bibliographical information about the text. The choice of extracts and the associations of text and place were made only on a case by case, and somewhat extempore, basis.

4 The prototype received minor funding awards from the English Literature department at the University of Edinburgh and also from a Knowledge Exchange impact grant, although it was mainly created through the voluntary efforts of staff and postgraduate students from English Literature, Informatics and Edinburgh College of Art. The prototype can be found at *Palimpsest*.

Figure 2.1 Palimpsest prototype screenshot of satellite version of the map.

Source: *Palimpsest*, created by the original Palimpsest Team using Imagery © 2015 DigitalGlobe, Getmapping plc, Infoterra Ltd & Bluesky, Landsat, The GeoInformation Group.

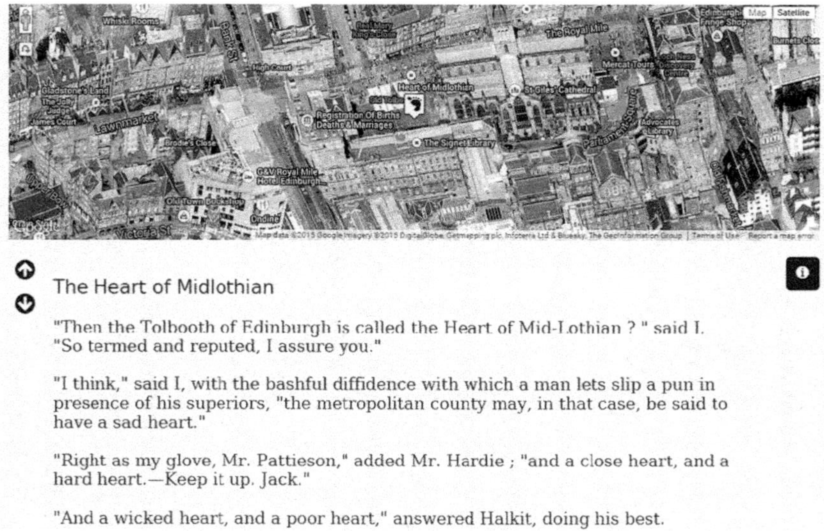

Figure 2.2 Palimpsest prototype screenshot with short text excerpt.

Source: *Palimpsest*, created by the original Palimpsest Team using Imagery © 2015 DigitalGlobe, Getmapping plc, Infoterra Ltd & Bluesky, Landsat, The GeoInformation Group.

The current AHRC-funded project promises a fuller and more systematic realisation of the prototype's promise, by combining a bottom-up gathering of large collections of digital texts with top-down filtering of these collections by literary and language technology scholars. *Palimpsest* is now an interdisciplinary research project involving literary critics and the Language Technology Group at the University of Edinburgh, the University of St Andrews Human-Computer Interaction research group and the Edinburgh-based national data centre, EDINA. So, whilst the project's purpose remains to explore the dimensions of literary Edinburgh, this is to be achieved through a radical change in methods: we are using natural language processing technology to text-mine literary works set in Edinburgh and make possible interactive visualisations of the results. By uncovering works that have sunk into obscurity, and by tracing the narratives told through place-names in Edinburgh across time, we will be able to reflect on existing critical models of the nature of Edinburgh and its literature. *Palimpsest* will provide a new means to reflect on, excavate and celebrate the sedimentary processes that have given our city its literary shape, within and beyond the conventional boundaries of the literary canon.

We are focusing on the literary use of place-names as an important dimension of a text's engagement with place within and across a dataset comprising thousands of extracts from Edinburgh texts, including genres such as novels, short stories, memoirs and travel narratives and dating between the sixteenth century and the twenty-first century. *Palimpsest* will offer a richly multifaceted way to explore the literary cityscape of Edinburgh. The interface will display place-name usage in these works on contemporary and historic maps, along with other graphic representations of the frequency and density of such usage, either across space or time ranges. This will also permit the visualisation of the collocations of place-name mentions, and we hope also to include some basic element of sentiment analysis allowing users to view the kinds of associations that place-names have gathered around them. Furthermore, it will be possible for users to set either restrictive or expansive criteria for the exploration of our dataset through this interface. Some of these criteria will draw on the metadata associated with the digitised texts we are using, so it will be possible to view the data alongside such aspects as author, genre and date of publication. Other criteria, though, will draw more fundamentally on grammatical properties and affective qualities extracted through text-mining, permitting users to explore the range of collocations associated with a selected set of Edinburgh toponyms. Combining both these approaches will give users a greater number of entry points, and passages through, the data.

In order to facilitate this, we are working with a large corpus comprising thousands of digitised 'Edinburgh-centric' literary works, which has been extracted from the hundreds of thousands of OCR'd and out-of-copyright literary texts in the collections of the Hathi Trust, Project Gutenberg, the

British Library, the Oxford Text Archive and the National Library of Scotland, alongside a limited set of copyright texts which have been made available to us with the kind permission of authors and publishers. The initial stage involves determining which texts in the collections involve significant Edinburgh settings and which are of some kind of literary interest – considering the defining parameters of both these factors required a perhaps surprising amount of discussion. The minimum requirement for the former was the use of the word Edinburgh (or a synonym) and at least one place-name; and, for the latter, that the text belong to a recognisably literary genre, such as the novel or short story, or otherwise have strong narrative or locodescriptive components – both memoirs and travel journals, for example, might well merit inclusion on these terms. A tool that assists with this first stage is the Edinburgh Geoparser, which extracts location names and geo-references them, and which has been customised into a loco-specific gazetteer of 1,200 place-name queries through the aggregation and weighting of OpenStreetMap, OSLocator, RCAHMS and Historic Scotland data in order to deal with geo-referencing fine-grained place-names in literary texts, including streets, buildings, or particular sites in the urban environment (such as the Scott Monument), local areas, or more general archaic, poetic, vernacular and dialect terms (such as Edenborough, Embra, Edina, 'Auld Reekie'). Literary scholars then examine the texts that have been identified in order to check their relevance for inclusion, and also to feed back information that helps to refine the Edinburgh Geoparser. We further aim to identify continuities and divergences in frequencies of place-name usages and associations, through a combination of the strengths of machine and close reading, and thus to produce a detailed map of the development of Edinburgh as a city of the literary imagination.

The principal output of the project will be the interactive website that will allow users to curate their own maps of geolocated Edinburgh literature, and to explore locations in the city through geolocated extracts from those works, with a mobile-friendly version useable on handheld devices. Other outputs will include the new version of the Edinburgh Geoparser enhanced with fine-grained location geo-referencing including the Edinburgh place-name gazetteer, with text-mining components that extract literary properties around locations, the text-mined XML output for the *Palimpsest* corpus and the XML schema. The web resources are to be made freely available to all users; the software will also be released as an open access resource.

Taxonomies of Toponyms

Place-names are ubiquitous in fiction and non-fiction narratives, of course, but the choice of the toponym as our primary marker of geocritical interest might seem surprising because the use of a toponym is often passed by unmentioned in critical literary analyses in a way the referencing of a real character is not. However, as Nicholas Horsfall has suggested, 'names are signposts in a child's imagination and thereafter in an adult's memory':

> my mental map of childhood London is not a visible map at all, but remains an elaborate tissue of homes, aunts, toyshops, dentists, friends' homes, favourite walks, cinemas. These private gazetteers of ours are perforce idiosyncratic and disorderly; they interrupt unpredictably and inconveniently those slightly more disciplined maps-in-the-mind which we acquire at school, and, if we are very lucky, from parents who read or, better still, recite to us the poetry they love. For that poetry (and all I say is just as true of song) will prove to contain *names*: 'silent on a peak in Darien', 'down in Demerara', 'at Flores in the Azores', 'silently rowed to the Charlestown shore', which will, many of them, sink into the mire of the not-explained and not-understood, only to emerge, many years on, gleaming, and weighed down with some relevant information. For names are also an integral part of our social memory. (305)

There is a long, modern tradition of locative writing that seeks to mobilise the powers of the place-name, from the chorographical endeavours of seventeenth-century writers such as Michael Drayton, through the Wordsworthian interest in 'Poems on the Naming of Places' to Seamus Heaney's reflections in 'The Sense of Place' on the way in which there is a tension between place as learned, literate and conscious or as lived, illiterate and unconscious, and now the contemporary topopoetics outlined, for example, by Moslund.[5] In this Heideggerian mould, which echoes to some extent the influential reading of Wordsworth developed by Jonathan Bate, toponyms evoke the understanding of language as naming, as a primordial calling forth:

> Precisely because naming things brings things forth or close, as if standing them up before our senses, the word opens toward a prelinguistic sensation that compares with the sensation of the thing itself. This is not a mimetic copy or a true reflection of the object of the thing-world but a triggering of the sensations of that object. (32)

Yet names can conceal as well as unconceal and the grounding relation to 'earth', that might be revealed in the poetry that most concerns Heidegger and his followers, can also be obscured. As Moslund explains:

> Our awareness of this ever-present and fundamental physical grounding or emplacement is lost on our daily communicative uses of language because of the silent workings of the senses, the taken-for-grantedness

5 Alan Gillis has ably situated the use of the place-name in modern and contemporary Irish and Scottish poetry in the context of this longer tradition.

> of place and the dominance of practical, discursive, or meaning-based uses of language. (34)

Since toponyms function in such practical, discursive and meaning-based uses of language, and such registers are taken up by narrative fiction, the names evoked by writing might participate in the overlaying of any awareness of such 'physical grounding'. Furthermore, although Moslund, for one, is happy to acknowledge the important role that names can play in topopoetic reading, he is far from conceding their primacy on other grounds, too. There is so much more, after all, to place in literature than the given names of localities. Indeed, as in Wordsworth's 'Poems on the Naming of Places', the act of naming can be counterposed to names accreted or inherited. Stories can be 'placed' without any resort to toponyms: whereas diegesis binds locus and narration together, a name is far from being essential to such articulations. Place, in other words, can be nameless.[6]

Outwith the kinds of Heideggerian emplacement envisaged by Moslund, toponyms play a distinctive set of roles that can be elaborated in terms developed from narratology. As Teresa Bridgeman says, narrative has most commonly been understood as fundamentally temporal rather than spatial. However, narratologists have recently redefined narrative in more open terms consistent with our own stress on diegesis, with Monika Fludernik emphasising experientiality, David Herman seeing it both as an object of interpretation and as a means of making sense of experience, and Marie-Laure Ryan thinking in terms of storyworlds and transmediality. Bridgeman describes changing notions of space: in realist nineteenth-century novels it appeared a stable and concrete phenomenon; in modernist works it became filtered through the perceptions of protagonists; and then in postmodernist fiction it became further destabilised, with different spaces multiplying and merging (56). Readers of literary texts require minimal spatio-temporal hooks to orient themselves: emotional engagement and associations are linked to spatial, as well as temporal parameters, with the dimensions of and movements in narrative worlds varying in scope, and with the proximity and distance between people often marked in topographical terms (60, 63). Bridgeman adds to Lakoff and Johnson's notions of the path and container, the portal (whether door or window), as a space through which the character and reader can be transported (55).

Marie-Laure Ryan's narratological categorisation of space evokes a rippling outward and inward of parameters, both into the world of the reader and into the counterfactual world of counterfactual characters. Ryan's

6 In the topographical poetics of Sean Borodale's works *Walking to Paradise* and *Notes for an Atlas*, to offer just one example, no place-names feature in the detailed, phenomenological presencing of the material world, although the texts are set in the Lake District and London respectively.

overlapping categories stretch from the 'spatial frames' and 'the setting', which are the surroundings of events and the socio-historico-geographical environment in a narrative; to 'the story space', which is the space relevant to the plot via the actions and thoughts of the characters; to 'the narrative world', which is the coherent world fleshed out by the reader according to their knowledge and experience; and, in addition, to 'the narrative universe', which encompasses both the actual world of the text and 'all the counterfactual worlds constructed by characters as beliefs, wishes, fears, speculations, hypothetical thinking, dreams, and fantasies'. These categories can be usefully reapplied in thinking through how place-names are operating. They can also be supplemented by reapplying some of the categories that emerge in relation to definitions of character, as for instance in Alan Palmer's *Fictional Minds*, which defines character in terms of grammatical person, literary device, speech position, semes, actant and non-actual individual (37–8). Such taxonomies of toponyms indicate the kinds of interrogations, both within a text and, more significantly and unusually, across large corpora of literary texts, that will become possible through *Palimpsest*.

For instance, adapting Palmer's schema would give us the following, and far from exhaustive, range of toponymic possibilities. Firstly, a grammatical place-name is construed as an entity of discourse. So, for example, in the following description by John Gibson Lockhart, Edinburgh is the subject of the sentence, whereas London is allotted only a subordinate clause and a lesser dignity:

> Edinburgh, even were its population as great as that of London, could never be merely a city. Here there must always be present the idea of the comparative littleness of all human works. Here the proudest of palaces must be content to catch the shadows of mountains; and the grandest of fortresses to appear like the dwellings of pigmies, perched on the very bulwarks of creation. (3–4)

Secondly, a structural place-name, unsurprisingly, plays a role or element in the story structure. In James Hogg's *The Private Memoirs and Confessions of a Justified Sinner*, the Black Bull (now gone and only vaguely geolocatable), alongside the unusually clement weather, provides a vital orientation point for the narrative of a riot and the torrent of motion that follows:

> The concourse of people that were assembled in Edinburgh at that time was prodigious; and as they were all actuated by political motives, they wanted only a ready-blown coal to set the mountain on fire. The evening being fine, and the streets thronged, the cry ran from mouth to mouth through the whole city. More than that, the mob that had of late been gathered to the door of the Black Bull, had, by degrees, dispersed; but, they being young men, and idle vagrants, they had only

> spread themselves over the rest of the street to lounge in search of farther amusement: consequently, a word was sufficient to send them back to their late rendezvous, where they had previously witnessed something they did not much approve of. (Hogg 23)

The movement of people and news flows out and back across the city, and the Black Bull is identified as a key organising locus for such movement – the crowds disperse from there, and then reform around it, precipitating the explosive intervention of a riot.

Thirdly, a place-name can also serve as a means of achieving an aesthetic, even a meta-aesthetic, effect. In Mary Shelley's *Frankenstein*, for example, Edinburgh locations appear in a brief flurry:

> I visited Edinburgh with languid eyes and mind; and yet that city might have interested the most unfortunate being. Clerval did not like it so well as Oxford; for the antiquity of the latter city was pleasing to him. But the beauty and regularity of the new town of Edinburgh, its romantic castle and its environs, the most delightful in the world, Arthur's Seat, St. Bernards Well, and the Pentland Hills, compensated him for the change and filled him with cheerfulness and admiration. (Shelley 162)

Though brief, this passage is a shorthand and spatialised reference to the Enlightenment and Romantic ideals organised by the city and its environs: with its natural beauty and orderly New Town architecture, Edinburgh indirectly signals the extent to which the disenchanted Frankenstein has become lost to all such pleasures, values and ambitions.

Lastly, a place-name may also stand for a thematic element, a semantic complex or macrocosm composed of semes and unified by a proper name. In his *Picturesque Notes*, Robert Louis Stevenson conjures the ways in which the Old Town of Edinburgh became imbued by its historical and literary past, as well as by its natural and architectural setting, with a sinister chill that fired dark tales in the imagination:

> So, in the low dens and high-flying garrets of Edinburgh, people may go back upon dark passages in the town's adventures, and chill their marrow with winter's tales about the fire: tales that are singularly apposite and characteristic, not only of the old life, but of the very constitution of built nature in that part, and singularly well qualified to add horror to horror, when the wind pipes around the tall lands, and hoots adown arched passages, and the far-spread wilderness of city lamps keeps quavering and flaring in the gusts. (Stevenson 14)

Stevenson then goes on to conjure up chilling tales, some now more familiar than others, of Deacon Brodie, Begbie the Porter and Burke and Hare;

however, he does not provide a more precise location within the 'lands' here invoked, the high-rise tenements of Edinburgh's Old Town.

So place-names fulfil a range of crucial diegetic functions. Their significance goes beyond this, though, as their real-world referentiality is asserted or assumed when place-names and the places they denote are taken from what Piatti and Hurni have called 'geospace' (219): that is, the extradiegetic or sometimes metadiegetic realm. Gabriel Zoran comments that 'despite the possibility of distinguishing between the space of the text and that of the world, one cannot point to any constant correlation between them', and yet place-names provide exactly such an explicit point of referential correlation (310). Such correlations, though, are inevitably intertextual as well as referential, mediated through precedent invocations of the same names. Thus, for example, Stevenson's passage down the High Street in his *Picturesque Notes* cannot avoid also referencing Walter Scott's fictive use of this location. 'Here, for example', says Stevenson, 'is the shape of a heart let into the causeway. This was the site of the Tolbooth, the Heart of Midlothian, a place old in story and namefather to a noble book' (10). The primarily Scots term 'namefather' here marks the intertextual processes through which names are bestowed and inherited. They accumulate down the generations: in the opening pages of her memoir, *Curriculum Vitae*, Muriel Spark declares how '[d]etails fascinate me. I love to pile up the details. They create an atmosphere. Names, too, have a magic, be they never so humble' (11). Spark describes a feeling of affinity with Stevenson arising from a sense of shared experience that comes from their both having lived in Edinburgh, whose place-names she recites in the incantatory mode evoked by Horsfall: 'The Braid Hills, the Blackford Hill and Pond, the Pentland Hills of Stevenson's poems, his "hills of home" were mine too' (*Curriculum Vitae* 35).

Diegetic Deformance

To focus on the place-name, then, is to attempt to isolate a distinctive set of forms that the creative investment in place can take. Such a move opens up the possibility of making the kinds of intervention best described by Jerome McGann and Lisa Samuels as acts of critical 'deformance': 'invasions or distortions of the documentary foundation of the artifact' (115). Deformance of this sort sets itself against the self-denying ordinance that McGann and Samuels associate with critical interpretation in its customary mode, which 'brackets off from attention crucial features of imaginative works, features wherein the elemental forms of meaning are built and elaborated', features 'that readers tend to treat . . . as preinterpretive and precritical' (115). To treat the isolation of place-name occurrences from other textual elements in a narrative as a way of activating critical possibilities is to invite a new and distinctive form of critical attention to place in literature. Within a particular text, it brings attention to bear on the kind of toponyms deployed – what kind of place is thus named – as well as on their functions. It renders visible what might otherwise remain unnoticed, giving an at least preliminary critical pertinence to measures such as sequence, frequency and density, which might well be thought, from most traditional standpoints, to be preinterpretive or precritical. Yet it is possible to envisage such considerations as offering a new way of mapping the diegetic cityscape of Edinburgh as it is forged both in individual texts and across the wider canon of narrative writing that takes the city for its setting. The place-name punctures the diegesis of a particular work, providing a nodal point at which any such diegesis is joined to any other in which the same place-name occurs. From this position outside the fictional city of any one text, considerations of sequence, frequency, or density embody a critical potentiality in a different mode: mapping the shape of the diegetic cityscape assumes a comparative, cumulative and historical dimension.

The Edinburgh that might be evoked through this intertextual and referential toponymic play undoubtedly belongs amongst those sites classed by Eric Prieto as 'the *hauts lieux* of the literary tradition: places that have a distinct cultural and topographical profile and that have given rise to a whole body of literature' ('Geocriticism, Geopoetics, Geophilosophy' 22). In fact, this profile is very well established and often rehearsed, yet *Palimpsest* will allow for more systematic examinations of these features of the literary history of the city, and the tracing of their emergence and changing associations over time. As a recent and characteristic literary-critical portrayal puts it:

> The poet Hugh MacDiarmid referred to Edinburgh as a 'mad god's dream'. It exemplified antisyzygy, his preferred creative term, meaning a 'zigzag of contradictions' . . . The Edinburgh I walk through each day is part Piranesi, part Peter Greenaway. I can't tire of its soaring bridges that never cross water, its Tetris blocks of Gothic tenements framed in classical Palladian arches, its tug-of-war between secret vennels and stately locked doors.
>
> Edinburgh's centre is riven, bifurcated: on one hand, the vertiginous, overlapping, haphazard, medieval Old Town, and on the other, the geometric, unfolded, planned, neoclassical New Town. (Kelly 18)

There is a sense, then, in which the topography of literary Edinburgh is both literal and metaphorical, heavily freighted or overdetermined in a manner that both shapes and constrains the efforts of those contemporary writers who use its names to define the space of their fictions. Ian Rankin has jointly credited both Stevenson and the instantiation in Edinburgh of the 'Caledonian antisyzygy' with inspiring his own tartan noir:

> I owe a great debt to Robert Louis Stevenson and to the city of his birth. In a way they both changed my life. Without Edinburgh's split nature Stevenson might never have dreamed up *Dr Jekyll and Mr Hyde*

> and without *Dr Jekyll and Mr Hyde* I might never have come up with my own alter ego Detective Inspector John Rebus.

By contrast, Irvine Welsh has handled the familiar diegetic cityscape with more than a degree of disenchanted scepticism, memorably voiced by Begbie in *Trainspotting*:

> These burds ur gaun oantay us aboot how fuckin beautiful Edinburgh is, and how lovely the fuckin castle is oan the hill ower the gairdins n aw that shite. That's aw they tourist cunts ken though, the castle n Princes Street, n the High Street. Like whin Monny's auntie came ower fae that wee village oan that Island oaf the west coast ay Ireland, wi aw her bairns. The wifey goes up tae the council fir a hoose. The council sais tae her, whair's it ye want tae fuckin stey, like? The woman sais, ah want a hoose in Princes Street lookin oantay the castle. . . . Perr cunt jist liked the look ay the street whin she came oaf the train, thoat the whole fuckin place wis like that. The cunts in the council jist laugh n stick the cunt n one ay they hoatline joabs in West Granton, thit nae cunt else wants. Instead ay a view ay the castle, she's goat a view ay the gasworks. That's how it fuckin works in real life, if ye urnae a rich cunt wi a big fuckin hoose n plenty poppy. (Welsh 115–6)

Books not only replicate hegemonic structures but also hold up the means to question them. Here, notably, the place-name 'West Granton' plays contrapuntally against Princes Street, the High Street and Edinburgh Castle, stretching and decentring the established topography. Yet, later in the same text, it only takes a trip away for a less belligerent insider to give voice to a partial re-enchantment with the city and its customary *haut lieu*. As Welsh's protagonist Renton puts it on his return: 'But when ye come back oot ay Waverley Station eftir bein away fir a bit, ye think: Hi, this isnae bad' (228).

Other writers have found differing ways to reflect on this sense of an abundantly and densely significant landscape. In her brief autobiographical note, 'Edinburgh-born', Spark describes her stay in the city of her birth in the spring of 1962, when she was attendant on her dying father. Resident, as she specifies, in the North British Hotel (now, in another mode of significance, the Balmoral) alongside Waverley Station, she surveys exactly the standard – and expansively named and framed – panorama to which Welsh alludes:

> From where I sat propped in the open window frame, I could look straight onto Arthur's Seat and the Salisbury Crags, its girdle. When I sat the other way round I could see part of the Old City, the east corner of Princes Street Gardens, and the black Castle Rock. In those days I experienced an inpouring of love for the place of my birth, which

> I am aware was psychologically connected with my love for my father and with the exiled sensation of occupying a hotel room which was really meant for strangers. (Spark, 'Edinburgh-born')

Later in the piece the topography of this landscape is fused with Edinburgh habits of mind and tongue, which Spark suggests are a crucial and distinctive part of the education furnished to the city's children. She focuses on 'nevertheless', 'this word of final justification', which comes to stand for an 'Edinburgh ethos' that has had a formative influence but that she can also hold at sufficient distance to make it the subject of her own enquiry. Such habits become, as the piece moves towards its conclusion, features of the landscape itself:

> The Castle Rock is something, rising up as it does from pre-history between the formal grace of the New Town and the noble network of the Old. To have a great primitive black crag rising up in the middle of populated streets of commerce, stately squares and winding closes, is like the statement of an unmitigated fact preceded by 'nevertheless'. (Spark, 'Edinburgh-born')

And the landscape's parental associations, its formative powers over the native, are made explicit in the final lines:

> When the shrill telephone in my hotel room woke me at four in the morning, and a nurse told me that my father was dead, I noticed, with the particular concentration of the fuddled mind, that the rock and its castle loomed as usual in the early light. I noticed this, as if one might have expected otherwise. (Spark, 'Edinburgh-born')

For all its evocation of an inextricable symbolism, however, this piece also manages to play with and through just such an assumption. The rock still stands even after the father has died, and in the surprise of this perception the assumption of symbolism is undone by the brute facticity of its own 'nevertheless'. Yet this is to reinforce exactly the meaning that has been drawn from the rock hitherto: its facticity was not quite as brute as had been suggested. In standing for 'unmitigated fact', that is to say, it couldn't simply be such a fact itself.

Elsewhere, Spark finds other ways to explore the scope and limits of Edinburgh's 'distinct cultural and topographical profile'. In a celebrated passage from *The Prime of Miss Jean Brodie*, Sandy and her schoolfellows are taken on a walk through the Old Town, where 'The Canongate, The Grassmarket, The Lawnmarket, were names which betokened a misty region of crime and desperation: "Lawnmarket man jailed"' (Spark, *The Prime* 32). This initial statement is followed by a passage of rich description of the scene in the Grassmarket, before one of the abrupt shifts of narratorial perspective for which the novel is rightly famous introduces a disruption of what had appeared to be secure associations:

> And many times throughout her life Sandy knew with a shock, when speaking to people whose childhood had been in Edinburgh, that there were other Edinburghs quite different from hers, and with which she held only the names of districts and streets and monuments in common. (Spark, *The Prime* 33)

The very names that had 'betokened . . . crime and desperation' only a few paragraphs earlier here thin to empty tokens: their intersubjective nature does little to undo Sandy's adult perception of fundamental human separateness. This points to a narratorial play with names, with the fact that they are shared and recognised, but also that such sharing can sometimes conceal profound differences of apprehension and understanding; or, from the other way round, that singular experiences are nonetheless brought together – not always comfortably – in the shared use of a name. Here, as elsewhere, we see how important the place-names of Edinburgh are to those writers wrestling with the city's densely significant topography, and how their reflective attention to such names generates critical, or at least metafictional, engagements with the referential and representative powers and functions of such names.

Topopoetic Prospects

Nevertheless, with a place as richly resonant as Edinburgh, our deformative emphasis on the place-name as a way of opening up topopoetic prospects might still be thought to face insuperable difficulties. As Robert J. Tally Jr. has argued of Westphal's ambitions for his geocritical method:

> it invariably raises the question of the corpus. How does one determine exactly which texts could, in the aggregate, reasonably constitute a meaningful body of material with which to analyze the literary representations of a given geographical site? That is, if the Dublin of James Joyce is far too limited, since it relies on the perspective of a *(sic)* only single author or a few of his own writings, then how many authors and texts representing Dublin would constitute a feasible and credible starting point for a geocritical study of the Irish capital? With certain cities, such as Paris, London, Rome, or New York, the almost mythic status of these places and the seemingly innumerable textual references to them render any geocritical analysis, at least those laying claim to a kind of scientific value, impossible.

As Westphal admits, 'to attempt to undertake a full-scale geocritical analysis of those hotspots would be madness'. A geo-centered method, if it aims truly to avoid the perception of bias, seems somewhat doomed from the start. ('Foreword' xii)

There are several issues here, not least the relation between 'scientific value' and the 'feasible and credible'. Of most relevance, however, is the claim of doom or impossibility, the madness in attempting to undertake 'full-scale geocritical analysis of . . . hotspots': the suggestion is that the sheer profusion, variety and mobility of textual engagements with certain places will disable any attempt at worthwhile geocritical study. Yet one might wonder at the feasibility and credibility of a geocriticism that had to shy away from, rather than engage with, such profusion. For this is an ineliminable aspect of the literary experience of the modern city, especially – in his *Picturesque Notes*, for example, Stevenson ascends Calton Hill to obtain a luminous prospect over the city that is uncomfortably faithful to just such profusion, variety and mobility. 'It is the character of such a prospect', he says, 'to be full of change and things moving. The multiplicity embarrasses the eye; and the mind, among so much, suffers itself to grow absorbed with single points' (Stevenson 31).

The problem, as Tally and Westphal see it, is simply that there is too much for any critic, and thus any criticism, to handle. The eye remains embarrassed: analysis – in the sense of the discernment of patterns, processes, similarities and contrasts, the creation of maps – will not be able to take place. Perhaps, though, such pessimism has not taken account of the ways that an encounter with such plenitude might be staged and how criticism itself might develop in the process. In bringing text-mining and georeferencing technologies to bear on exceptionally large collections of digitised texts in the pursuit of its own deformative intervention, the *Palimpsest* project seeks to demonstrate that there are indeed ways of engaging feasibly, and fruitfully, with such profusion. This is the possibility that critics such as N. Katherine Hayles, Franco Moretti and Stephen Ramsay have been claiming for modes of digital textual analysis, whether we think of these as 'hyper reading', 'distant reading', or 'algorithmic criticism'. In this connection, it is worth noting that Ramsay has himself drawn on McGann's and Samuel's notion of deformance to argue for the continuity of computational and critical analysis against a prevalent view that would oppose them, and that McGann and Samuel were themselves seeking to register some of the critical possibilities opened up by the use of digital forms of enquiry (32–57). Yet this has other implications, especially for our sense of what might be the appropriate form for such criticism. Our multiversal map of literary Edinburgh will not just reproduce a single, and already familiar profile – there will be no simple, flat surface on which all our data can be arrayed at once, and the whole apprehended in a single, monocular vision. In allowing the apprehension of many patterns of toponym usage, within and across an unprecedented range of canonical and non-canonical sources, the critical possibility of our visualisations will not be exhausted in any one query or image. On the one hand, this will make our dataset and interface a resource for critics who wish to treat it precritically, as it were – as a kind of index, concordance or gazetteer. Yet, beyond that, the project seeks to explore the possibility that just this spatialised, mappable, diegetic cityscape, this mobile, retraceable, generative surface, might constitute a multiform and open act of critical fidelity to a place – in Stevenson's words – old in story.

Works Cited

Bachelard, Gaston. *The Poetics of Space.* Boston: Beacon Press, 1969.

Bate, Jonathan. *Romantic Ecology: Wordsworth and the Environmental Tradition.* London: Routledge, 1991.

Borodale, Sean. *Walking to Paradise.* Grasmere: The Wordsworth Trust, 1999.

———. *Notes for an Atlas.* Dartington: Isinglass, 2002.

Bridgeman, Teresa. 'Time and Space'. *The Cambridge Companion to Narrative.* Ed. David Herman. Cambridge: Cambridge University Press, 2007/2010. 52–65.

Campbell, Donald. *Edinburgh: A Cultural and Literary History.* Oxford: Signal Books, 2003.

Casey, Edward S. *Getting Back into Place: Toward a Renewed Understanding of the Place-World.* Second Edition. Bloomington: Indiana University Press, 2009.

Certeau, Michel de. *The Practice of Everyday Life.* Trans. Stephen Rendall. Berkeley: University of California Press, 1988.

Cresswell, Tim. *Place: A Short Introduction.* Oxford: Blackwell, 2004.

Edinburgh City of Literature. 2015. Web. 20 March 2015. <http://www.cityofliterature.com/>.

Entrikin, Nicholas. *The Betweenness of Place: Towards a Geography of Modernity.* Houndmills: Macmillan, 1991.

Fishkin, Shelley Fisher. '"Deep Maps": A Brief for Digital Palimpsest Mapping Projects (DPMPs, or "Deep Maps")'. *Journal of Transnational American Studies* 3.2 (2011): 1–33.

Fludernik, Monika. *Towards a 'Natural' Narratology.* London: Routledge, 1996.

Foster, Allan. *The Literary Traveller in Edinburgh: A Book Lover's Guide to the World's First City of Literature.* Edinburgh: Mainstream Publishing, 2005.

Gillis, Alan. 'Names for Nameless Things: the Poetics of Place Names'. *Modern Irish and Scottish Poetry.* Eds, Peter Mackay, Edna Longley and Fran Brearton. Cambridge: Cambridge University Press, 2011. 204–21.

Hayles, N. Katherine. *How We Think: Digital Media and Contemporary Technogenesis.* Chicago: Chicago University Press, 2012.

Heaney, Seamus. 'The Sense of Place'. *Preoccupations: Selected Prose, 1968–1978.* London: Faber & Faber, 1980. 131–49.

Herman, David. *Storytelling and the Sciences of the Mind.* Cambridge, MA: MIT Press, 2013.

Hogg, James. *The Private Memoirs and Confessions of a Justified Sinner.* 1824. Ed. Ian Duncan. Oxford: Oxford University Press, 2010.

Horsfall, Nicholas. 'The Poetics of Toponymy'. *Literary Imagination: The Review of the Association of Literary Scholars and Critics* 4.3 (2002): 305–17.

Kelly, Stuart. *Scott-land: The Man Who Invented a Nation.* Edinburgh: Polygon, 2010.

Lakoff, George and Mark Johnson. *Metaphors We Live By*. Chicago: University of Chicago Press, 1980.

Lefebvre, Henri. *The Production of Space*. Trans. D. Nicholson-Smith. Oxford: Blackwell, 1991.

A Literary Atlas of Europe. 2015. Web. 20 March 2015. <http://www.literaturatlas.eu/en/>.

Lockhart, John Gibson. *Peter's Letters to his Kinsfolk*. New York: Goodrich, 1819/1820.

Lownie, Andrew. *The Edinburgh Literary Companion*. Edinburgh: Birlinn, 2005.

Mapping the Lakes: A Literary GIS. N.d. Web. 20 March 2015. <http://www.lancaster.ac.uk/mappingthelakes/>.

McGann, Jerome, with Lisa Samuels. 'Deformance and Interpretation'. *Radiant Textuality*. By Jerome McGann. New York: Palgrave Macmillan, 2001. 105–35.

Modernist Versions Project. N.d. Web. 20 March 2015. <http://web.uvic.ca/~mvp1922/>.

Moretti, Franco. *Atlas of the European Novel, 1800–1900*. London: Verso, 1998.

———. *Distant Reading*. London: Verso, 2013.

Moslund, Sten Pultz. 'The Presencing of Place in Literature: Toward an Embodied Topopoetic Mode of Reading'. *Geocritical Explorations: Space, Place and Mapping in Literary and Cultural Studies*. Ed. Robert T. Tally Jr. New York: Palgrave Macmillan, 2011. 29–43.

Our Town Stories: Edinburgh. 2015. Web. 20 March 2015. <http://www.ourtownstories.co.uk/#overview/>.

Palimpsest: Literary Edinburgh. N.d. Web. 20 March 2015. <http://www.ed.ac.uk/literatures-languages-cultures/english-literature/research-activities/palimpsest>.

Palmer, Alan. *Fictional Minds*. Lincoln: University of Nebraska Press, 2004.

Piatti, Barbara and Lorenz Hurni. 'Editorial: Cartographies of Fictional Worlds'. *The Cartographic Journal* 48.4 (2011): 218–23.

Prieto, Eric. *Literature, Geography, and the Postmodern Poetics of Place*. New York: Palgrave Macmillan, 2011.

———. 'Geocriticism, Geopoetics, Geophilosophy and Beyond'. *Geocritical Explorations: Space, Place and Mapping in Literary and Cultural Studies*. Ed. Robert T. Tally Jr. New York: Palgrave Macmillan, 2011. 13–27.

Ramsay, Stephen. *Reading Machines: Toward an Algorithmic Criticism*. Urbana: University of Illinois Press, 2011.

Rankin, Ian. 'Creating Rebus: Edinburgh's Favourite Detective, Edinburgh – the Jekyll and Hyde city'. Web. 20 March 2015. <http://www.ianrankin.net/creating-rebus/>.

Ryan, Marie-Laure. 'Space'. *The Living Handbook of Narratology*. Eds, Peter Hühn et al. Hamburg: Hamburg University. Web. 20 March 2015. <http://www.lhn.uni-hamburg.de/article/space>.

Shelley, Mary. *Frankenstein: or, The Modern Prometheus*. 1818. Ed. M. K. Joseph. Oxford: Oxford University Press, 1969.

Spark, Muriel. *Curriculum Vitae: A Volume of Autobiography*. Manchester: Carcanet, 1992.

———. 'Edinburgh-born'. *The New Statesman* 64 (1962), 180.

———. *The Prime of Miss Jean Brodie*. London: Penguin, 1965.

Stevenson, Robert Louis. *Edinburgh: Picturesque Notes*. 1878. London: Seeley, Jackson and Halliday, 1879.

Tally, Robert T. 'Foreword: A Geocriticism of the Worldly World'. *The Plausible World: a Geocritical Approach to Space, Place and Maps.* By Bertrand Westphal. Trans. Amy Wells. New York: Palgrave Macmillan, 2013. xi-xvi.

———, (ed.). *Geocritical Explorations: Space, Place and Mapping in Literary and Cultural Studies.* New York: Palgrave Macmillan, 2011.

Tuan, Yi-Fu. *Space and Place: The Perspective of Experience.* Minneapolis: University of Minnesota Press, 1977.

Welsh, Irvine. *Trainspotting.* 1993. London: Vintage, 2011.

Westphal, Bertrand. *Geocriticism: Real and Fictional Spaces.* Trans. Robert T. Tally. New York: Palgrave Macmillan, 2011.

———. *The Plausible World: a Geocritical Approach to Space, Place and Maps.* Trans. Amy Wells. New York: Palgrave Macmillan, 2013.

Wilson, John. *Noctes Ambrosianae.* Vol. 1. Edinburgh: William Blackwood, 1864.

Zoran, Gabriel. 'Towards a Theory of Space in Narrative'. *Poetics Today* 5.2 (1984): 309–35.

3 Geographical Text Analysis

Digital Cartographies of Lake District Literature

Ian Gregory and Christopher Donaldson

Introduction

This chapter introduces an interdisciplinary approach to the digital geographical analysis of digital literary corpora. It does so by presenting a case study undertaken as part of Lancaster University's *Spatial Humanities: Texts, GIS, and Places* project.[1] Combining corpus-based approaches, automated geo-parsing techniques and geographic information systems (hereafter GIS) technology, this study investigates literary responses to the landscape of the English Lake District (Figure 3.1). The focus of this investigation is a custom-built, 1,500,000-word georeferenced corpus of Lake District literature. This corpus consists of eighty digitised texts, ranging in date from 1622 to 1900. A historically representative sample of writing about the Lakes region, the corpus comprises a variety of canonical and non-canonical works, including the guidebooks of Lakeland luminaries, such as those of William Wordsworth and Harriet Martineau, as well as more ephemeral publications, such as *Black's Shilling Guide to the English Lakes*. In engaging with this resource, our aim is to exemplify how a hybrid corpus- and geographic-based methodology – which we label geographic text analysis – can be used in conjunction with more traditional forms of close reading and contextual analysis to understand how literary landscapes, such as the Lake District, were perceived and represented in the past.

GIS and the Geographies of Digital Literary Corpora

GIS and 'Macroanalysis'

GIS are a foundational form of digital geospatial technology that has been integral to innovation in the social sciences since the 1970s. In the humanities the application of GIS is a much more recent phenomenon, and is, in part, a result of the rapid growth of humanities computing and

1 The research leading to these results has received funding from the European Research Council (ERC) under the European Union's Seventh Framework Programme (FP7/2007–2013) / ERC grant 'Spatial Humanities: Texts, GIS, Places' (agreement no. 283850).

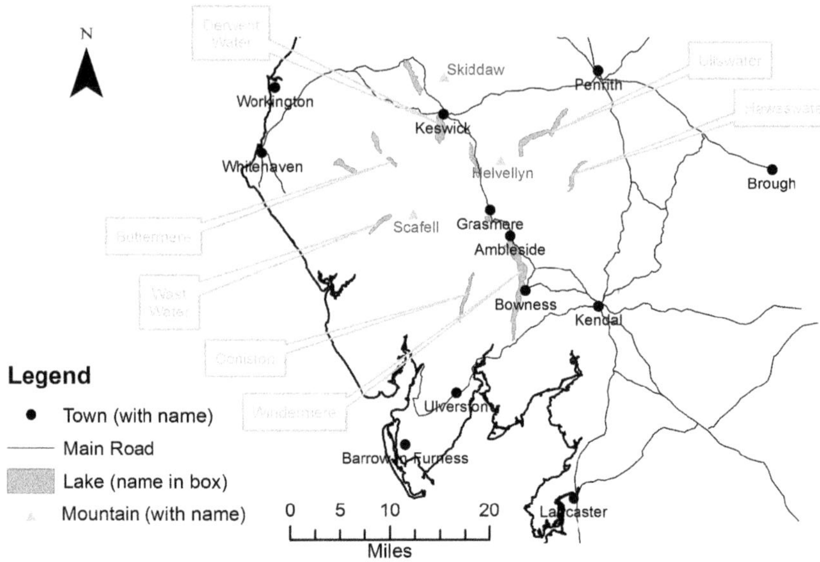

Figure 3.1 The major geographical features of the Lakeland region.
Source: The Authors; created using ArcGIS © 2015.

the proliferation of the digital humanities over the past twenty years. More broadly, the emergence of humanities GIS is also a consequence of the widespread adoption of geographical perspectives, approaches and techniques (or the 'spatial turn') across the arts and sciences (see Tally). Within literary studies, one of the more recent disciplines to take such a turn, the adoption of geographical principles and practices has chiefly been driven by pioneering research projects, such as ETH Zurich's *A Literary Atlas of Europe*, Trinity College Dublin's *Digital Literary Atlas of Ireland* and the University of Edinburgh's *Palimpsest: Literary Edinburgh* – to name only three. Taking inspiration from the ground-breaking work of Franco Moretti, Matthew L. Jockers and the Stanford Literary Lab, each of these projects has endeavoured to show how geospatial technologies can transform the way we engage with the geographical and spatial dimensions of individual literary works as well as those of large literary corpora.

Underpinning these endeavours is a conviction in the value not only of geographical thinking, but also – more specifically – of maps as 'analytical tools' for displaying information derived from literary works in ways that 'bring to light relations that would otherwise remain hidden' (Moretti, *Atlas* 3). Maps, in so many words, are valuable because they can serve the literary scholar both as instruments for generating abstract representations of particular aspects of specific works and, furthermore, as a means of compiling information from multiple works and of collectively assessing and comparing them. This latter process of aggregate analysis, which has variously been called 'distant reading' (Moretti, 'Conjectures') and 'macroanalysis' (Jockers), represents a major advance for the discipline of literary studies, since, as Jockers explains, it enables literary scholars to make substantial use of the 'massive digital-text collections' now available to them and, in the process, to launch the discipline fully into the digital age:

> Today, in the age of digital libraries and large-scale book-digitization projects, the nature of the evidence available to us has changed, radically. Which is not to say that we should no longer read [individual] books . . . but rather to emphasize that massive digital corpora offer us unprecedented access to the literary record and invite, even demand, a new type of evidence gathering and meaning making. (Jockers 7–8)

Put simply, instead of engaging with only a handful of outstanding or exemplary works, literary scholars should create new knowledge about those works by studying them in relation to the larger corpora – whether construed historically (in terms of period), formally (in terms of genre, type or mode), tropologically (in terms of specific themes and motifs) or otherwise – into which they can be assembled. Geographical text analysis, the methodology introduced in the following pages, aims to facilitate just this sort of approach by augmenting the traditional methods of textual analysis employed in literary studies with techniques from geography and corpus linguistics (see Gregory et al.).

Geoparsing

The first step in performing geographical text analysis is geoparsing. Geoparsing involves identifying and extracting place-names from the corpus under analysis and assigning each place-name to a coordinate-based location. Geoparsing can, of course, be performed manually, but when working with a large corpus there are obvious incentives for automating the procedure. In this case study we have used a customised version of the Edinburgh Geoparser: an open-source, automated georeferencing tool that consists of two interlinked components. The first of these components is a 'geo-tagger', which uses Named Entity Recognition (NER) technology to identify and extract place-names (including named settlements, landmarks and geological formations). The second component is a 'geo-resolver', which allocates coordinate data to the extracted place-names using digital gazetteers (Grover et al). Once the corpus has been geoparsed in this fashion, the georeferenced place-names it contains can be extracted, along with their relevant co-text (the text to the left and right of the place-name), and imported into a GIS application where it can be displayed and analysed.

It should, of course, be emphasised that automated geoparsing is not an error-proof process. Place-names are surprisingly complex for software to process automatically. Errors can occur for a variety of reasons. In some cases the software may simply fail to recognise that a word really is a place-name and this will result in it being omitted. There is also the potential for errors of inclusion stemming from the difficulty that even state-of-the-art software has in disambiguating between place-names, personal names and toponymic titles (such as the Bishop of Carlisle or the Duke of Devonshire). There is, moreover, the additional difficulty of disambiguating between places with the same name. As a result, when automatically geoparsing a historical text corpus, it is important to remember that the raw output produced will almost invariably contain oversights and inaccuracies that the researcher will need either to account for or, ideally, to identify and correct. In order to avoid these problems, which stem from an over-reliance on technology, we have developed an iterative method of implementing, reviewing and correcting the results from the geoparsing process (see Rupp et al.).

The Geoparsed Lake District Corpus: Initial Visualisations and Observations

In total, there are almost 40,000 instances of mappable place-names in the corrected Edinburgh Geoparser output for the Lake District corpus, which means that place-names account for roughly 2.6 per cent of all the tokens in the corpus.[2] Of these nearly 40,000 place-names, 96 per cent refer to locations in the UK, 88 per cent of which are in northwestern England or southwestern Scotland: in other words, within and around the greater Lakeland region. Notably, only some 60 per cent of these locations are within the boundaries of the modern Lake District National Park: a finding which reflects the fact that early literary accounts of the Lakes – such as John Dalton's influential *Descriptive Poem, Addressed to Two Ladies after their Return from Viewing the Mines at Whitehaven* (1755) – are often as concerned with the periphery of the region as its centre. This finding, furthermore, goes some way towards confirming the contention, expressed elsewhere, that the Lake District, though ostensibly bounded, is geographically extensive (see Cooper; Nicholson).

Once a corpus has been geoparsed and the geoparser output has been corrected and assembled in a GIS layer, one can begin performing a geographical text analysis by assessing the spatial dimensions of the geography that the corpus contains. The most elementary way to do this is by creating

2 In corpus linguistics each occurrence of a particular word within a corpus is called an *instance*. The units that comprise a corpus are, moreover, called *tokens* rather than words. This is because a token can be a word, a numeral or a punctuation mark.

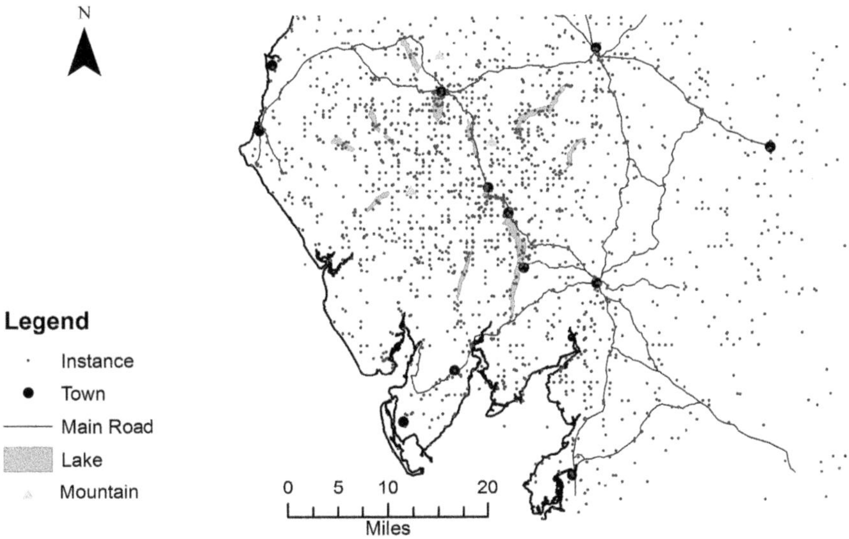

Figure 3.2 A dot-map of place-name instances taken from the Lake District corpus.
Source: The Authors; created using ArcGIS © 2015.

a dot-map in which the referenced place-names are converted into point-data and displayed as dots on the map-interface of the GIS. As shown in Figure 3.2, dot maps are a simple way of representing georeferenced data; however, they are of limited value for interpreting a large corpus of qualitative sources such as our collection of Lake District writing. This is in large part because they represent each place marked on the map in the same way and, as a result, tend to 'flatten-out' datasets instead of highlighting the variations they contain. Take a quality like the frequency of references to a specific place, for example. Dot-maps are inadequate as a means of representing frequency because, when displaying point data, GIS applications superimpose multiple place-marks in the same location (see Fotheringham et al.). Consequently, a dataset may contain dozens of references to a particular place and only one reference to another, but, on a dot-map, these places will appear in exactly the same way. An additional problem with dot-maps – and one of particular relevance for this study – is that they tend to imply an accuracy of location that may be misleading if the geographical entity in question extends over a significant extent of space. Think, for instance, of a mountain, lake or estuary.

In order to overcome these limitations it is useful to employ an analytical technique such as *density smoothing*, which is a common method for simplifying and displaying point patterns. Performing a density smoothed analysis involves calculating the number of points that occur near to each location on the study area (see Lloyd). The results can be presented using a 'heatmap' in which areas of the higher density are shown with darker shading. For the present purposes, it suffices to say that the application of this technique results in maps that are both easier to evaluate and that do not misleadingly imply that each place-name in the corpus corresponds to a single, precise location.

To this end consider Figure 3.3, which displays the distribution and density of the places that are referenced in the corpus and which are located in and around the greater Lakeland region. Studying these maps enables us to draw a number of initial observations. The most significant of which being that, whereas the dot-map (Figure 3.2) suggests that these places are spread around the Lake District, the darker shading on the density map (Figure 3.3) indicates a different underlying pattern. Instead of being more or less evenly distributed, the geography of the corpus is shown here to be marked by areas of greater and lesser density, with clusters of references forming in specific localities. These include the areas near Skiddaw and Keswick, as well as areas south and west of this including Borrowdale, Buttermere and, to a lesser extent, Scafell. Moving eastwards, one notices other significant clusters stretching from Thirlmere and Helvellyn, and Ullswater.

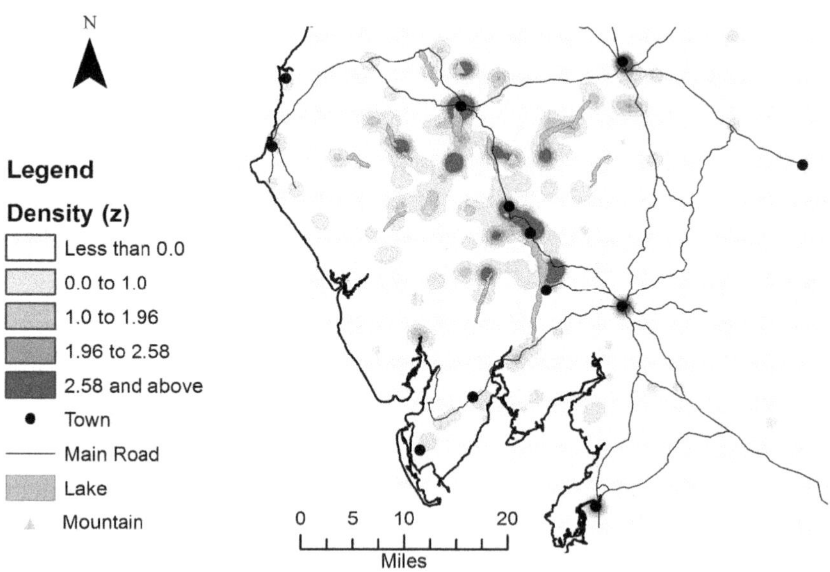

Figure 3.3 A density-smoothed distribution of place-names in the Lake District corpus.[3]
Source: The Authors; created using ArcGIS © 2015.

3 A z-score of 0 is the mean density of the dataset, 1.0 is one standard deviation above the mean, and 1.96 and 2.58 are expected to represent 5% and 1% of the values (two-tailed) in the dataset.

Further south, there are the clusters around Grasmere, Ambleside, Langdale, Windermere and Coniston. There are also clusters centred on the larger settlements to the east of the Lake District including Lancaster, Kendal and Penrith (see Figure 3.1 for orientation).

That these localities are the ones most frequently mentioned in the corpus stands to reason. Each, after all, figures prominently enough in the literary and cultural history of the region to remain integral to our conception of Lake District heritage today. Yet, one must be mindful that density maps still need to be interpreted with care. Specifically, one must bear in mind that the density map depicts a generalisation of the pattern displayed in the dot-map, and that, in certain cases, difficulties of disambiguation complicate our ability to distinguish whether a cluster is centred precisely on the spot it should be. For instance, with locations such as Coniston and Windermere, a village and a town which share their names with nearby lakes, one must be mindful that it is difficult, if not impossible, for an automated process, to determine which is being discussed. Notwithstanding such complications, density smoothing *does* allow us to identify that – collectively – the texts in the corpus contain a disproportionate number of references to places within specific parts of the Lake District. Concomitantly, it also allows us to perceive the existence of geographies of absence: areas which are either mentioned infrequently, such as the ones around Haweswater and Shap Fell, or which are ignored all together, such as the ones to the north of Bassenthwaite and to the south of Ennerdale. If, as seems reasonable, we are willing to use the frequency of place-name references within the corpus as an index for the amount of interest and attention received by any given area, these preliminary observations mean that even within the centre of the region, the works in the corpus pay the most attention to a handful of key locations and pay far less attention – or altogether neglect – several others.

Comparing Descriptions of Different Locations

Viewing Figure 3.3 gives us a sense of the various places mentioned in our corpus of Lake District writing. It helps us, moreover, to discern the number of times these places are mentioned and thus, by extension, the amount of attention given to each. This, in turn, prompts us to investigate why certain places receive more attention than others. It also encourages us to find out what is being said about them. Examining these sorts of issues is integral to geographical text analysis. In order to do this, however, it is necessary to combine GIS-facilitated spatial analysis with complementary methods from corpus linguistics, such as collocation analysis: a basic approach for identifying words that are frequently paired with specific named entities, such as place-names.

Skiddaw and Scafell

As a way of illustrating this process, let us compare the words that are frequently associated with two locations that are the sites of clusters in Figure 3.3: the area around Skiddaw and the one around Scafell. With peak elevations of 931m and 964m respectively, Skiddaw and Scafell are amongst the highest and most iconic mountains in the Lake District. This similarity aside, however, they are located in markedly different parts of the region and, as a result, have markedly different associations. Whereas the latter rises above the sloping, slate-rich fells in the north of the district, near the popular tourist town of Keswick, the former rises amongst the steep volcanic group in the centre of the region, near the head of the less-accessible and, notably, less-frequented valley of Wasdale. Performing collocation analysis of the place-names *Skiddaw* and *Scafell* enables us to ascertain whether or not – and if so, how – these differences are manifest in the accounts of the Lake District that comprise our corpus. In order to perform this analysis, however, one must first define what one means by Skiddaw and Scafell. This is not as simple a task as it may seem, since there are several named locations around each mountain that have names of their own but are nonetheless part of the same massif. Scafell Pike, the highest summit in England, which is one kilometre northeast from Scafell, but part of the same range, is one excellent example; as is Lingmell, which is an outlying shoulder of the mountain. In brief, to search for Scafell alone would be inadequate, as it would fail to take such contiguous formations into account.

One way of compensating for this would be to perform a collocation analysis not of Skiddaw or Scafell alone, but of all the place-names found within a defined radius of each summit. Arguably, however, this approach is unsatisfactory because the distance of the radius would be entirely arbitrary, and neither mountain range is circular in shape. A preferable approach, and the one adopted here, is to perform density smoothing using a bandwidth determined by the formula presented by Fotheringham et al., and to define clusters using the resulting density smoothed pattern. Performing this analysis reveals a strong cluster of place-name references around Skiddaw and a weaker one around Scafell (shown in Figure 3.3). These clusters can be taken as indicative of the area of interest that corresponds to each mountain. For the Skiddaw cluster this gives us a list of nearly 700 place-names which contain the word Skiddaw, such as Skiddaw Fell, and a handful of variant spellings, such as Skiddow. The place-names that comprise the Scafell cluster are slightly more varied and include not only variant spellings, such as Scawfell, but also the names of other nearby landmarks, such as Great End, Styhead Pass, Styhead Tarn, Mickledore and Lingmell. Whether or not these places can be said collectively to constitute the location of Scafell depends on the nature of the research being undertaken. In this case study, we have decided to accept them as indicative of the extent of the Scafell range and to explore the consequences.

Initial Findings

In the first instance, simply counting and comparing the number of place-name references that occur in each cluster indicates key differences between the two. The cluster around Skiddaw contains 691 references; the one around Scafell includes only 533, which suggests that the former is a more prominent location in the corpus than the latter. This, in itself, may be unsurprising given that Skiddaw towers above Keswick, which is one of the more famous tourist resorts in the Lake District. At the same time, however, the difference between these two figures is not as great as one might expect it to be. (Given that more tourists flocked to Keswick than any other settlement in the region during the period represented in the corpus, one might fairly expect Skiddaw to be mentioned even more frequently than it is.) Turning to the texts themselves, it is striking to note that whereas locations within the Skiddaw cluster are mentioned in sixty-one of the eighty works in the corpus, those within the Scafell cluster are only referenced in thirty-two.

All such inferences must be tentative, but these findings would seem to indicate that although fewer writers discuss the area around Scafell, those who do seem to mention the places that comprise it fairly frequently. Carrying this reasoning one step further, one might posit that although Scafell figures in fewer accounts than Skiddaw, the writers who mention the former devote a significant amount of attention to it. Intriguingly, consulting the corpus corroborates this contention, as it reveals that four works – Edward Baines's *A Companion to the Lakes* (1829), C.N. Williamson's 'The Climbs of the English Lake District' (1884) and the 1853 and 1900 editions of *Black's Shilling Guide to the English Lakes*[4] – account for most of the 533 references to the area around Scafell. Each mentions place-names in the Scafell cluster more than fifty times. (By contrast, only one work in the corpus, Baines's *Companion,* references places in the Skiddaw cluster more than fifty times.) It, of course, might be argued that the significance of the number of references per text depends greatly on the length of the texts in question. Here, however, a statistical comparison of the number of place-names per million words in each work across the corpus reveals a similar pattern: namely, that whereas the Scafell cluster has four works with over 250 place-name instances per million words, the Skiddaw cluster has only one. This finding suggests that although the area around Scafell is mentioned less frequently within the corpus than the area around Skiddaw, when mentioned it is discussed in more detail.

Collocation Analysis of Skiddaw and Scafell

Establishing the relative amount of attention paid to locations within the Skiddaw and Scafell clusters is one thing; exploring how they are being portrayed is another. This is where collocation becomes relevant. As noted

4 The 1900 edition of *Black's Shilling Guide* reproduces, with only slight modifications, the text of the 1853 edition.

Table 3.1 A selection of statistically significant collocates around place-names in (and around) the Skiddaw and Scafell clusters; these collocates are words occurring within 10 words of the search-terms (place-names). Only words with statistically significant t-scores and a minimum collocation frequency of 5 have been included.

Categories	*Skiddaw only*	*Both clusters*	*Scafell only*
Judgements of appearance	Picturesque, majesty, majestic, awful, grand, beauty, beautiful, mighty, interesting	Fine, lofty, magnificent, vast, towering	Rugged, steep
Physical features	Village, trees	Mountain(s), hill, top, summit(s), ridge, precipices, rock(s), crag(s),	Peak, chimney, chasm, cliffs
Weather	Sky, weather, sun, cloud(s), snow, mists		
Transport	Walk		Path, climbers, climbing

above, collocation analysis is a standard method within corpus linguistics for identifying words that appear unusually frequently or in close proximity to – that is to say, collocate with – one another. Evaluating collocation frequencies involves using statistical measurements (such as t-scores, the measurement we use here)[5] to compare how often each word occurs near the search-term in relation to the number of times it appears in the corpus as a whole.[6] Performing a collocation analysis on the basis of proximity requires the researcher to determine how many word tokens on either side of the search-term constitutes a position near the search-term. Here we have adopted a bandwidth of 10 word tokens as a measure of proximity.

In order to determine the kind of language used to describe locations within the Skiddaw and Scafell clusters, we used the names of the places contained in each as search-terms and recorded each of the statistically significant collocates (Table 3.1). For the Skiddaw cluster these included a range of complex aesthetic terms, such as *picturesque*, *beauty* and *beautiful*, *majesty* and *majestic* and *awful*. For the Scafell cluster, by contrast, the collocates comprise mainly words indicative of scale, size and physical appearance, such as *vast*, *lofty*, *steep*, *rugged* and *towering*. Intriguingly, whereas *vast*, *lofty* and *towering* are also significant collocates in the Skiddaw cluster, *rugged*

5 A t-score is a statistical test that compares two samples. In this case it allows us to compare the frequency of the word near to place-names with its frequency in the corpus as a whole.

6 For more on collocation see Barnbrook et al. The collocation analysis described here was performed using AntConc 3.2.4 (Anthony).

and *steep* are not; nor for that matter are *cliff*, *chasm* and *peak*, all of which collocate with locations in the Scafell group. This, of course, can be said to make sense: the fells around Skiddaw, as noted above, are rounded and rise above a famous tourist resort, whereas those around Scafell are more escarped and remote. What is intriguing, however, is that this confirms what might be expected: that – at the level of semantics – the works in the corpus respond to this difference.

Keyword Querying the Geography of the Corpus

From the foregoing analyses we can draw two preliminary conclusions: firstly, that the area around Scafell is described in more detail (but in fewer works) than the area around Skiddaw; and secondly that, in general, the word most frequently associated with Skiddaw and Scafell can be seen to correspond to the physical and geographical differences between the two mountains. These findings affirm the merits of the hybrid corpus- and geographic-based approaches showcased in this chapter. Crucially, however, they do not exhaustively demonstrate their potential. Collocation analysis can, after all, work in the opposite direction. In addition to helping to determine the keywords that collocate with a given location, it can assist us in identifying the locations that collocate with any given search term. When we combine GIS and collocation, moreover, we can also explore the distributions of those locations and their relation to one another. With this in mind, consider the following examples, which are based on the distribution of the place-names that collocate with two of the key terms identified in the previous section: *awful* and *steep*.

Collocations with Awful

Figure 3.4 displays the distribution of the places that collocate with *awful* within the corpus.[7] As this map indicates, although the term *awful* collocates with Skiddaw, it is even more strongly associated with locations near Derwentwater and Borrowdale (the lake and the valley south of Keswick) and Ullswater and Patterdale (the lake and village southwest of Penrith). This pattern would seem to indicate that although the works in the corpus occasionally describe mountains as awful, the term is more often associated with locations within some of the region's lower-lying valleys. Reading the works that contain the word *awful* supports this impression. In her famous account of her journey through the Lakes in 1794, for example, the novelist Ann Radcliffe describes Ullswater as 'bounded on one side by the precipices of Place Fell, Martindale Fell, and several others equally rude and awful that

7 *Awful*, in this context, should be understood in the eighteenth-century sense of the word: 'That which strikes with awe, or fills with reverence' (Johnson 139).

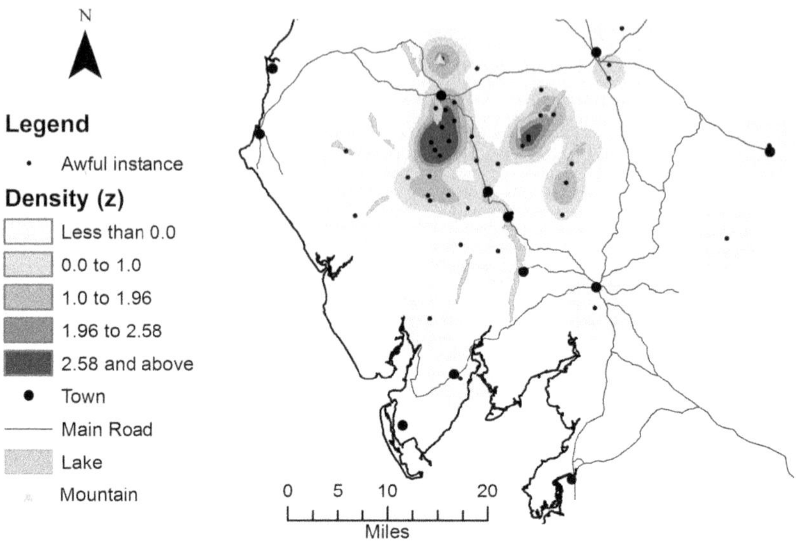

Figure 3.4 Place-name instances that collocate with *awful*.

Source: The Authors; created using ArcGIS © 2015.

rise from its edge' (258). Similarly, Charles Cooke's *Tourist's and Traveller's Companion* of 1827, which incorporates passages from Radcliffe's account, reports that the village of Patterdale is 'bounded by the precipices of Place Fell, Martindale Fell, and several others equally rude and awful, that rise from its edge' (74). Descriptions of Borrowdale, near Keswick, apply *awful* in a similar way. When describing a boat trip on Derwentwater, for instance, Edward Baines mentions passing 'Wallow Crag, whose awful precipice towers over the wood [that] spread around its base' (121). Radcliffe, for her part, makes note of 'the awful rocks, that rise over the fall of Lowdore' (349) when describing her tour through Borrowdale.

In each of these cases, we find writers using the word *awful* to describe mountains and cliffs viewed from a position of lower elevation. This suggests that, in the period represented by the corpus, the sense of awestruck wonder denoted by *awful* implied the appearance of something towering above the viewer. This impression is further supported by other instances of the word within the corpus. Thus, in Thomas West's account of the vale of Keswick, we find reference to 'the skirts of Skiddaw, which raises here in awful majesty his purple front' (97). Elsewhere, Cooke refers to 'towering Skiddaw wrapped in awful shade' (83). The list of examples goes on and on. In fact, the only notable exception to this trend is found in Baines's *Companion*, which offers the following account of a journey above Langdale: 'At a fearful depth beneath the summit lies the Stickle Tarn on one side, . . . and at the other side the more awful depth of Great Langdale . . . and descended an easier path to the level of Angle Tarn, which lies at an awful depth beneath the precipitous summit of Bowfell' (247). Given that Baines is the only writer in the corpus who uses *awful* in this way, this exception can be viewed as useful in that it underscores a general rule.

In addition to implying a particular vantage point, *awful* also seems to be a word that only a particular subset of writers use. The word only turns up 155 times in the corpus, and more than half of these occurrences are found in five works. Two of these works (West's *Guide to the Lakes* and Radcliffe's *Journey*), were published in the late eighteenth century; the other three (Baines's *Companion*, Cooke's *Companion* and John Robinson's *Views of the Lakes*) were published between 1827 and 1833. In each of these cases, except the latest text (Robinson's *Views*), *awful* collocates with a place-name each time it occurs. In the other four works which contain several instances of the word, it collocates with a place-name at least fifty per cent of the time. This suggests that, within the corpus *awful* mainly serves as a term for landscape description but only does so for a relatively limited period of time: approximately 50 years.

Collocations with Steep

Figure 3.5 indicates the distribution of the locations that collocate with *steep*. As noted above, this word is more strongly associated with the area around Scafell than the one around Skiddaw. Yet, as Figure 3.5 suggests, *steep* is more often used to describe the valleys than the mountains of the Lake District. The localities that most commonly collocate with *steep* include Buttermere, Borrowdale, Keswick, Coniston, Ullswater and Patterdale, as well as the area around Grasmere, Ambleside and the Langdales. Examining the co-text surrounding these collocates reveals that *steep* most often occurs in the description of roads, routes and directions of travel. Thus, in Charles MacKay's *Poetry and Scenery of the English Lakes*, the tourist is advised that the 'road from Buttermere, through Newlands to Keswick, leads by a very steep ascent' (161). Likewise, Samuel Leigh's *Guide* tells us that '[l] eaving Coniston Waterhead, the tourist ascends a steep hill' (17). Harriet Martineau's *Complete Guide*, for its part, warns that the 'descent to all the Ambleside inns is steep' (42).

Intriguingly, although roads are frequently described as *steep* in this manner, mountains are only rarely. Investigating the place-name collocates in the Scafell cluster illustrates this. Here, the only place-name within the cluster that collocates with *steep* is Sty Head Pass, the mountain road between Wasdale and Borrowdale. Surveying the whole of the Lakes region one notices the cluster of collocates near Helvellyn (the third highest peak in the district). Here again, however, inspection of the co-text reveals that these collocates occur in passages that describe journeys below the mountain. Hence, William Green's *New Tourist Guide* (1819) informs us that the

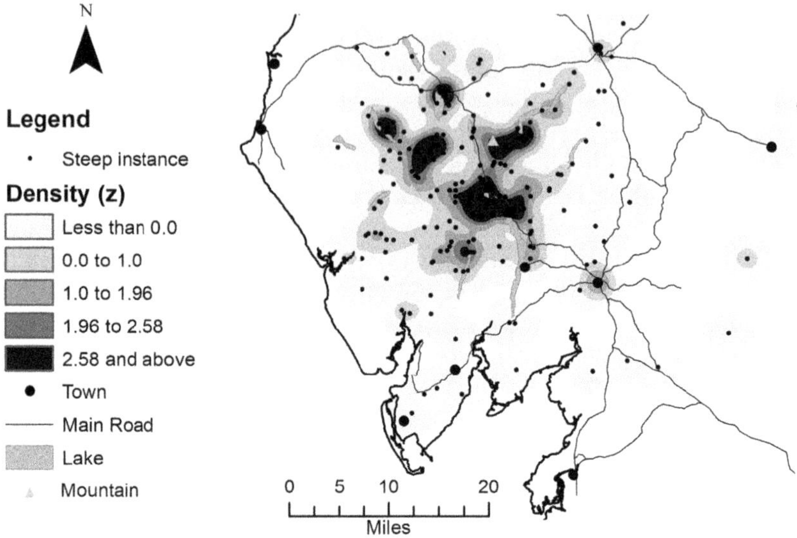

Figure 3.5 Place-name instances that collocate with *steep*.
Source: The Authors; created using ArcGIS © 2015.

road beside 'the lake passes . . . under the steep and shaggy brow of Helvellyn' (423). In sum, like *awful*, the word *steep* seems to imply a specific context. Whereas the former tends to imply a position in relation to a site of higher elevation, the latter tends to indicate a consideration or discussion of travel.

Calibrating Keyword Collocations against the Geography of the Corpus

One potential criticism of the maps in Figures 3.4 and 3.5 is that they fail to account for the fact that some place-names appear in the corpus more often than others and are, therefore, likely to collocate with queried keywords queries more frequently. As a result, one might claim that the foregoing maps tell us more about the number of times individual place-names are mentioned in the corpus than they do about the actual distribution of collocations with the search-term in question. When examining the cluster around Keswick in Figure 3.5, for instance, one might wonder whether this pattern appears because Keswick frequently collocates with *steep* or whether it occurs because Keswick is a place-name that recurs throughout the corpus and is, thus, more likely to collocate with any search term. Equally, one might wonder if those locations that appear to collocate with the search-terms only a few times – such as Workington and Whitehaven – appear to be less significant because their names are not mentioned as often in the corpus. In order to address these sorts of queries we need to augment the approaches to geographical text analysis outlined above by using more complex methods of spatial analysis. In effect, we need to use statistical scans to measure and compare the distributions from Figures 3.4 and 3.5 with the one displayed in Figure 3.3. In doing so, we need to identify both those locations where there are significantly more collocations in Figure 3.4 or 3.5 than would be expected, on the basis of Figure 3.3 (hereafter called hot-spots), *and* those locations where there are significantly fewer collocations in Figure 3.4 or 3.5 than would be expected on the basis of Figure 3.3 (hereafter called cold-spots). This can be done using a test called Kulldorf's Spatial Scan Statistic.[8]

Figure 3.6 shows the results of applying Kulldorf's Spatial Scan Statistic to evaluate the distribution of place-name collocations with *awful*. The map displayed here is more complicated than the ones above; but, effectively, it shows us that the technique has identified two hot-spots: one on the eastern side of Derwentwater and another stretching along the eastern shore of Ullswater towards Haweswater and Kentmere. The first of these two clusters indicates that locations and landmarks along the precipitous south-eastern edge of Derwentwater do, in fact, collocate with *awful* more regularly than might be expected given the number of times they are mentioned in the corpus. In contrast, although other areas – such as the one around Skiddaw and the one extending from Borrowdale to the Langdales – also collocate with *awful*, the spatial scan suggests that this has more to do with the number of times they are mentioned in the corpus than it does with a particularly strong association with the search-term.

The second cluster identified in Figure 3.6 requires even more careful interpretation. As with the eastern side of Derwentwater, the area around Ullswater comprises a number of locations (including Patterdale, Place Fell and Martindale Fell) that are mentioned several times in the corpus. This might lead us to infer that these locations often collocate with *awful* simply because they appear so frequently. However, the spatial scan indicates that many of the names of these locations do, in fact, collocate with *awful* more regularly than might be expected. Thus, the observation that it is frequently described as *awful* is not simply a consequence of this place being commonly mentioned in the corpus.

Figure 3.7 displays the output of the spatial scan of the place-name collocations with *steep*. Here the pattern is more complex than the one produced by the spatial scan of collocations with *awful*. One noticeable difference between the two is the presence of cold-spots (that is, places that collocate less frequently with the term than would be statistically expected). Clusters of cold-spots are found throughout the peripheries of the central Lakeland

8 See Kulldorf; in this analysis the scan was implemented using SatScan (Kulldorf, *SatScan*).

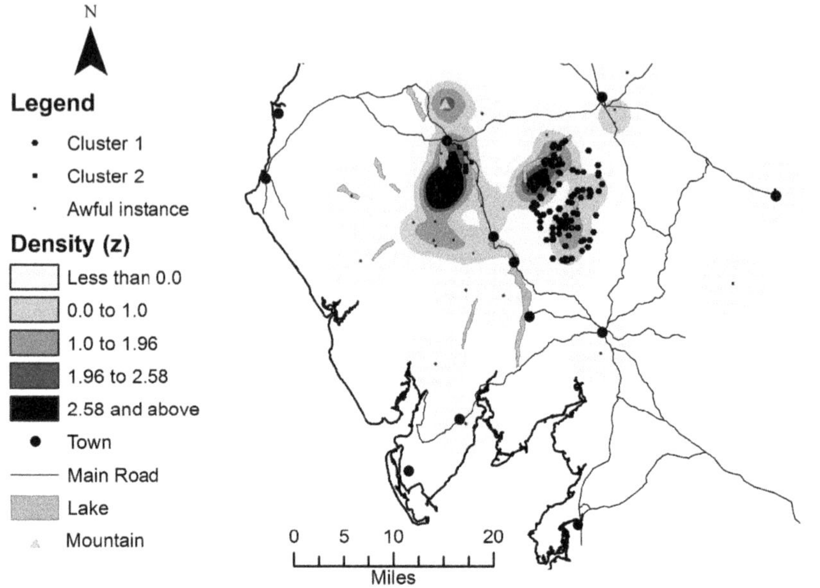

Figure 3.6 Spatial scan statistic of place-name instances that collocate with *awful*.
Source: The Authors; created using ArcGIS © 2015.

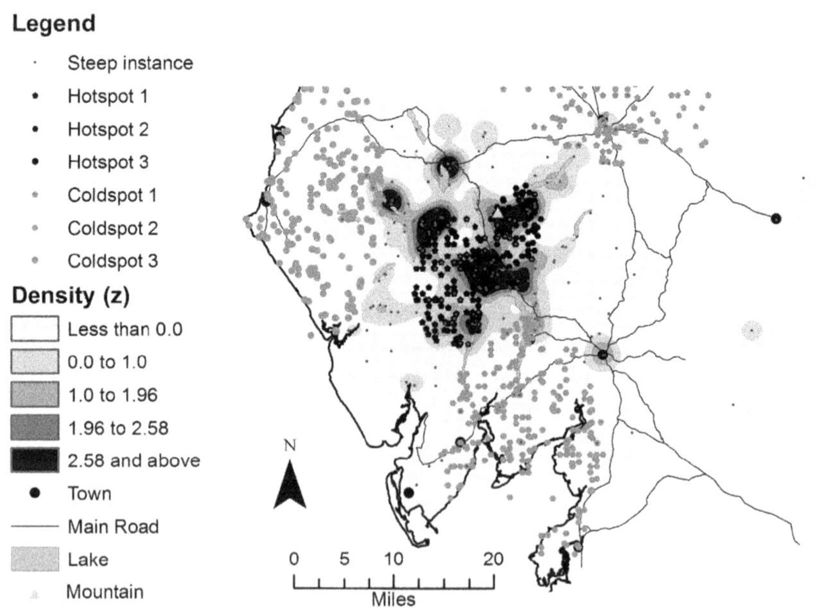

Figure 3.7 Spatial scan statistic of place-name instances that collocate with *steep*.
Source: The Authors; created using ArcGIS © 2015.

fells, from the Cumbrian coastal plain eastwards to the Eden Valley and southwards to the southern Lakes. Upon reflection, this might be said to make sense: each of these areas is, after all, distinguished by relatively level terrain and thus, one suspects, would be unlikely to be regarded as steep. In contrast, the distribution of hot-spots indicates that only a handful of key localities in the upland centre of the region – including the areas around Patterdale, Borrowdale, Coniston and the Langdales – collocate with *steep* more often what one would statistically expect. Panning out, moreover, one notices an intriguing difference between the clusters identified by the density-smoothed surface and those identified by the spatial scan. These include the areas around Ambleside, Keswick and Buttermere. Each of these settlements is frequently mentioned in the corpus. Each, furthermore, is a popular tourist destination from which steep roads radiate. Yet, for all that, the spatial scan indicates that although they often collocate with *steep*, this has more to do with the number of times they are mentioned than with a significant association with the word *steep*.

Comparing the patterns in Figures 3.6 and 3.7 furnishes us with some significant insights into the geography of the corpus, specifically because it shows us the place-names associated with *awful* and *steep* do not overlap. They do occasionally come close to one another: the fells east of Derwentwater are *awful*, whereas Borrowdale, just to the south, is *steep*; and whereas Patterdale is *steep*, the fells east of Ullswater are *awful*. In general, however, the two terms are associated with different localities. This finding is intriguing since it illustrates how we can explore in detail the ways in which different parts of the Lake District are perceived and represented. Noticing this pattern in the descriptive semantics of the corpus invites us to assess whether or not it correlates with patterns in the physical geography of the Lakes region itself. In order to do this, however, it is first necessary to incorporate additional contextual data within the GIS environment.

Incorporating Contextual Data: Digital Terrain Models

One of the most useful, yet underrated, features of GIS technology is its ability to integrate georeferenced data from different sources. For the purposes of this case study, this means that we can add extra contextual data – such as topographic relief data – to enhance our analyses. The standard of incorporating relief data is by employing digital terrain models (DTMs), which use pixel-data to represent the surface of the earth (Burrough and McDonnell). Consider Figure 3.8, which displays a DTM for the Lake District on which the place-name collocates with *awful* have been superimposed.

One benefit of examining point-data, such as the collocates displayed in Figure 3.8, against the backdrop of a DTM, is that it enables us to examine how their distribution maps onto the contours of the terrain. A key feature to consider here is elevation. Once a layer displaying point-data has been

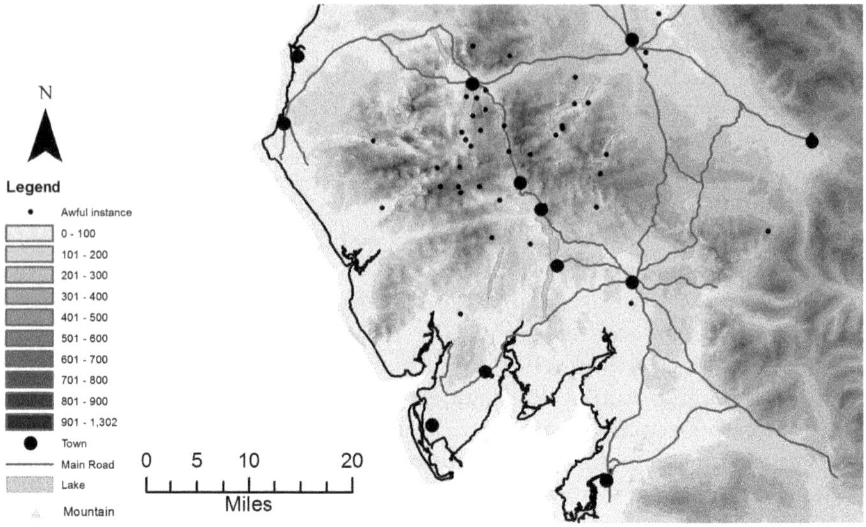

Figure 3.8 A Digital Terrain Model of the Lake District with *awful* place-name collocations.[9]
Source: The Authors; created using ArcGIS © 2015.

superimposed on a DTM, one can merge information from the two data tables and, in doing so, assign each item of point-data a height above sea level. One must, of course, remember that the accuracy of the location of individual points may vary. Nevertheless, using the DTM to assign each point an elevation is an efficient and reasonably reliable way of assessing whether items of data – such as collocations with *awful* and *steep* – occur at a range of different heights or whether they tend to cluster around particular altitudes. With this in mind, have a look at the graphs displayed in Figure 3.7, which document the different elevations of the places whose names collocate with *awful* and *steep*. In both cases the elevation of all the locations mentioned in the corpus (represented by a dotted line) have been included for comparison.

As Figure 3.9a indicates, whereas most of the places in the corpus are below 200m, those that collocate with *awful* tend to be located at moderate to high altitudes (between 200m and 600m), and are unusually common between 300 and 600m when compared to the corpus as a whole. Places that collocate with *steep* follow a similar pattern in that they become unusually common above 300m. This suggests that whereas the highest parts of the Lake District (those above 600m) are sometimes associated with *steep*, they tend not to be associated with *awful*. Cold-spots for *steep* are, unsurprisingly, commonly found in low areas. In sum, then, this analysis confirms the pattern described in the previous sections: namely, that both *steep* and *awful* are

9 The coast line and the DTM do not match up completely because of differences in tidal definitions.

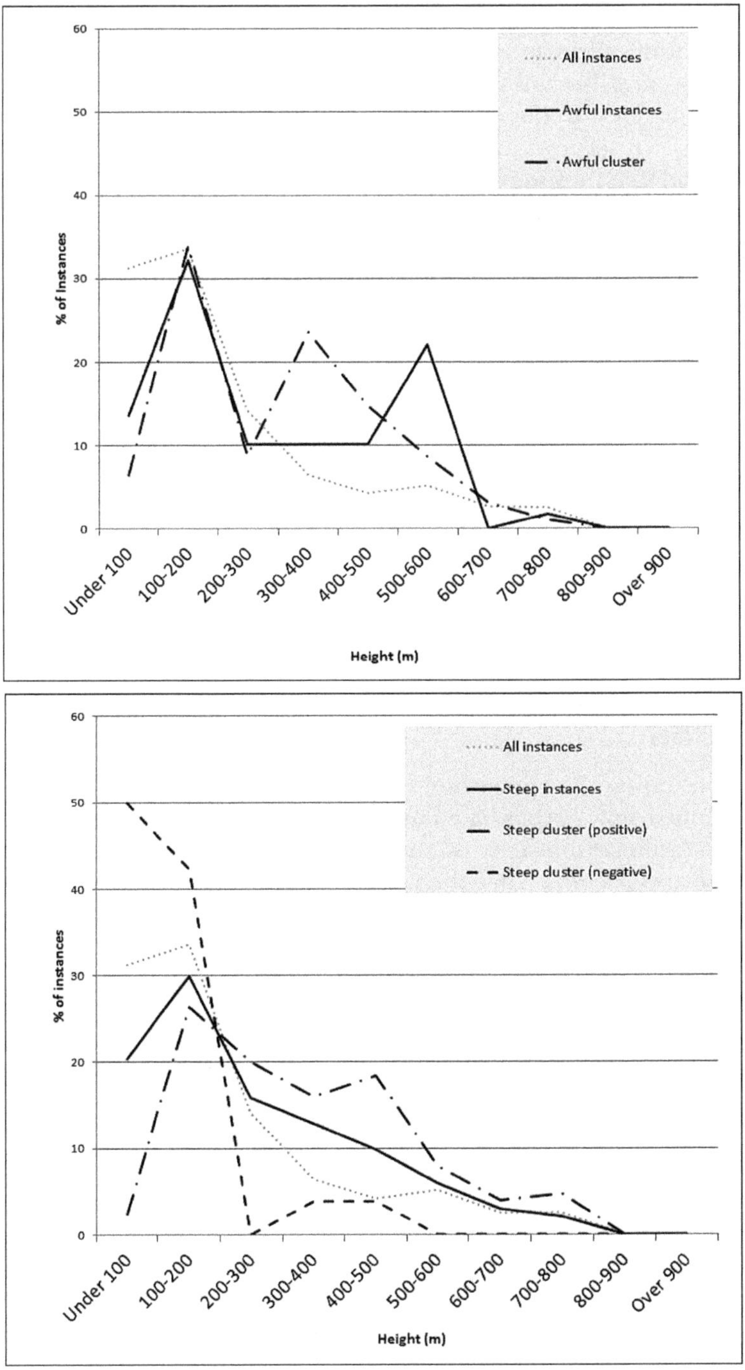

Figure 3.9 The heights of the instances of (1) *awful* and (2) *steep* instances compared to all instances in the corpus.

Source: The Authors.

associated with valleys and mountain passes. High mountains are sometimes described as *steep*, but this tends to be obscured because mountainous places are mentioned less frequently than locations in the surrounding valleys.

The previous sections indicated that these two terms were used in different ways and that they were associated with different geographies. Although *awful* occurs relatively infrequently, when it does appear it is typically used to describe high landmarks viewed from a point of lower elevation. The eastern sides of the Derwentwater valley and of Ullswater are localities associated with *awful* particularly often, as is Harter Fell at the southern end of Haweswater. Places in and around Borrowdale also frequently collocate with *awful*; however, this is at least in part because they are mentioned so frequently in the corpus. *Steep*, by contrast, tends to be particularly associated with roads and the places they connect. Once the popularity of these places is taken into account, however, passes such as Honister and Wrynose still stand out as significant collocates with *steep*. Some of the higher fells occasionally collocate with *steep*; although this is not immediately apparent because they are not mentioned very frequently in the corpus. As with *awful*, *steep* frequently occurs in passages that describe the high slopes of valleys and hillsides seen from a lower altitude. These valleys and hillsides are, however, in different places than the ones associated with *awful*.

Conclusions

In demonstrating the application of geographical text analysis, this chapter has illustrated how corpus-based and geographical approaches can be integrated to facilitate the study of any corpus. Our discussion has focused on how these approaches enable the researcher to explore the underlying geography of a corpus by identifying, extracting and displaying its geographical information; to apply collocation analysis to examine how the corpus thematises this geographical information; to assess how specific concepts and terminology from the corpus relate to its underlying geography; and to incorporate additional contextual data to enhance his or her analysis. In general, this chapter has indicated that the summaries produced by geographical text analysis – namely: maps, tables and graphs – are effective tools for identifying geographical patterns within texts. It should be noted, however, that these tools are not capable of providing definitive answers about those patterns. In effect, they pose questions and point the researcher towards the parts of the corpus that will likely contain the answers to them. In this chapter, these tools have been applied in an exploratory way and many potential questions they raise have not been explored. A more applied paper will focus more on comparing preselected places, search-terms, times, writers or genres to explore the basic question of how different words and uses of language are associated with different places. This work remains to be done; however, the potential of geographical text analysis as a method for aiding our understanding of the geographies within texts should be clear in that it can provide a framework within which we can improve our understanding of the geographies within texts.

Works Cited

Anthony, Laurence. AntConc Homepage. *Laurence Anthony's Website.* 2014. Web. 20 March 2015. <http://www.laurenceanthony.net/>.

Baddeley, M.J.B. (ed.) *Black's Shilling Guide to the English Lakes.* 22nd ed. London: Adam & Charles Black, 1900.

Baines, Edward. *A Companion to the Lakes.* 2nd ed. London: Hurst, Chance & Co., 1830.

Barnbrook, Geoff, Oliver Mason and Ramesh Krishnamurthy, *Collocation: Applications and Implications.* London: Palgrave Macmillan, 2013.

Black's Shilling Guide to the English Lakes. Edinburgh: Adam & Charles Black, 1853.

Burrough, Peter A., and Rachael A. McDonnell, *Principles of Geographical Information Systems.* 2nd ed. Oxford: Oxford University Press, 1998.

Cooke, Charles. *The Tourist's and Traveller's Companion to the Lakes.* London: Sherwood, Jones & Co., 1827.

Cooper, David. 'The Poetics of Place and Space: Wordsworth, Norman Nicholson and the Lake District'. *Literature Compass* 5 (2008): 807–21.

Fotheringham, A. Stewart, Chris Brunsdon and Martin Chartlon, *Quantitative Geography: Perspectives on Spatial Data Analysis.* London: Sage, 2000.

Green, William. *The Tourist's New Guide.* Vol. 1. Kendal: Lough & Co., 1819.

Gregory, Ian, David Cooper, Andrew Hardie and Paul Rayson. 'Spatializing and Analysing Digital Texts: Corpora, GIS and Places'. *Spatial Narratives and Deep Maps.* Eds, David Bodenhamer, John Corrigan and Trevor M. Harris. Bloomington: Indiana University Press, 2015. 150–78.

Grover, Claire, et al. 'Use of the Edinburgh Geoparser for Georeferencing Digitized Historical Collections'. *Philosophical Transactions of the Royal Society A* 368 (2010): 3875–89.

Jockers, Matthew L. *Macroanalysis: Digital Methods and Literary History.* Urbana-Champagne: University of Illinois Press, 2013.

Johnson, Samuel. *A Dictionary of the English Language.* Vol. 1. London: Strahan, 1755.

Kulldorf, Martin. 'A spatial scan statistic'. *Communications in Statistics: Theory and Methods* 26 (1997): 1481–96.

———. *SaTScan,* 2015. Web. 20 March 2015. <http://www.satscan.org/>.

Leigh, Samuel. *Leigh's Guide to the* Lakes. London: Leigh, 1830.

Lloyd, Christopher D. *Local Models for Spatial Analysis.* Boca Raton, FL: CRC Press, 2007.

MacKay, Charles. *The Scenery and Poetry of the English* Lakes. London: Longman, 1846.

Martineau, Harriet. *A Complete Guide to the English Lakes.* Windermere: J. Garnet, 1855.

Moretti, Franco. *Atlas of the European Novel, 1800–1900.* London: Verso, 1998.

———. 'Conjectures on World Literature'. *New Left Review* 1 (2000): 54–68.

Nicholson, Norman. *Greater Lakeland.* London: R. Hale, 1969.

Radcliffe, Ann. *A Journey Made in the Summer of 1794.* Vol. 2. 2nd ed. London: Robinson, 1795.

Robinson, John. *Views of the Lakes in the North of England.* London: Whittaker & Co., 1833.

Rupp, C.J., et al. 'Dealing with heterogeneous big data when geoparsing historical corpora'. *Proceedings of the 2014 IEEE Conference on Big Data* 2015: 80–83.

Tally Jr., Robert T. *Spatiality.* Abingdon: Routledge, 2013.

West, Thomas. *A Guide to the Lakes.* London: Richardson and Urquhart, 1778.

Williamson, C.N. 'The Climbs of the English Lake District', *All the Year Round,* 35 (1884), 77–82, 110–14.

4 Mapping Fiction

The Theories, Tools and Potentials of Literary Cartography

Barbara Piatti

Literary Geography and Literary Cartography: A Brief Introduction

As all lovers of literature know, there are hundreds, if not thousands, of places that have been transformed because they have provided the setting for a memorable work of fiction. Alphonse Daudet's colourful descriptions of Fontvieille, for instance, will be forever connected in the minds of his readers with the countryside of Provence. The works of the Brontë sisters, for their part, have enriched our appreciation of the wild, windswept moors of West Yorkshire. Similarly, for many readers, the cityscape of Dublin remains imaginatively overdetermined by the writings of James Joyce. On the other side of the Atlantic, writers ranging from John Steinbeck to Paul Auster have left marks on locations that are now commonly, even canonically, perceived through the filter of their works. A comprehensive list of instances of this sort of layering of fiction over actual places would fill an atlas. The point is that, in enumerating specific examples, one gradually becomes aware that the physical geography of our world is enveloped by a geography of fiction.

Scholars interested in this geography of fiction have long been exploring the relation of literary spaces and actual places (see Piatti; Döring). With the recent emergence of digital literary cartography, however, efforts to visualise and analyse this relation have greatly intensified. Situated at the interface of literary studies, digital cartography and spatial analysis, this genuinely interdisciplinary field of research is concerned not only with addressing the practical and methodological challenges that attend efforts to map the settings of literary works, but also – and even more fundamentally – with expanding the knowledge and insights that literary mapping can afford.

Literary Geography and Literary Cartography

Literary geography and literary cartography can be said to constitute two distinct yet interrelated areas of enquiry (Piatti and Hurni, 'Cartography of Fictional Worlds'). So, it may be useful to begin by drawing a few distinctions between them. Literary geography, as suggested above, is generally concerned with the geographical analysis of literary works. As a product of the late nineteenth and early twentieth centuries (see Sharp; Nagel), the practice of literary geography is rooted in the appreciation of the way works of fiction engage with knowable places. The efficacy of this engagement, of course, depends on the frame of reference that a work assumes in negotiating the relationship between the material world it inhabits and the fictional world it creates. Some works (such as, say, those of Steinbeck or the Brontës) are strongly tethered to actual locations, whereas others (such as, for example, the novels of C. S. Lewis or Lewis Carroll) drift more freely between the actual and imaginary. In either case, the goal of literary geography is both to explore how works of literature negotiate their relationship to the space of the external world and to consider how this negotiation influences the way those works are read and received. On the other hand, literary cartography, as the practice of mapping works of literature, can best be thought of as a subfield of literary geography. Whether in mapping a single work and its spatial elements, or in mapping several works and their attributes, literary cartography typically relies on abstract symbology and quantitative methodologies to render the spatial information contained in works of literature visible as a geospatial overlay. Literary geography and literary cartography can thus be linked in a hierarchical fashion: whereas literary geography is a field of study, literary cartography is one of the cardinal interpretive practices of that field.

When considering the development of literary cartography, it becomes apparent that certain types of literary works are more amenable to spatial representation than others. For instance, whereas the mapping of nineteenth-century realist novels provides, in most cases, convincing and valuable results, the mapping of much postmodern fiction is far more challenging and poses far more questions and problems than solutions. In a more general sense, it has to be acknowledged that literary cartography is not the only critical procedure which facilitates further thinking about literary geography, since, as Franco Moretti has emphasised, some of the greatest studies ever written on space and narrative do not include a single map (*Graphs, Maps, Trees* 35).

Referentiality, Ambiguity and Uncertainty

One of literary cartography's traditional starting points is the assumption that a large part of fiction refers to the mappable world, or what – to invoke Edward Soja's term – might be called 'first space' (Soja). Works of fiction can refer to the physical world in a variety of ways, including the use of identifiable toponyms or the inclusion of topographically specific descriptions. As Malcolm Bradbury observes in his *Atlas of Literature*, '[a] very large part of our writing is a story of its roots in a place: a landscape, region, village, city, nation or continent' (Bradbury 7). In addition to such realistically constrained spaces, however, literary works are also able to create limitless spaces, including imaginary realms comprised of purely invented cities, countries, continents and entire stellar systems. Both H.P. Lovecraft's *Dreamlands* and the desert planet setting of Herbert Frank's *Dune* spring to mind. Between these two extremes of fictional geographies – the topographically accurate and the entirely imagined – one finds literary works in which various degrees of transformed settings, spaces and places appear. An excellent example of this is Robert Harris's counterfactual novel, *Fatherland* (1992), which envisions a world in which the Allies have lost the Second World War and which, accordingly, presents the reader with the familiar, but at the same time fundamentally altered, space of Berlin, here transformed into Hitler's Reichshauptstadt. Such a partly real, partly invented place can be described and critically analysed, of course, but, at the same time, its geographies can also be displayed as a cartographic product (Piatti and Hurni, 'Mapping the Ontologically Unreal' 334–6).

The ambiguity of a literary work is often regarded as a sign of its intrinsic value, since uncertainty opens the way for interpretation and critical debate. When it comes to literary cartography, however, such ambiguity can be problematic (Bodenhamer et al. 171). Although the advanced techniques of digital cartography, especially cartography aided by geographic information systems (GIS) technology, are able to cope with a high degree of uncertainty, literary cartography has to deal with more than one uncertainty factor, both in its primary materials and on a methodological level. Works of literature do not, after all, always provide specific information about the topographical and geographical dimensions of their settings. Indeed, as Anne-Kathrin Reuschel and Lorenz Hurni have clarified, there are no less than five sources of uncertainty that one must confront when mapping fictional spaces: the artistic freedom of the author; the variety of different terms used to describe places and spaces; the vagueness of certain geographical concepts; the different ways that readers interpret the geographical information found in literary texts; and, finally, the fact that most forms of cartographic visualisation are too rigid to represent the sort of 'fuzzy' geographies that many literary works contain (Reuschel and Hurni 298–9). An additional source of spatial uncertainty might be added to this list as the production of a literary map can also serve to complicate – rather than resolve – the literary geographies embedded within a text. Franz Kafka's fragmentary novel *Amerika* (1927), for example, contains many inaccurate topographical references. The fictional space of this novel can be mapped and laid over a contemporary satellite map of New York, but the production of this layered map simply serves to prompt further critical questions. Was Kafka unable to access detailed and accurate information about the geography of the city? Or was his seemingly erroneous portrayal of New York an intentional creative strategy? Creating a map of such a teasingly 'slippery' literary geography (to use Sally Bushell's term) is not essential for the prompting of such questions; rather, it helps to foreground them. In any case, as will become evident in the following pages, even a scientific approach to literary mapping – such as the database-supported digital mapping of literature – must take such ambiguities and uncertainties into account.

Criticism of studies in literary cartography

Literary cartography is also a contested critical approach, and some literary scholars remain sceptical about the application of GIS, and other kinds of quantitative cartographical technologies, in the study of literature. One of the most commonly expressed criticisms is that such approaches endorse a too 'homogeneous conception of mapping' (Hewitt 158) that both fails to account for the uniqueness of different literary works and lacks the sensitivity required to engage with the kinds of 'slippery' or equivocal geographic information that works of literature often contain (Bushell). Another common criticism is that the mapping of literary works relies on a process of abstraction, quantification and isolation that significantly reduces their semantic content. It is important for researchers who are engaged in literary cartography to keep these criticisms in mind. At the same time, however, it is important for those who criticise literary cartography to be aware that cartographers do not typically insist that the findings of their mapping experiments are binding interpretations. Rather, they present their results as provocations to intellectual curiosity and as ways of generating questions to guide inquiry. Literary mapping, then, is only one stage in a process of interpretation. Exciting literary-cartographical visualisations do not always provide final answers, but they invariably do support researchers in shaping their interpretations.

It is, moreover, to be regretted that many of the scholars who have expressed scepticism about the merits of a literary mapping project have stopped short of making applied contributions to advance the current state of the art. Instead, they criticise the maps developed by other scholars, voicing their opinions about how literary maps should or should not appear, but without thereby producing maps of their own. It would be a major advance in the development of literary cartography if literary cartographers and literary critics could join forces in a more concerted way than has occurred in the past. This would provide interdisciplinary opportunities both for the exchange of knowledge and for the sharing of best practice. More particularly, this sort of collaboration could lead to further self-reflexive consideration of the practice of literary cartography including the potential for developing new systems of symbology.

A Literary Atlas of Europe

Having provided a basic definition of literary geography and literary cartography, I now want to turn to a concrete example of literary mapping in the digital age. Accordingly, in the following pages I propose to showcase some of the findings from *A Literary Atlas of Europe*, a prototype literary mapping project that I initiated and led with the assistance of colleagues at ETH Zurich between 2007 and 2014. Collectively, the *Literary Atlas of Europe* project comprised a team of specialists in literary studies, cartography, computer science and graphic design. Working together, this interdisciplinary team conducted research on the literary geographies of three distinct regions: the Alpine region surrounding Lake Lucerne and the Gotthard Pass in Switzerland; the coastal region of North Frisia, which is, for the most part, located in Germany; and the city of Prague, the capital of the Czech Republic. The literary geographies of these three regions were examined using a corpus of extracts from works of fiction published over the past three hundred years. By the end of the project, in 2014, the project team had spatially analysed extracts from nearly three hundred different works.

The initial phase of this analysis involved differentiating between the various types of spaces present in the literary works we examined. In order to do this, we identified five key spatial categories. The first of these categories was setting: in other words, the location in which the action represented in the narrative occurred. The second category, zone of action, was specified in order to account for narratives that combine several different settings. The third category, projected spaces and places, was developed in

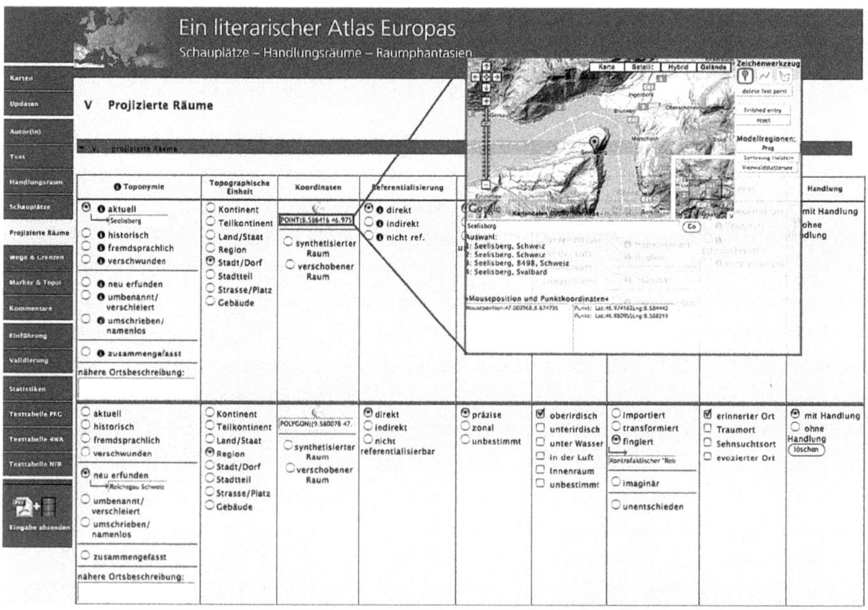

Figure 4.1 Extract of the online template for the data entry in *A Literary Atlas of Europe*.

Source: © ETH Zurich, Institute of Cartography and Geoinformation.

order to account for those locations that characters dream about, desire or remember. The fourth category, topographical markers, included places that, though mentioned in a work, do not play a significant role as a setting. Finally, a fifth category, paths and routes, was created in order to account for the movement of characters between different locations (Piatti et al.). The project team catalogued the spaces corresponding to these categories in each of the works considered, and we created visualisations of these spaces using digitisation tools that allowed us to draw points, lines and polygons on webmaps such as OpenStreetMap and Google Maps.

During this process, however, we encountered a number of problems. Whereas some of the works of fiction we considered were clearly anchored in existing regions, villages and cities, others were far more difficult to localise. In the latter case, only careful reading – along with detailed investigations of possible 'real-world' counterparts of the fictional setting – enabled us to identify a probable position or zone on the map. Vanished places (that is, actual locations mentioned in the works considered, but which are no longer extant) were another category that required additional research. Working on the premise that literary cartography deals with the referentiality between first space (that is, as noted above, the mappable world) and fictional space, the research team had to decide whether a named location has disappeared from the first space or was simply the invention of the author. The consideration of these and other issues ultimately led to the creation of more complex maps with a higher, and at the same time more condensed, degree of information.

To clarify the process, then, the team members with expertise in literary studies conducted careful readings of the extracts compiled by the project, and they entered the relevant spatial information for each into a customized online database (see Reuschel et al.). This database allowed the project team to record information about each of the five spatial categories we designated (setting, zone of action, projected spaces/places, topographical markers and paths and routes), which could then be queried using checkboxes and dropdown menus with pre-determined definitions. In addition, the database also allowed the team to enter further written explanations and comments in a text field. Once the information was entered into the database, the cartographers could start their work, which chiefly involved designing and developing a series of interactive literary-geographical maps that integrated tailor-made background maps and innovative subject-specific symbology. For example, in order to distinguish between settings and projected spaces/places, two different colour scales were used. Whereas the scale red – orange – yellow was used to indicate settings, the scale purple – blue – cyan was used to indicate projected spaces/places. The darker the colour, therefore, the stronger the correlation to first space of the mappable world (Reuschel et al.). The maps created, then, variously represented the geography both of individual works, and of multiple works by individual authors, as well as particular topics and attributes.

Maps and Visualisations from *A Literary Atlas of Europe*

Having introduced *A Literary Atlas of Europe*, I now want to present and discuss some of the maps produced by the project team. These include maps detailing journeys and movements, zones of action and projected spaces/places. The selection intends to demonstrate one of the key critical convictions of the project: namely, that literary maps and spatial visualisations can help to generate research questions to guide further critical inquiry.

Journeys and Movements

Plate 3 shows the hidden spatial dynamics of Friedrich Schiller's *William Tell* (1804), which is set amidst Alpine scenery near Lake Lucerne. Although dramatic works have generally been excluded from the corpus of extracts compiled by the project, this text has been chosen to serve as an example partly on account of its canonicity and partly because it is exceptionally rich in geographical and topographical detail, containing more than 150 references to specific landmarks and locations. The play is, furthermore, diegetically enriched by descriptions of the journeys made by characters: whether they are past trips (indicated in purple), planned excursions (indicated in green) or journeys that take place within the story and which are witnessed and reported by other characters (indicated in yellow). In depicting these different journeys, the map displayed in Plate 3 evinces how central the flow of ideas, thoughts and information between locations is to the thematics of Schiller's drama. What is more, in doing so, it calls attention to a spatial dimension within the work that is otherwise only indirectly apparent in the text of the drama itself.

Following on from this, Plates 4 and 5 display the routes of characters moving through two canonical fictionalised regions: the coastal landscape of North Frisia and the city of Prague. When comparing the two visualisations, one immediately notices that, whereas Prague appears to be a space of movement through and across, North Frisia is largely an enclosed space: one in which movement mostly takes place within. Interestingly enough, looking at Plate 4, one notices that only a few journeys (green) lead from North Frisia to the outside world. This is contrary to what one might expect in a coastal region, where the sea might facilitate movement to and from other areas. Instead, the map suggests that North Frisia, as a narrated space, is characterised by a static rather than a dynamic profile: a space wherein most characters either remain fixed or move within small circles. The literary mapping of Prague, by contrast, presents a strikingly different picture. Here, a complex overlay of paths radiate from, over or through the city, which appears as a central node connecting the Czech capital with the rest of the world and vice versa. Within Prague, characters are constantly on the move. The black, dense clusters of lines represent the movements within the city, indicating 'hot spots' – or areas of particularly intense activity – on

Wencelas Square and Charles Bridge. The green, red and yellow lines indicate journeys where Prague is either a stopping point within a longer trip or the origin or terminus of an excursion.

If you compare the two maps side by side some intriguing questions suggest themselves: How historically consistent are these observations? Was there a period when the fictional Prague was less dynamic? Did the narrated North Frisia once have greater connectivity with the surrounding territories, including Denmark to the north or the City of Hamburg to the south? Did it sometimes even appear on a world map? Answers to these questions might be found through further engagement and careful reading; but the maps produced by the project team clearly serve an important intermediary function by drawing these spatial issues to our attention.

Zones of Action

The next map (Plate 6) shows an extract of model region North Frisia. Whereas the settings of Theodor Storm (1817–88), the author of the famous novella *The Dikemaster* (1888), are labelled in green, red is used to indicate the settings and zones of action of other contemporaneous writers, including Detlev von Liliencron (1844–1909) and Klaus Groth (1819–99). The map – a bivariate map, featuring two different topics – immediately reveals that Storm somehow managed to occupy the territory near the town of Husum (the large green dot in the middle) all by himself. It seems as if other writers sought different settings – further to the west, for example – where the world of the North Frisian Islands begins. Is it that no other writer dared to 'enter' Storm's fictionalised area? Or, alternatively, did they just not care to do so? These are the sort of questions raised by studying a literary map, and they are questions that can only be answered conclusively by returning to the works themselves and to accompanying materials such as reviews, diaries and letters. The phenomenon of such blocked zones, it should be noted, is not unique to Northern Frisia. Similar observations can be made in the case of Schiller's *William Tell*, as the literary geography around Lake Lucerne is dominated by the play and other adaptations of the legend that inspired it.

Projected Spaces

Projected spaces/places are an exceptionally fascinating literary-geographical category. The term, as mentioned above, refers to those spaces or places that characters think of, remember, desire or imagine. They are, accordingly, spaces or places situated in geographical locations that are often distant from the characters who invoke them. These projected locations often perform specific functions in the narratives in which they appear, as they add extra layers of meaning to the geography of a narrated world. Consider, for example, the role that projected space plays in Gustave

Flaubert's *Madame Bovary* (1856). In this novel, the beautiful Emma lives with her husband, whom she soon comes to detest, in a small provincial village in Normandy. Feeling isolated and bored, Emma frequently daydreams of Paris, using a map of the city as a prompt to her imagination. With this aid she plunges herself, figuratively, into the vibrant urban space of the metropolis:

> She bought a plan of Paris, and with the tip of her finger on the map, she walked about the capital. She went up the boulevards, stopping at every turning, between the lines of the streets, in front of the white squares that represented the houses. At last she would close the lids of her weary eyes, and see in the darkness the gas jets flaring in the wind and the steps of carriages lowered with much noise before the peristyles of theatres. (Flaubert 51)

In terms of the maps produced by the *Literary Atlas of Europe* project, the projected space that springs from Emma's map finds a counterpart in our visualisations of Storm's works. In Plate 6, Storm appeared as the dominant author of a small region of near the town of Husum. When we map the setting and projected spaces of Storm's stories (Plate 7), however, a different picture comes into focus. Instead of a map of North Frisia alone, here we have a map of the wider world that includes projected spaces in Hong Kong, India, North America, South Africa and the Virgin Islands. This revelation of the global dimension of Storm's works may seem surprising, since he is conventionally regarded as a realistic author whose fiction only concerned a small region in Northern Germany.

Through these maps, the interdisciplinary work of the *Literary Atlas of Europe* project clearly demonstrates that visualising literary geographies requires abstract symbolisation. There is a danger that the use of pictograms or icons might lead to a touristic trivialisation of the textual content. For cartographers, therefore, the principal challenge is to develop an abstract – but at the same time optically and cartographically useful – symbology that can be transferable to other kinds of geographics and that, at the same time, can be implemented in a webmap service.

From Individual Maps Towards a Literary-Geographical System

The maps presented above were created to assist with the analysis of specific works, authors and literary geographies. Significantly, however, the *Literary Atlas of Europe* project team also developed methods for generating visualisations of groups of works using statistical queries. During this process, we developed a prototype for a large-scale literary geographical information system in the form of a flexible toolbox. Some of the main functionalities we developed for this toolbox include the capacity to move/zoom from the local/regional level to a continental – or even global – view, thereby providing easy access to maps – and, by extension, imaginative geographies – on different scales. Via mouseover, the user is provided, on every scale, with information regarding the relevant settings and projected spaces/places represented on the map. As demonstrated above, most of the visualisations we have created project the literary geographies of works of fiction onto modern background maps. Crucially, however, the majority of the works considered by the project were set in the past: in Prague around 1900; in North Frisia in the late nineteenth century; and around Lake Lucerne from 1477 to 2005. In order to accommodate this, the geographical system we developed also allows the user to select historical maps for each model region.

Consider, for example, Figure 4.2, which displays the literary geography of Paul Leppin's 'Das Gespenst der Judenstadt' (1914), a story which is set in the former 'Judenstadt' (Jewish Ghetto) in Prague. The background map displays only the grid of the streets and some building blocks; but the historical map, provided on the left side of the screen, offers an abundance of details that can be explored through a virtual magnifying glass. By studying the representation of the historical – and now vanished 'Judenstadt' – with its narrow streets and maze-like backyards, the user obtains an even deeper appreciation of the urban space depicted in Leppin's narrative.

Ultimately, however, the major advantage of a whole databased system is that the user is able not only to filter data at will, but also to compare maps alongside one another and, furthermore, to use the same sample selection

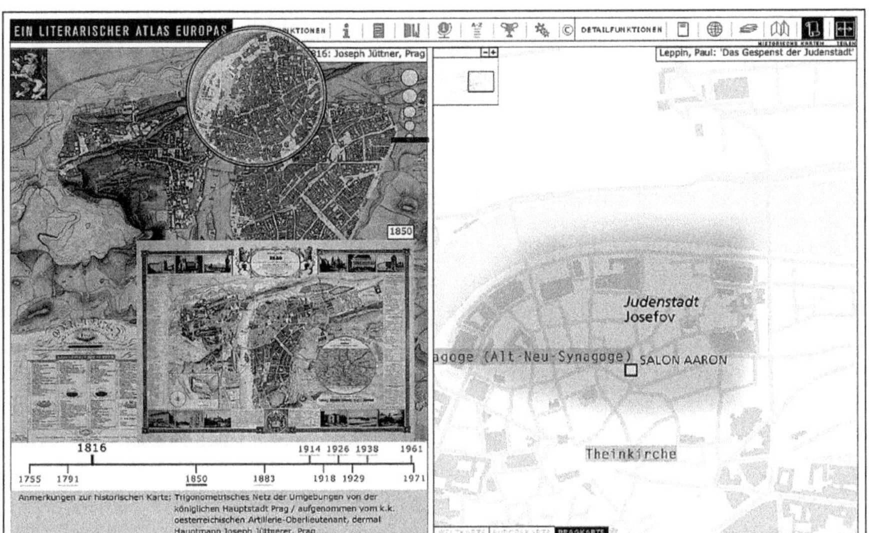

Figure 4.2 Extract from the scenario for *A Literary Atlas of Europe.*

Source: © ETH Zurich, Institute of Cartography and Geoinformation.

Figure 4.3 Literary gravity centres and peaks around Lake Lucerne 1800–2006.
Source: © ETH Zurich, Institute of Cartography and Geoinformation.

of data to create different visualisations and representations from maps in 2D and 3D to diagrams and charts. The data could be displayed as a simple map, in the way Storm's territory is presented in Plate 6, but the system also allows an output in a 3D-visualisation. Figure 4.3, for instance, shows the gravity centres of fictional settings around Lake Lucerne, from 1800 until 2006 in 3D. The green peaks indicate places with a high density of fictional accounts. In other words, then, a literary-geographical system offers many options and it is the user's choice which mode of visualisation best serves his or her needs.

An Outlook: The Limits and Potentials of Literary Cartography

Thanks to innovations in digital mapping technology, literary cartography stands at the brink of a new era. For a long time, there were far more ideas regarding literary mapping than mapping solutions available. The rich possibilities of a digital, interactive animated cartography, with database support, have only recently begun to be explored. GIS technologies – that is, digital, interactive and animated mapping – seem to be increasingly able to facilitate the visualisation of literary space. They offer a range of options such as cartographic layering and the side-by-side comparison of multiple maps. What is more, they allow the user both to toggle between micro- and macro-cartographies and to call up additional information in the form of texts, diagrams and information graphics. Hence, it is not surprising that, over the past twenty years, literary scholars have increasingly come to perceive the potentials of mapping-based research. But in order for literary scholars to conduct such research, expertise from the field of geographic information sciences is needed. Through the development of more sophisticated geovisualisation techniques, more literary scholars might be tempted to revisit key issues of literary geography and to examine what elements of literary space can be mapped. Equally, literary scholars also need to question what elements are ultimately unmappable and should, therefore, be appreciated as such. They might also assess whether different types of cartographic visualisations are needed for different literary periods or genres. Is there, for example, a need to develop different mapping strategies for Victorian and Modernist novels? For works of prose and poetry? For urban and rural settings?

Addressing these questions requires literary scholars and digital cartographers to work together, and this collaboration will require both groups to give up a certain amount of intellectual autonomy. This, of course, will be a highly demanding task. Some literary-geographical maps will simply support and underline facts that could also be retrieved from pure readings. In this case they will function as – often revealing – *illustrations*. Even more exciting, of course, will be those maps in which new spatial patterns appear: those maps that make what is otherwise invisible visible. In such instances, a map will become a true *tool of interpretation*. But, in all cases, it is crucial to accept that the maps in the field of literary geography will rarely provide final answers, especially since their production process is connected with a range of uncertainties.

One demand within literary studies is to find a way to standardise criteria and visualisation methods to be able to compare results from different investigations and different projects. Ungern-Sternberg has put it as follows: 'It is important and should be possible to develop cartographic standards that ensure the comparability of such undertakings', therefore 'we have to find and agree on criteria, scales and methods of how to transform the literary rhetoric of space into a pictorial system' (Ungern-Sternberg 244). Hence, future mapping approaches and visualisations should serve as a contribution towards one possible common 'pictorial language' for literary cartography. The field of literary mapping is a dynamic and growing one, and a series of completed or ongoing projects follow similar or related goals to *A Literary Atlas of Europe*, such as *Viennavigator*, *Mapping St Petersburg* and the *Cultural Atlas of Australia*. The most important prerequisite for being able to compare and link such projects, their collected data and their results, will be the development of a transferable typology and symbology. For the scientific community, this could help create new research topics and generate new questions. For now, then, the approaches of literary cartography clearly have the potential to stimulate progress in two areas. Firstly, they can provide detailed profiles of single fictionalised spaces (or model regions) based on the maps of individual works. Secondly, they can simplify the comparison between such model regions. Such approaches could become future, innovative chapters in a cartographically supported and hence spatially organised history of literature.

It is important, of course, for all literary map-makers to reflect on the actual meanings of the twin terms – literary geography and literary cartography – with which this chapter began. At present, a twofold answer can be provided. First and foremost, the process of mapping literary texts facilitates a richer, deeper understanding of the (imaginary) geographies of such texts. The second part of the answer goes beyond the purely academic interest. That is to say, one could even dare to speak of a general benefit. A literary-geographical reading can change our understanding not only of books, but also of the world in which we live. Through literary geography, after all, we learn more about the production of places, their historical layers, their meanings, functions and symbolic values. If places emerge from a combination of real elements and fictional accounts, then the tools and means of literary mapping can open up new dimensions in terms of 'reading' a landscape or cityscape. UNESCO, the United Nations agency responsible for promoting international collaboration through education and culture, offers us an example. Amongst the types of landscapes worthy of UNESCO's coveted World Heritage label are so-called 'associative cultural landscapes'. These landscapes, according to UNESCO, can be awarded World Heritage status 'by virtue of the powerful religious, artistic or cultural associations of the natural element rather than material cultural evidence, which may be insignificant or even absent' (UNESCO). This definition – a layer of cultural associations without direct material evidence – is perfectly applicable to fictionalised cities, regions and landscapes, which are subject to literary geography and literary cartography. Against this background, literary-geographical methods can provide powerful analytical tools for exploring the boundaries between fiction and reality.

In literary geography and literary cartography there is a need for flexibility, for freedom to experiment, to imagine and even to take risks. The range of maps developed by literary mapping projects is growing and impressive: from hand-drawn maps to automatically rendered ones (Bär and Hurni). Yet, in spite of this diversity, the best literary maps continue to share a common ground in that they stem from research that regards literary mapping as a practice that returns the reader to the text with new questions and ideas.

Works Cited

Bär, Hans Rudolf, and Lorenz Hurni. '*Improved Density Estimation for the Visualisation of Literary Spaces*'. The Cartographies of Fictional Worlds. *The Cartographic Journal* 48.4 (2011): 309–16.

Bodenhamer, David J., Trevor M. Harris and John Corrigan. 'Spatial Narratives and Deep Maps: A Special Report'. Deep Mapping and the Spatial Humanities. *International Journal of Humanities and Arts Computing* 7.1–2 (2013): 170–75.

Bradbury, Malcolm. *The Atlas of Literature.* London: De Agostini Editions, 1996.

Bushell, Sally. 'The Slipperiness of Literary Maps: Critical Cartography and Literary Cartography'. *Cartographica* 47:3 (2012): 149–60.

Cultural Atlas of Australia. 2012. Web. 20 March 2015. <http://australian-cultural-atlas.info/CAA/index.php>.

Döring, Jörg. 'Zur Geschichte der Literaturkarte (1907–2008)'. *Mediengeographie. Theorie – Analyse – Diskussion.* Eds, Jörg Döring and Tristan Thielmann. Bielefeld: transcript, 2009. 247–90.

Flaubert, Gustave. *Madame Bovary.* Trans. Eleanor Marx-Aveling. The Electronic Classics Series. Web. 20 March 2015. <http://onlinebooks.library.upenn.edu/webbin/gutbook/lookup?num=2413 >.

Hewitt, Rachel. 'Mapping and Romanticism'. *Wordsworth Circle* 42.2 (2011): 157–65.

A Literary Atlas of Europe. 2015. Web. 20 March 2015. <http://www.literaturatlas.eu/en/>.

Mapping St Petersburg Experiments in Literary Cartography. 2011. Web. 20 March 2015. <http://www.mappingpetersburg.org/site/>.

Moretti, Franco. *Graphs, Maps, Trees. Abstract Models for a Literary Theory.* London: Verso, 2005.

Nagel, Siegfried Robert. *Deutscher Literaturatlas. Die geographische und politische Verteilung der deutschen Dichtung in ihrer Entwicklung nebst einem Anhang von Lebenskarten der bedeutendsten Dichter.* Wien: C. Fromme, 1907.

Piatti, Barbara. *Die Geographie der Literatur. Schauplätze, Handlungsräume, Raumphantasien.* Goettingen: Wallstein, [2008] 2009.

Piatti, Barbara, and Lorenz Hurni. 'Mapping the Ontologically Unreal: Counterfactual Spaces in Literature and Cartography'. *The Cartographic Journal,* 46.4 (2009): 333–42.

———. Piatti, Barbara and Lorenz Hurni, 'Editorial: Cartographies of Fictional Worlds'. *The Cartographic Journal.* 48.4 (2011): 218–23.

Piatti, Barbara, Anne-Kathrin Reuschel, Hans Rudolf Bär, William Cartwright and Lorenz Hurni: 'Mapping Literature: Towards a Geography of Fiction'. *Cartography and Art, Lecture Notes in Geoinformation and Cartography.* Eds, William Cartwright, Georg Gartner and Antje Lehn. Heidelberg: Springer, 2009. 177–92.

Reuschel, Anne-Kathrin, and Lorenz Hurni. 'Mapping Literature: Visualisation of Spatial Uncertainty in Fiction'. *The Cartographic Journal* 48.4 (2011): 79–94.

Reuschel, Anne-Kathrin, Barbara Piatti and Lorenz Hurni. 'Modelling Uncertain Geodata for the Literary Atlas of Europe'. *Lecture Notes in Geoinformation and Cartography, Understanding Different Geographies.* Eds, Karel Kriz et al. Wien: Springer, 2013: 135–57.

Sharp, William. *Literary Geography.* London: Pall Mall, 1904.

Soja, Edward W. *Thirdspace: Journeys to Los Angeles and Other Real-and-Imagined Places.* Oxford: Blackwell Publishers, 1996

UNESCO. 2011. Web. 20 March 2015. <http://en.unesco.org/>.

Ungern-Sternberg, Arnim von. 'Dots, Lines, Areas and Words. Mapping Literature and Narration (With some Remarks on Kate Chopin's "The Awakening")'. *Cartography and Art, Lecture Notes in Geoinformation and Cartography.* Eds, William Cartwright, Georg Gartner and Antje Lehn. Heidelberg: Springer, 2009. 229–52.

Viennavigator. N.d. Web. 20 March 2015. <http://viennavigator.metaspots.net/front/desktop>.

5 Bloomsday's Big Data

GIS, Social Media and James Joyce's *Ulysses*

Charles Travis

Introduction

As James Joyce once famously boasted to his friend, Frank Budgen, his aim in writing *Ulysses* (1922) was 'to give a picture of Dublin so complete that if the city one day suddenly disappeared from the earth it could be reconstructed out of my book' (Budgen 69). Ironically, it is now Joyce's 'Big-Data' novel (eighteen episodes, 740 pages, 265,000 words and a lexicon of 30,030 terms, titles and expressions) that is being digitally repurposed and mapped onto the streets of the twenty-first century city during its annual Bloomsday celebrations. Thanks to the digital revolution and the proliferation of social media networks, virtual performances of Joycean Dublin occur spontaneously not only on 16 June, but over the course of the year on local and global scales.

In a phenomenological sense, *Ulysses* has always served as a literary geographical information system-cum-time-machine to transport tenacious readers willing to navigate the novel's kaleidoscopic stream-of-consciousness into the balmy environs, odours, ambience and slow bustle of an Edwardian summer's day in Dublin. Arriving at 8.00 in the morning on 16 June 1904, readers simultaneously experience dawn rising over the Irish Sea from a Martello tower in Sandycove, whilst enjoying a breakfast of kidneys, scented faintly with the tang of urine, from Duglatz's Butchers on the north-side of the Liffey. After negotiating the labyrinth of city streets, libraries, museums, parks and graveyards, eating a gorgonzola sandwich at Davy Byrne's pub and escaping the nationalistic raves of the Citizen in another, the torrid heat of Bella Cohen's brothel and the unwanted attention of a pair of British soldiers at midnight, readers of *Ulysses* soon drink Epps Soluble Cocoa and discuss Science and Art at 1.30 in the morning on 17 June, whilst gazing at the 'heaven-tree of stars' (Joyce, *Ulysses* 819) from the garden of a house at 7 Eccles Street in Dublin's north-inner city.

Joyce consulted *Thom's Directory* whilst working on *Ulysses* and 'wrote the *Wandering Rocks* with a map of Dublin before him on which he traced in red ink the paths . . . and calculated to a minute the time necessary for his characters to cover a given distance of the city' (Budgen 124–5). In

1904, the year in which Joyce set his novel, the only messaging service, aside from couriers and telegraphy, was an efficient postal service which could deliver letters across the city several times a day, allowing individuals to engage in multiple correspondences from sunrise to sunset. Today, with the aid of GPS-enabled smartphones and social media platform communications, it is not only possible to communicate instantaneously but also to download virtual maps and guides such as *Joyceways*: a selective, erudite app, albeit stricken with common Google Map errors, which will be discussed in section 2 below. On a more sophisticated level, it is possible to use Global Positioning Satellite (GPS) devices, open source geographic information systems (GIS) and online database structures to conduct digital-hermeneutic mappings to correlate specific sites mentioned in *Ulysses* with social media network-enhanced Bloomsday celebrations at local and global scales.

This chapter outlines the creation and implementation of a Humanities Geographic Information System/Social Media (hereafter, HumGIS/SM) model of *Ulysses* drawing on the cartographical and literary tools Joyce himself employed to conceptualise and publicise his novel. By spatially contextualising *Twitter* social network feeds (*Flickr* and *YouTube* will also be discussed) with the cartographic structure of Joyce's novel, digital literary mapping methods can be engaged to plot a repurposing narrative of *Ulysses* as it 'blooms' spontaneously across the streets of Dublin and over the span of the globe. This model cross-pollinates literary, historical and neo-geographical tropes and techniques to explore the relations between text, time, social media network language activity and digital ecosystem productions of social and cultural space. The emergence of these types of digital systems and relations promises to not only change the ways in which scholarship is expressed, but more profoundly the manner in which it is measured, perceived, assessed and consumed.

Literature, the Digital Revolution and Social Media

In the past decade, the confluence of mobile telecommunications, online geo-spatial platforms and social media networks has disseminated augmented realities and e-lexicons into the sphere of mass-produced popular fiction, in addition to influencing and shaping authors' writing styles and depictions of place. In particular, there seems to be a geo-techno-cultural convergence in the public consumption of mass market 'pulp'/'airport' fiction crime thrillers and the ubiquitous use of iPad, laptop and smartphone technology. Paul Hoggart once observed that 'if you want to get under the skin of a foreign culture, skip the guided tours of cathedrals and art galleries, and read its crime thrillers' (Hoggart); and whether styled 'highbrow' or mass market, it can be seen that the literary tropes of contemporary mass market fiction and the techno-tropes of augmented reality devices are cross-pollinating each other's cultural spheres in interesting and unpredictable ways to re-purposely shape and convey perceptions and experiences of place. For example, the techno-culture of augmented reality has become strongly apparent during the last decade in works of popular crime thriller novelists Michael Connolly and Lee Child. In Connolly's novel, *The Gods of Guilt* (2013), a GPS 'low-jack' tracking device allows the Lincoln Lawyer Mickey Haller to play out a cat and mouse scenario in which iPhone records, videos and online escort services become plot devices to map the CCTV urban geography of a sinister city of cameras. In Lee Child's novels, the one-man wrecking crew and de-mobbed Military Policeman Jack Reacher embodies a walking smartphone. Child's geospatially-referenced dialogue structure shapes Reacher's voice, as locations, vectors, directions, travel time, encyclopaedic knowledge and philosophical references embed themselves in his observations about the places in which he finds himself, whilst waging one-man wars in the desolate small town and urban spaces of the American heartland. Both Connolly and Child have produced bodies of work in which their narratives and characters witness societies shifting from emergent telecommunications, open-source, social media technology to the use of geospatially augmented smartphone technology realities. The geo-technology trend is also showing its presence in contemporary 'literary fiction'. The Dutch born narrator of Joseph O'Neill's *Netherland* (2008) employs *Google Earth* as a vicarious device to remedy the home sickness colouring his life in Manhattan's Chelsea Hotel. In O'Neill's *The Dog* (2014) his main character employs *Google StreetView* to virtually roam the Eastern European hometowns of the escorts he consorts with as a philanthropic trustee in Dubai.

Geo-web technologies are increasingly used to market and virtually enhance the reader's experience of not only Connolly and Child's novels but also those of Ian Rankin, Stieg Larsson and other authors, whose works have been digitally mapped, including Charles Bukowski, Agatha Christie, Ursula K. Le Guin and Lisa Jackson. Indeed, many contemporary authors have dedicated web pages and apps which, when downloaded to GPS-activated smartphones, provide digital tour guides of the sites and settings of their novels, allowing readers to engage in social media facilitated performances of literature, narrative and place, which in turn facilitate the posting of their augmented reality experiences on *Twitter, Flickr* and *YouTube.* As a result, new digital ecosystems are emerging as literature and its tropes are infusing and influencing geospatial technology-aided social productions of space and vice-versa.

Samuel Beckett, Joyce's countryman and protégé, once asked: 'must literature alone be forever left behind on worn out paths abandoned long ago by music and painting?' (Cochran 92). The same question might be asked, today, about academic criticism of literature and the arts. The emergence of these new types of digital ecosystems triangulated by social media, and the indices of literature and history, promises not only to change the ways in which such scholarship is expressed but more profoundly the manner in which it is received, analysed and experienced. Although it has been argued that social media influences behaviours and perceptions, it can also be argued that deeper cultural forces contextualise and mediate such technological agencies (Long and Travis). From a certain perspective, Joyce's *Ulysses* can be seen to have subverted British imperial cartography with a kaleidoscopic narrative that escaped the nets of colonial latitude and longitude. Today, that same novel reaches across space and time to mediate language and cultural activity, on what have been described as 'neoliberal' (Leszczynski) geo-techno-cultural social media networks, during the annual Bloomsday celebrations. However, can such social media performances subvert their own technology, as Joyce's work subverted the cartography which helped shape it?

Joyce's Cartography

Joyce composed *Ulysses* in self-imposed exile from Ireland, re-imagining Dublin from fragments of memory and personal correspondence, whilst the First World War detonated national cataclysms across the world. In the novel he laments, through Stephen Dedalus, that 'history . . . is a nightmare from which I am trying to awake' (Joyce, *Ulysses* 42). *Ulysses* takes place largely within the consciousness of his principal characters: the twenty-two year old Dedalus and thirty-eight year old advertising salesman, Leopold Bloom. In considering how to create a HumGIS/SM model of *Ulysses*, despite the current convergence of mass-market fiction and digitally augmented realities of place, it was recognised that the relations between accessible geographic information and the construction of fictional narratives is a long-standing and significant one. A century ago, urban information was commonly compiled in city gazetteers supplemented by foldout maps. The 1904 edition of *Thom's Official Directory of the United Kingdom of Great Britain and Ireland*, employed by Joyce whilst writing *Ulysses*, is one example. The map allowed Joyce to 'remotely sense' the layout of Dublin's streets, districts, pubs, churches, houses and neighbourhoods as he plotted – from his various locations in Trieste, Zurich and Paris – the journeys of Dedalus and Bloom as they crossed the city on the 16 June 1904. It has long been maintained that the trajectories of Joyce's characters across the city were based upon Homer's depiction of the 10-year journey of Odysseus and his crew after the end of the Trojan War. The Greek hero's travels across the Mediterranean and Aegean seas depicted in the *Odyssey* form the basis for the eighteen episodes in *Ulysses*. The episodes transposed from the Homeric epic upon the streetscapes of Edwardian Dublin follow the journeys of Dedalus and Bloom (as well as those of other key figures) around the city from 8.00 on the 16 June, to 2.00–4.00 the next morning. Both men are experiencing alienation in their daily lives: Dedalus is in mourning for his mother and troubled by his refusal to pray at her deathbed; Bloom, a Jew in a largely Catholic country, carries the burdens both of the recent death of his son Rudy and of knowing that his wife Molly is cuckolding him with a *bon-vivant* named Blazes Boylan. Their separate peregrinations across the different districts and neighbourhoods of Dublin over the course of the day are joined during the phantasmagorical 'Nighttown' section of *Ulysses* in Bela Cohen's brothel. The pair then walk to Bloom's house on 7 Eccles Street where, after a short visit, Dedalus goes his own way. The novel closes with a monologue capturing Molly Bloom's thoughts and feelings about her husband as she falls into the arms of Morpheus and passes into a dreamscape of desire and longing.

The 1904 *Thom's* edition of *Dublin and Environs*, consulted by Joyce when writing *Ulysses*, reflected the 'official' statistical records of 1903 and displayed the city's twenty wards, supplemented with a directory, street lists, tradesman's catalogues and census counts. The map and its directory allowed Joyce to conceive mental schemas upon which to weave his kaleidoscopic 'verbal representation of Dublin' (Hart and Knuth 13). Both Eric Bulson and Jon Hegglund have already suggested how this creative engagement with cartography can be connected with the discourses of imperial power and colonial subjugation that resonate throughout Joyce's writings on Ireland. Although I am aware of the special place that *Ulysses* occupies within this critical context, here I am more interested in the ontological implications for literary mapping in the digital age found in Joyce's creative blending of discursive and cartographical tools. Budgen recalled that 'to see Joyce at work on the Wandering Rocks' episode of the novel 'was to see an engineer at work with compass and slide-rule, a surveyor with theodolite and measuring chain, or more Ulyssean perhaps, a ship's officer taking the sun, reading the log and calculating current drift and leeway' (Budgen 123). Joyce's literary navigation of Dublin was also informed by the visual techniques of Cubism, Futurism and Dadaism. Budgen pointed out that 'the multiplicity of technical devices is proof that Joyce subscribed to no limiting aesthetic creed and proof also that he was willing to use any available instrument that might serve his purpose' (198). One can only imagine what might have occurred if Joyce, in addition, to *Thom's* map, had enjoyed access to the neogeographical survey methods enabled by today's GIS and Global Positioning Systems devices, as well as by *Google Maps*, *Fusion* and *Earth*, and by smartphone technology and social-networking platforms such as *Twitter*, *Flickr* and *YouTube*. (I coin the neologism *tweetflickrtubing* to describe this methodological integration of social media platforms, geospatial technology and the study of Joyce's work as it intersects with language, history and place.) Would he have creatively repurposed the cornucopia of information from the city's digitised Big Data streams to plot his novel? All such conjecture must be tentative, but it is tempting to think that Joyce's artistic sensibilities might have inclined him to have enjoyed playing with digital mapping and social media technology. Remembering a conversation with Joyce about the *Cyclops* episode in 1919, Budgen recalls the following exchange:

> 'Does this episode strike you as being futuristic?' said Joyce.
> 'Rather cubist than futurist,' I said.
> 'Every event is a many-sided object. You first state one view of it and then you draw it from another angle to another scale, and both aspects lie side by side in the same picture[.]' (Budgen 156–7)

When Joyce created the labyrinthine routes of Dedalus and Bloom crossing Dublin on 16 June 1904, it is as though he anticipated artistic mapping techniques made possible now not only by neogeographical applications, but also by Immersive/Experiential GIS and *Second Life* virtual platforms. The former approximates a *Star Trek* 'holo-deck' experience by projecting three-dimensional GIS-generated models of physical and historical landscapes upon the walls, floor and ceiling of an enclosed space to give the viewer the sense of being embodied in another landscape, place or period. The latter, provides the user with a virtual and interactive avatar that allows users to navigate digitally sculpted environments.

Translating *Ulysses* for GIS and Social Media Mapping

The HumGIS/SM model of *Ulysses* inspired by this imaginative counterfactual exercise draws specifically upon the 1920 schema Joyce created for his Italian critic and translator Carlo Linati to whom he wrote: 'in view of the enormous bulk and the more than enormous complexity of my damned monster-novel it would be better to send . . . a sort of summary-key-skeleton schema' (Ellman 187). The *Linati Schema* was one of three versions Joyce drafted during and after the completion of his novel. One other was the 1921 schema designed for the American academic Stuart Gilbert, who published the first definitive study of Joyce's novel in 1930. The schemas served a few purposes for Joyce. The first partial-schema he drafted aided him (along with *Thom's* map) in plotting *Ulysses*; but the 1920 and 1921 schemas were presciently employed to outline the novel's sub-structure for selected critics, such as Linati and Gilbert, in order to facilitate the critical decoding of his work for the emerging academic audience of the early twentieth century. Joycean disciples soon embraced this 'mathematical catechism' (Joyce, *Letters* 159) and established a literary cult whose exegetical outputs have devastated vast forests and consecrated the greasy tills of the annual Bloomsday pilgrimage. Lastly, the schemas were used to market his sprawling book, structuring its notoriously convoluted narrative into a more comprehensible form, particularly for American publishers, but also for Hollywood filmmakers, as after its initial publication (and thanks in no small part to the notoriety it attained from the ensuing obscenity trial) interest was raised by the cinema mogul Samuel Goldwyn to produce a film based upon the novel. In any event, Joyce was rewarded with a high profile review in the *New York Times* and an appearance on the cover of *Time* magazine.

From a GIS perspective, the schemas can be seen as proxy attribute databases, containing a cornucopia of relevant information which, when geocoded to the actual sixteen sites in Dublin where Joyce set the eighteen episodes of *Ulysses,* provide templates for a spatialised literary analysis of his novel. Both the 1920 and 1921 schemas list the eighteen Homeric episodes, but have individual variations and slight chronological differences. The *Linati Schema* (1920) designates place, title, time, colour, people, science/art, meaning, technic, organ and symbols. In contrast the *Gilbert Schema* (1921) is shorter, listing title, scene, hour, organ, colour, symbol, art and technic. The schemas can be employed to cartographically visualise and connect the eighteen interweaving spatial narrative strands composing the fabric of Joyce's novel to the sixteen actual sites in Dublin. The HumGIS/SM model can then synchronise the resulting layers of images, words and vectors into a contrapuntal, multi-dimensional digital hermeneutic, providing the means to 'reconnect the representational spaces of literary texts not only to material spaces they depict, but also reverse the moment' (Thacker 63). With HumGIS/SM techniques we can further operationalise Andrew Thacker's perspective of literary geography by integrating the socio-spatial dialogics of Bloomsday social media posts and streams into a multitude of contemporary reconnections and reversals intersecting the material spaces of Dublin and the textual spaces of *Ulysses.*

Creating the HumGIS/SM Model

To create the HumGIS/SM model of *Ulysses,* a spreadsheet (see Figure 5.1) was populated with information from the *Linati Schema* (since it contains

	Titolo	*Ora*	*Colore*	*Persone*
1	Telemachus	8–9	oro, bianco	Telemaco Antinoo { Mentor Pallas I Proci Penelope (Madre)
2	Nestore	9–10	marrone	Nestore Telemaco Pisistrato Elena
3	Proteo	10–11	azzurro	Proteo Menelao Elena Megapente Telemaco

LOCATION	PLACE	TITLE	TIME
53.288665, -6.113593	Sandycove	Telemachus	8am - 9am
53.273294, -6.109999	Dalkey	Nestor	9am - 10am
53.332151, -6.200680	Sandymount Strand	Proteus	11am - 12am
53.357701, -6.264365	Eccles Street	Calypso	8am - 9am
53.343500, -6.249671	Westland Row	Lotus Eaters	9am - 10am
53.372631, -6.276880	Glasnevin Cemetary	Hades	11am - 12pm
53.348136, -6.261678	Abbey Street	Aeolus	12pm - 1pm
53.340286, -6.255029	National Museum	Lestrygonians	1pm - 2pm
53.341077, -6.254525	National Library	Scylla & Charybdis	2pm - 3pm
53.326725, -6.228801	RDS Ballsbridge	Wandering Rocks	3pm - 4pm
53.345883, -6.269784	Ormond Quay	Sirens	4pm - 5pm
53.349474, -6.269108	Capel Street	Cyclops	5pm - 6pm
53.326936, -6.208234	Sandymount Strand	Nausicaa	8pm - 9pm
53.340265, -6.245988	Holles Street	Oxen of the Sun	10pm - 11pm
53.351828, -6.254225	Mabbot Street	Circe	11pm - 12am
53.348309, -6.254961	Butt Bridge	Eumaeus	12am - 1am
53.357701, -6.264365	7 Eccles Street	Ithaca	1am - 2am
53.357701, -6.264365	7 Eccles St. Bedroom	Penelope	Infinity

Figure 5.1 Fragment of the Carlo Linati Schema (1920).

Source: © The Author; Ellman 1974.

more comprehensive data than the *Gilbert Schema*) and geo-coded according to decimal degree designations of latitude and longitude to sixteen Dublin sites described by Joyce in the eighteen Homeric episodes of the novel. Both the 1904 *Thom's Map of Dublin and Environs* and Ian Gunn and Clive Hart's *James Joyce's Dublin* guided the use of the 'what's here' function of the *Google Maps* app and GIS 'ground-truthing' methods (whereas the former displays decimal degree coordinates and location information and images, the latter monitors site visits in order to reconcile cartographically calculated positions with 'field' sourced geospatial data: a necessity with *Google Maps* as its coordinate designation and location descriptions can be mismatched and erroneous). Subsequently, sites were identified and then geo-coded either at the beginning, end or centre of each episode's spatial range, since various characters' movements and positions in *Ulysses* (such as in the *Wandering Rocks* episode) occur simultaneously and often at multiple locations within and beyond the geographical and temporal boundaries of each episode designated by the *Linati Schema*. The geo-coded *Excel* database was then imported into *Google Fusion*, and site locations visualised through its *Google Maps* function (see Figure 5.2). The database was also converted

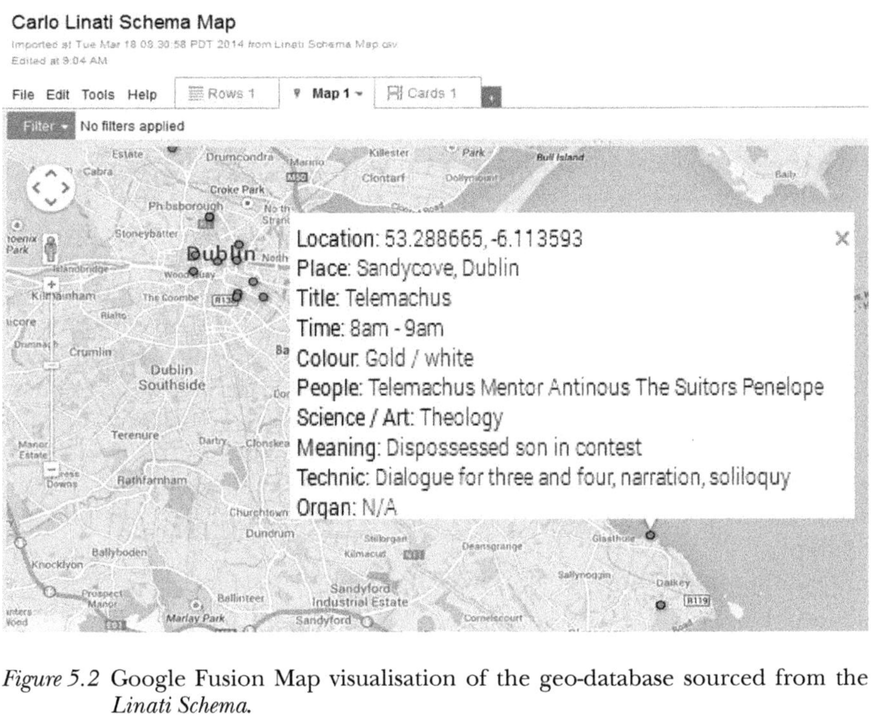

Figure 5.2 Google Fusion Map visualisation of the geo-database sourced from the *Linati Schema*.

Source: © The Author; Google Fusion & Google Maps 2014.

to a CSV file and imported into the *ArcGIS Online* platform and integrated with a live social media map layer.

Integrating social network feeds from *Twitter* (as well as *Flickr* and *YouTube*) within the HumGIS/SM model of *Ulysses* created from the *Linati Schema* database provides a way to plot Joyce's convergences of language, narrative and place as they mingle with the Babel-like 'e-semiologies' produced by the intersecting axes of Homeric episode sites and live social media post locations on Bloomsday. With this approach it is possible to visualise the geo-linguistic landscape of *Ulysses*, and how it is repurposed through twenty-first-century social media translations not only in the streetscapes of Dublin but also on a global scale.

Mapping Bloomsday

The HumGIS/SM survey of Bloomsday 2014 was performed remotely due to my attendance at a Global Humanities conference in Hannover, Germany on 16 June. Employing 'Bloomsday' as a keyword in the *Twitter* (as well as the *Flickr* and *YouTube*) search engine, I conducted hourly surveys (from the Central-Hotel Kaiserhof and the Herrenhausen Palace) using my laptop computer and a cloud-based digital mapping service hosting my database visualisations (MacBook Pro and ArcGIS Online). Over the course of the day, I watched as Joycean enhanced reality performances of language and place proliferated across the HumGIS/SM digital model of Dublin and created clusters of activity across the globe. The predominant activity in Dublin occurred on *Twitter*, with 56 tweets recorded. But random surveys on local and global scales taken before and after 16 June 2014 revealed that *Flickr* and *YouTube* posts exhibited the most prolific activity, with hundreds, if not thousands of individual feeds coalescing into a Bloomsday 'data squall'. In addition, tweets were often hyperlinked to other websites, photo images and video-clips, forming waxing and waning constellations of digitally connected networks and ecosystems. Bloomsday social-media posts suggest that *Ulysses*, rather than being just a solitary reading experience of a select academic cadre, has transformed into a collective spectacle of performance and social production, generating a global digital ecosystem obviously rooted in Dublin, but possessing world-wide electronic tentacles where awareness of Joyce's novel, internet penetration, social media platforms, smartphones and tablets are all available and are all in play.

The Bloomsday survey was divided into two categories – local and global – and weighed by place and chronology respectively. The first survey focused on *Ulysses*' Homeric episode sites in Dublin specifically; the second survey focused on a global chronology of Bloomsday *Twitter* activity, including Dublin-based outliers (posts not included in cluster formations) not located near any of the sites identified in the first survey. *Flickr* and *YouTube* posts were monitored as well, and did appear spontaneously on Bloomsday, although not as frequently as tweets. Photographs and videos with time-stamps for 16 June tended not to be spontaneously posted but, rather, were subsequently uploaded to these two visually themed platforms during the following weeks. In general, *Flickr* and *YouTube* reflected annually aggregated social media activity indicating that Bloomsday digital-ecosystems seem to proliferate on a spontaneous basis, illustrating a sustained interest and virtual interaction with the Joycean dimensions of Dublin.

Dublin Episode Site Tweets

Social media activity on 16 June 2014 in Dublin related to Bloomsday celebrations indicated that the majority of the *Twitter* activity clustered around eight of the sixteen sites in Dublin linked to the eighteen episodes in *Ulysses*. Tweet time-stamps did not necessarily correspond with the *Linati Schema*'s chronology. It seems that the variable of place dominated the variable of chronology in determining the time of the day in which people engaged in social media activity on Bloomsday in Dublin. On a descending scale of activity, tweets were posted from around the following sites:

1 **The National Museum & Davy Byrne's Pub (*Lestrygonians*, LS 13.00–14.00)**

- 12.50 Nassau Street / Spar Shop / Porterhouse: 'There are a lot of people in Dublin wearing funny hats. It must be Bloomsday'.
- 14.29 Duke Lane Upper: 'Bloomsday in Dublin'.
- 17.15 Duke Lane Upper: 'There are about 50 quite elderly ladies getting rowdy in Davy Byrne's'
- 17.25 Davy Byrnes Pub: 'Bloomsday at Davy Byrnes can't be beaten'.

2 **Joyce Martello Tower at Sandycove (*Telemachus*, LS 8.00–9.00)**

- 06.46 Sandycove: 'A Grand day for a skyte 'round Dublin #*Bloomsday*'
- 13.48 Sandycove: 'Not even Marilyn Monroe forgets #*bloomsday*. Sun is shinning (sp) and it's a day where we remember one of Ireland's most famous writers'.
- 19.43 Sandycove: 'Warm sunshine marrying over the sea . . . & the views are epic'.

3 **Dalkey (*Nestor*, LS 9.00–10.00)**

- 08.39 Barnhill Rd., Dalkey: 'It's *Bloomsday*!!!!!'
- 12.08 Dalkey Dart Station: 'Bloomsday! Readjoyce Readjoyce'
- 21.43 Johnston Rd., Dalkey: V. Woolf: 'Joyce – an undelivered genius, whom one can't neglect . . . but must help them out, at considerable pains to oneself. Ouch'.

4 **Glasnevin Cemetery (*Hades*, LS 11.00–12.00)**

- 12.25 Gravedigger's Pub, Glasnevin: 'Nice way to start Bloomsday & our 1st Wedding Anniversary . . . thanks to all our team & customers!'
- 20.32 Gravedigger's Pub, Glasnevin: 'Still Bloomsday!'

5 **The National Library (*Scylla and Charybdis*, LS 14.00–15.00)**

- 14.00 Dawson Street: 'Bloomsday reading @Hodges_Figgis kicking off now'.
- 14.09 Dawson Street: 'Mary Morrissy about to read from her story in #Dubliners100 @Hodges_Figgis'

6 **Eccles Street (*Calypso*, LS 8.00–9.00)**

- 20.54 Dorset Lane: 'Am I obliged to drink because its Bloomsday??'

7 **Westland Row (*Lotus Eaters*, LS 9.00–10.00)**

- 21.44 Lincoln Place: Happy Bloomsday! 'Where Leopold Bloom bought his lemon soap'.

8 **Sandymount Strand (*Proteus*, LS 22.00–23.00)**

- 11.19 Iveragh Court: 'So it is . . . yes . . . the 16th . . . yes . . . of . . . Ju ne . . . yes. I'll do . . . yes . . . to . . . Sandymount . . . yes . . . and walk the strand . . . yes . . . and . . . yes . . . '.

Davy Byrne's Pub on Duke Street and Joyce's Martello Tower in Sandycove (see Figure 5.3) are respectively associated with the *Lestrygonians* and *Telemachus* episodes of *Ulysses*. These sites were made famous on 16 June 1954 when Patrick Kavanagh, Flann O'Brien and Anthony Cronin inaugurated Dublin's first official celebration by intending to recreate Dedalus' journey in a horse-drawn carriage. The literary sojourn commenced in Sandycove, meandered from Joyce's Martello Tower, and stopped at a few public houses on its way into the city. Either due to acrimony, boredom, drink or a combination of the three, the ill-fated experiment imploded at Davy Byrne's Pub on Duke Street, the site of Leopold Bloom's Gorgonzola Cheese lunch during the *Lestrygonians* episode.

Images of throngs standing around Byrne's pub (see Figure 5.4), sourced by the 2014 Bloomsday survey, suggest that Joyce rightly identified the bodily organ associated with this episode as the 'Oesophagus'. Although the National Library was geo-coded as the literary centroid of *Lestrygonians*, the survey suggests that, because of the social cluster indicated by the number of *Twitter*, *Flickr* and *YouTube* posts from the pub, the episode location should be re-sited to Davy Byrne's.

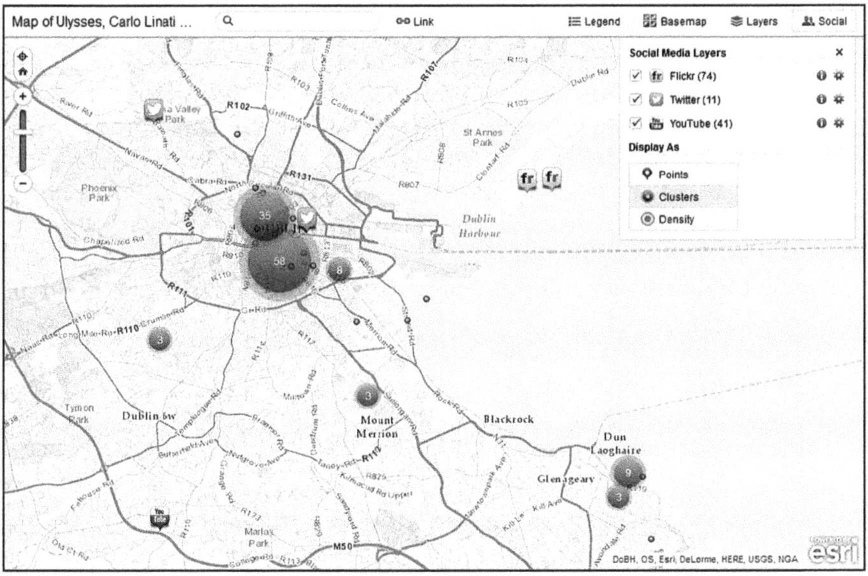

Figure 5.3 Bloomsday Social Media Concentrations.

Source: © The Author; ESRI with Permission.

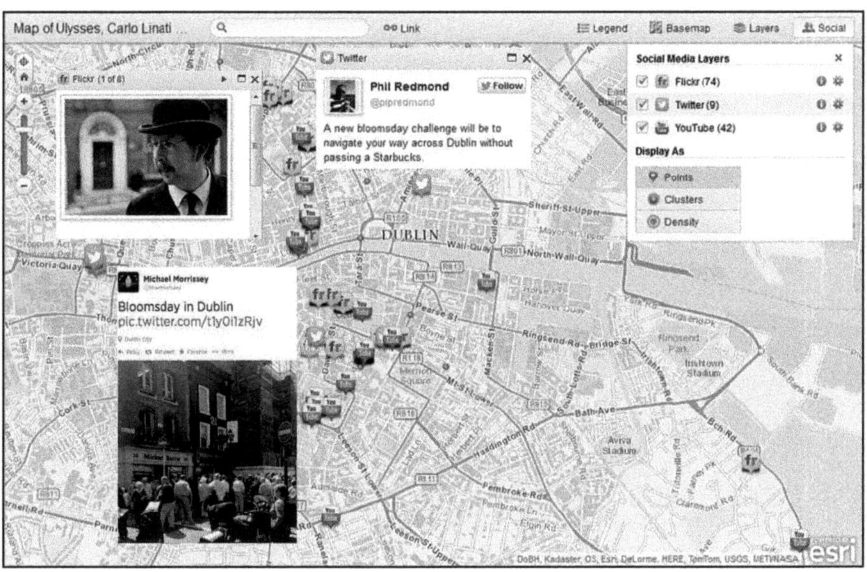

Figure 5.4 *Tweetflickrtubing* Bloomsday in the twenty-first century.

Source: © The Author; ESRI with Permission.

Global Chronological Tweets (GMT)

Global Bloomsday tweets proliferated across Europe, the Middle East, Asia, Australia-Pacific and the Americas (see Figure 5.5). The tweets listed below have been organised by the chronology of Homeric episodes listed in the *Linati Schema*. 'Orphan' tweets (those corresponding outside or to the hourly periods of 19.00–20.00 and 21.00–22.00, and thus not included in the schema) were either listed separately, with preceding episodes, or in the *Penelope* episode, whose temporal aspect encompasses 'Infinity'.

The first Bloomsday tweet originated in the southern hemisphere:

- 01.39 Melbourne, Australia: '@Blueboys2010 yes dear boy, but that's 110 years of controversy'.

I Telemachus and Calypso (8.00–9.00)

- 08.22 Mary's Lane & Church Street, DUBLIN: 'Morning all, and a very happy Bloomsday! I have my finest Edwardian frock on, pipe ready and kidney for lunch perhaps?'.
- 08.27 Mumbai, India: 'I'm taking a step a sinkapace forward on neats leather creaking'.
- 08.33 Singapore, Malaysia: 'Happy #Bloomsday to all in Dublin'.

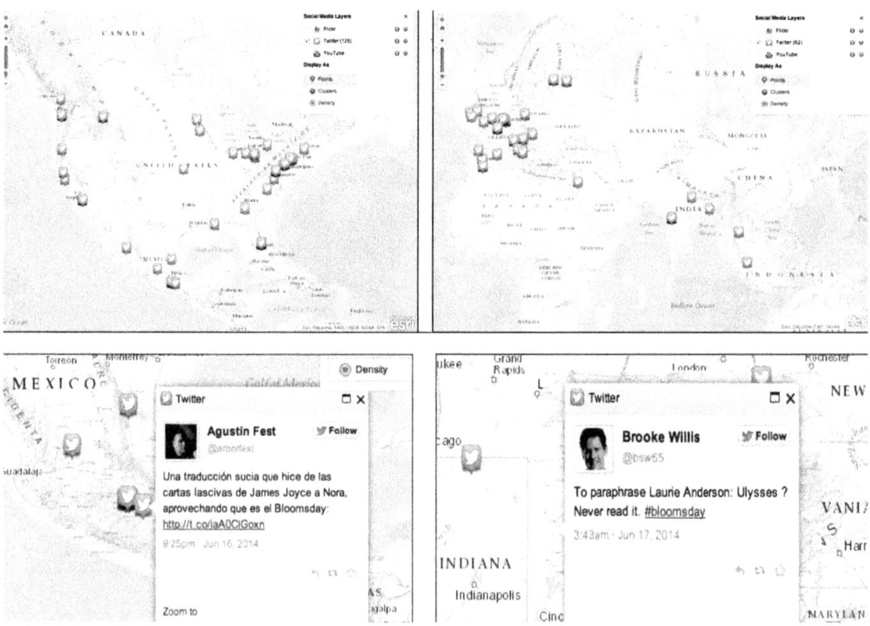

Figure 5.5 Global Bloomsday Tweets 16 June 2014.

Source: © The Author; ESRI with Permission.

- 08.33 Antwerp, Belgium: 'A happy #Bloomsday to all my Joycean friends, and everyone who is attending the #krommerun Joyce symposium!'.

II Nestor & Lotus Eaters (9.00–10.00)

- 09.05 Rome, Italy: 'Buon Bloomsday at tutti [Happy Bloomsday to all]!'.

III Proteus (10.00–11.00)

- 10.14 Longwood Ave. & Windsor Terrace, DUBLIN: 'Finally, I get around to using my Bloomsday Survival Kit- (it was a Christmas present from my daughter [. . .] !)'.
- 10.23 Jerusalem, Israel: '"Stately, plump Buck Mulligan came from the stairhead. . . . ineluctable modality of the visible. . . . . yes" Happy Bloomsday'.

IV Hades (11.00–12.00)

- 11.32 Lullymore Terrace, Dublin: 'Hey @colmtobin, are you pretending to *read* it or pretending to *have read* it?'.
- 11.54 Lower Baggot Street & Grand Canal, DUBLIN: 'Nothing like a combo of World Cup and bloomsday to hear people who don't like either Joyce or football talk about both'.
- 11.56 Brookwood, DUBLIN: 'I might break out Ulysses again in honour of Bloomsday –try to make it past page 15 this time. 7th time's a charm!!!!'.

V Aeolus (12.00–13.00)

- 12.00 Meredith, Australia: 'Enjoy Bloomsday, the original 24'.
- 12.29 South Circular Road & Griffith College, DUBLIN: 'Happy #Bloomsday #Bus ride #bloomsday2014 @Leonard's Corner'.
- 12.35 Antwerp, Belgium: 'Ahh Bloomsday. What Lovely things will I eat on you today. (Gorgonzola)'.
- 12.44 Biella, Northern Italy: 'I want to give a picture of Dublin so complete that if the city suddenly disappeared it could be reconstructed out of my book'.

VI Lestrygonians (13.00–14.00)

- 13.20 Kolkata (Calcutta) India: 'Look what I just picked up for about ∈ 1.50 on a street corner in Calcutta'.
- 13.25 Clontarf Road & Hollybrook Road, DUBLIN: 'Today's Bloomsday Google Doodle is genius'.
- 13.45 Baggot Street Upper, DUBLIN: 'Happy Bloomsday guys!'.

VII Scylla & Charybdis (14.00–15.00)

- NO TWEETS

VIII Wandering Rocks (15.00–16.00)

- 15.15 Sydney, Australia: ‘Happy #Bloomsday folks . . . Going to dark bed there was a square round Sindbad the Sailor roc’s auk’s egg’.
- 15.54 North Great George’s Street, DUBLIN: ‘I’m at The James Joyce Centre @jamesjoycecentr (Dublin) w/3 others’.

IX Sirens (16.00–17.00)

- 16.07 Glenageary, DUBLIN: ‘Two Bean coffee shop’s take on Bloomsday’.
- 16.00 Tymon Park & Kennington Road, DUBLIN: ‘Sorry did I imagine it or was someone going on about Bloomsday yesterday like it was the joke of the sesh lol help me out here’.

X Cyclops (17.00–18.00)

- 17.12 Helsinki, Finland: ‘Bloomsday (@O’Malley’s)’.
- 17.25 Rue De Vaugirard, Paris, France: ‘Sunny skies, fresh apricots and attempt number 27 to finish this exhaustingly brilliant novel: Happy #Bloomsday!’.
- 17.56 Durazno, Uruguay: ‘Eu: Hoje é Bloomsday! Namorado IRLANDES: isso tem alguma coisa a ver com flores? Er . . . Ok. [Me: Today is Bloomsday! Boyfriend IRISH: does this have anything to do with flowers? Er . . . okay.]’.
- 17.59 Washington, D.C. USA: ‘Plus I’m pretty sure Joyce would love hashtags’.
- 19.37 Antwerp, Belgium: ‘Also in Ulysses: “In the midst of death we are in life”. & “Rattle his bones. Over the stones. Only a pauper. Nobody owns” ’.

XI Nausicaa (20.00–21.00)

- 20.44 Camden, New Jersey, USA: ‘high in the effulgence symbolistic’.
- 021.02 Trinity College, DUBLIN: ‘Absolutely beautiful, blue sky #Bloomsday evening in Dublin’.
- 21.19 Léon, Spain: ‘Al igual que hacia leopold bloom, hoy he cenado dominos pizza y ahoar me voy a ver la ultima de xmen [The same as leopold bloom used to do, today I had dinner at dominos pizza and now I’m going to watch the last of the xmen movies] #bloomsday #home-najes’.
- 21.25 Mexico City, Mexico: ‘Una traducción sucia que hice de las cartas lascivas de James Joyce a Nora aprovechando que es el Bloomsday [Making the most of Bloomsday, here is a naughty translation I did of the lascivious letters to Nora by James Joyce], http://t.co/jaA0ClGoxn’.

XII Oxen of the Sun (22.00–23.00)

- 22:15 Goiás, Brazil: 'sim eu disse eu quero Sim [yes I said I want Yes]'.

XIII Circe (23.00–00.00)

- 23.09 Dartmouth, Cornwall, UK: 'spent Bloomsday pm with a pint of porter talking pomes in the company of . . . gents, 'twas a pleasure'.
- 23.54 Bronx, NY, USA: '@OverheardDublin: "Is #Bloomsday kinda like Paddy's Day for posh people?" '.
- 23.54 Theodore Roosevelt Island, Washington D.C., USA: '10 points to the best tweet to combine #bloomsday and #USAvGHA'
- 23.56 Toledo, Spain: 'Demuestra por algebra que el nieto de Hamlet es el abuelo de Shakespeare y qu él mismo es el espectro de su propio padre [He proves *by algebra that Hamlet's grandson is Shakespeare's grandfather* and that *he himself* is the ghost of his own *father*]'.

XIV Eumaeus (00.00–1.00)

- NO TWEETS

XV Ithaca (1.00–2.00)

- 1.41 June 17, 2014 Interstate 96, Delaware USA: '#Bloomsday Ulysses matters. Greatest novel of Twentieth Century. Exploded the Victorian novel. And the struggle to get back home never ends'.

XVI Penelope (Infinity)

- 2.32June 17, 2014, Washington, D.C.: 'Picking up #Ulysses again. Maybe I'll get more than 100 pages before utter frustration!'.
- 3.43Jun 17, 2014, Cleveland, Ohio, USA: 'To paraphrase Laurie Anderson: Ulysses? Never read it'.

An analysis of the Dublin and global tweet streams suggests that a comparison can be drawn between Joyce's writing style and the use of language on Twitter. In *Ulysses* Joyce perfected his free indirect discourse and stream-of-consciousness techniques (the latter influenced by Edouard Dujardin's 1887 novel *Les Lauriers sont coupés*) to imitate the various ways in which the human mind 'speaks' to itself, through complex fluid patterns, random interruptions, incomplete thoughts, half words and tangents (Norris and Flint 126). Tweets, which are limited to 140 characters, seem to mirror Joyce's writing techniques in conveying both focused and random thoughts.

For instance, one can imagine a tweet-map of Paddy Dignam's funeral cortege in the *Hades* episode, featuring Bloom, Mr. Power and Simon Dedalus (Stephen's father) 'tweeting' random thoughts, and observations as their creaking carriage travels across the streetscape of 1904 Edwardian Dublin from Sandymount, across the River Liffey to Glasnevin Cemetery. If one reads the episode, which alternates between external exposition and internal dialogues, it could be construed as a series of 'tweets'. (Dan Dawson's speech in the morning paper; Ben Dollard's 'The Croppy Boy'; Molly Bloom's Concert Tour; tramlines for cattle and funerals; the Gordon Bennett race in Germany; a collection for the Dignam children; the joking grave diggers in *Hamlet.*) Consequently, the Bloomsday 2014 tweets, refracted through the Joycean textual prism, can be seen in the same light, as people crossing the post-Celtic Tiger city (or tweeting from global locations) re-enact, and celebrate *Ulysses* one hundred and ten years later. The pluralities and synchronicities of the isolated thoughts and perceptions of Joyce's characters in the novel's text, juxtaposed with Bloomsday tweets in 2014, are collated and mapped by HumGIS/SM model to unify the digital dimensions of such places which Mark Graham and Matthew Zook contend are 'fractured along a number of axes such as location, language, and social networks' and characterised by 'correspondingly splintered representations customized to individuals' unique sets of abilities and backgrounds' (78).

Interpreting the Bloomsday GIS 'text-machine'

Digital humanities techniques of *deformance* and *ergodicity* serve as means to interpret and parse the digital text produced by the HumGIS/SM model survey, and to explore the various spatial and narratological juxtapositions that exist between the *Linati Schema,* Dublin city and Bloomsday social media ecosystem map layers. William A. Kretzschmar Jr. states 'when we choose to represent information on a map, we are engaged in scientific modelling' (2013). Alternatively, cartographical historian J.B. Harley observes that maps are 'cultural texts' due to their social construction, and argues that 'text' is a better metaphor to employ than 'mirror of nature' to situate and engage their functions. 'By accepting their textuality' Harley concludes, 'we are able to embrace a number of different interpretative possibilities' (2001, 159). *Deformance* bridges the creative tension between these two ontological perspectives on mapping by deliberately mis-reading and re-ordering the form and function of a text (such as reading a poem backwards or in the case of this chapter, parsing *Ulysses'* eighteen episodes through the prism of a social media-enabled GIS). By applying *scientia* to *poeisis* in GIS, *deformance* translates one type of ontological discourse to another, and seeks to explain unitary and unique phenomena, rather than establish general sets of rules or laws (McGann and Samuels 127). The result comprises a form of *ergodic* literature, not defined by its specific medium, but in the way it functions as a GIS 'text-machine'. Employed together *deformance* and *ergodicity* form types of narrative labour whose meaning must be created through the interactions of a GIS 'text-machine' author/coder/analyst- and its mapping subject(s), thereby creating the possibility of variety of spatial and narrative readings and interpretive possibilities (Aarseth, 1997; Staley, 2007).

By applying these interpretative techniques in GIS one can map and create non-linear and converging intertextual narratives that disrupt quotidian and epochal chronologies of time and space. Plotting the Dublin and global tweets together creates a trans-local, multi-spatial and temporal word-map that identifies people wearing funny hats because it is Bloomsday; elderly ladies getting rowdy in Davy Byrne's Pub; a wedding anniversary celebration in Glasnevin; and a Spanish tweeter hailing the day with a Domino's Pizza and the latest *X-Men* film feature. Individuals in Dublin, Paris and Washington D.C. resolve to attempt reading *Ulysses* again, whilst a tweeter in Uruguay mentions Bloomsday to her Irish boyfriend who asks if the day has anything to do with flowers. A few literary-minded types post Joycean lines from the novel, whilst two individuals from Dublin get suited up in Edwardian fashion items to face the day. One tweeter, reflecting on the day after a night out, wonders if Bloomsday was a joke brought up in a drinking session. A post from Mexico City advertises the tweeter's dirty online translations of Joyce's 'lascivious' letters to his wife, Nora Barnacle. The celebration of *Ulysses* converges with perhaps a larger global event to provoke a tweet which states that there is 'Nothing like a combo of World Cup and bloomsday [sic] to hear people who don't like either Joyce or football talk about both'. One wry observer from the Bronx asks if 'Bloomsday is Paddy's Day for posh people?' And two more tweets from the USA proclaim 'I'm pretty sure Joyce would love hashtags', and 'To paraphrase Laurie Anderson: Ulysses? Never read it'.

Conclusion

Digitally mapping and parsing the 'Big-Data' of *Ulysses* and Bloomsday allows us to reflect upon ways to reconceptualise and reconfigure literary criticism in the digital age as a creative, integrative and technology-enhanced mode of analysis and interpretation. Whether the semiotic import of the e-signs on the Bloomsday map/digital text symbolises a Homeric episode from *Ulysses*, or conveys a *Twitter* (or even *Flickr* or *YouTube*) post repurposing the novel in a different time-space, when visually and dialogically parsed through the lens of a HumGIS/SM model, they appear 'on the surface two worlds, one sensible, one intelligible, but deeper down a kaleidoscope of socially constructed appearances'; and they create 'a game of ontological transformations in which theory-laden observations are translated first into patches of colour, then into strings of words, finally into purposeful action' (Olsson 1999, 141; Olsson 2007, 80). In a sense, digital mapping and literature converges on the scale of language. Whether it is code being written to translate Cartesian mathematics into a geospatial technology software or open-source platform, or nouns, verbs, adjectives and adverbs being assembled into stream-of-consciousness or free-indirect-speech narratives which manifest into the novel *Ulysses*, both are forms of language, albeit with different ontological and epistemological orientations.

One can perceive a correlation between language, text and computing by viewing *Ulysses* as Joyce's staging ground for *Finnegans Wake* (1939), whose hyper-spatial use of language anticipated the linguistic and coding realms emerging today. Jacques Derrida once contentiously observed that Joyce's writing style was decades in advance of a 1000th-generation computer 'besides which the current technology of computers and micro-computerfied archives and translating machines remain a *bricolage* of a prehistoric child's toys' (Derrida 147). Indeed, Derrida concludes, these 'mechanisms are of a slowness, incommensurable with the quasi-infinite speed of the movements on Joyce's cables' (147). It can be argued that, in the twenty-first century, humanities computing, mapping and analysis methods are coming into synchrony with the discursive 'Big Data' speeding through 'Joyce's cables'. The HumGIS/SM model illustrates only one way out of many to plot the intractable relationship between *Ulysses* and current Bloomsday social media spatial productions and performances. The model illustrates the potential for the integration of GIS and neogeography as an avenue through which to consider other philological, linguistic, artistic and technical rewritings and parsings of history and place offered by digital literary criticism.

In the second volume of *Remembrance of Things Past* (1913) Marcel Proust observes: 'the only true voyage of discovery . . . would be not to visit strange lands but to possess other eyes, to behold the universe through the eyes of another, of a hundred others, to behold the hundred universes that each of them beholds . . . we really do fly from star to star' (657). By meditating on Proust's observation, we can see as the hundreds of posts from sites such as *Twitter*, *Flickr* and *YouTube* bubble up from the digital froth, how the HumGIS/SM model acts as a prosthesis allowing us to possess other eyes, behold the solipsistic universes and fly from acute instances of postmodern social media 'stardom': Andy Warhol's description of fame compressed from 15 minutes to 15 seconds. By focusing social media feeds through the geo-spatially-enabled prism of Joyce's *Linati Schema,* we can employ digital mapping theory and techniques to illustrate how *Ulysses*' 'Big-Data' symbiosis of language and place influences twenty-first century social and spatial productions and practices over ninety years after its publication.

Works Cited

Aarseth, Espen. J. *Cybertext: Perspectives on Ergodic Literature.* Baltimore: Johns Hopkins University Press, 1997.

Budgen, Frank. *James Joyce and the Making of 'Ulysses'.* Oxford: Oxford University Press, 1972.

Bulson, Eric. 'Joyce's Geodesy'. *Journal of Modern Literature* 25.2 (2001–2): 80–96.

Cochran, Robert. *Samuel Beckett: A Study of the Short Fiction.* New York: Twayne, 1991.

Derrida, Jacques. 'Two Words for Joyce'. *Poststructuralist Joyce.* Eds, Derek Attridge and Daniel Ferrer. Cambridge: Cambridge University Press, 1984.

Ellman, Richard. *Ulysses on the Liffey*. Boston: Faber & Faber, 1974.

Graham, Mark, and Matthew Zook. 'Augmented Realities and Uneven Geographies: Exploring the Geo-linguistic Contours of the Web'. *Environment and Planning A* 45.1 (2013): 77–99.

Harley, J.B. *The New Nature of Maps: Essays in the History of Cartography*. Ed. P. Laxton. Baltimore, MD: John Hopkins University Press, 2001.

Hart, Clive, and Leo Knuth. *A Topographical Guide to James Joyce's Ulysses*. Colchester: A Wake Newsletter Press, 1975.

Hegglund, Jon. 'Ulysses and the Rhetoric of Cartography'. *Twentieth Century Literature* 49.2 (2003): 164–192.

Hoggart, Paul. 'Kenneth Branagh is Kurt Wallander'. *The Times of London Online* 29 November 2008. Web. 20 March 2015. <http://www.thetimes.co.uk/tto/arts/tv-radio/article2445514.ece>.

Joyce, James. *Ulysses*. 1922. London: Penguin, 1992.

———. *The Letters of James Joyce*. Vol. 1. Ed. Stuart Gilbert. London: Faber & Faber, 1957.

Kretzschmar, William A. Jr., 'GIS for Language and Literary Study'. *MLA Studies Commons. Literary Studies in the Digital Age, An Evolving Anthology*. 2013. Web. 20 March 2015. <https://dlsanthology.commons.mla.org/gis-for-language-and-literary-study/>.

Leszczynski, Agnieszka. 'On the Neo in Neogeography'. *Annals of the Association of American Geographers* 104.1 (2014): 60–79.

Long, Tim, and Charles Travis. 'A Relay of Joy: An Artist and a Geographer Reflect Upon Cybernetic Assemblies and an Embodied Media Geography of Spätkapitalismus' *Aether: The Journal of Media Geography* 6 (2010): 66–87.

McGann, Jerome, with Lisa Samuels. 'Deformance and Interpretation'. *Radiant Textuality*. By Jerome McGann. New York: Palgrave Macmillan, 2001. 105–35.

Norris, David, and Carl Flint. *Introducing Joyce*. Cambridge: Icon Books, 2000.

Olsson, Gunnar. *Abysmal: A Critique of Cartographic Reason*. Chicago: The University of Chicago Press, 2007.

———. 'Constellations'. *Sistema Terra* 8.1–3 (1999): 141–60.

Proust, Marcel. *Remembrance of Things Past (À la recherche du temps perdu)*. Vol. 2. Trans. C.K. Scott Moncrieff and S. Hudson. Hertfordshire: Wordsworth Editions Ltd., 2006.

Staley, D.J. 'Finding narratives of time and space'. *Understanding Place: GIS and Mapping Across the Curriculum*. Eds, D.S. Sinton and J.J. Lund. Redlands, CA: ESRI Press, 2007. 35–48.

Thacker, Andrew. 'The Idea of a Critical Literary Geography'. *New Formations*. 57 (2005–6): 56–73.

Part II

Mapping Practices

Places, Writers and Readers

6 Mapping Fiction

Spatialising the Literary Work

Sally Bushell

> The true ground, the 'it', is everywhere and nowhere. It can be located on no map.
>
> (J. Hillis Miller, *Topographies* 52)

How do we move through and negotiate the experience of the literary work of art? What do we understand by 'literary place' or 'space' and what are the implications of this for the mapping or spatialising of literary texts in digital media? This chapter will begin to address such questions as well as raise some of the difficulties involved in applying concepts of mapping to literature. Unlike other chapters in this collection it does not emerge from a current digital project but, instead, considers ways forward for future projects in the field of Literary Studies. Throughout the chapter Thomas Hardy's novel, *The Return of the Native*, is used as an example to ground ideas.

Mapping Literary Place and Space

A map is the most effective, visual, means of presenting spatial and geographical information in a way that enables us to negotiate and make sense of things. It provides us with a spatial and intellectual confidence that extends beyond ourselves and it enables 'Gestalt' understanding of an environment beyond immediate visual perception.[1] Without it, or if we get lost, our experience of place will be radically different. 'Mapping', on the other hand, concerns an internalised cognitive process that *may* apply to the use of a map, but is also relevant to many other areas of life, where a sense of location or situatedness might be important. For *literary* mapping, the secondary activities contained in mapping – viewing the world from multiple

1 A gestalt theory of visual perception (originating in Germany in the 1920s) argues that the mind will organise related visual images into a whole in order to make sense of them. See, for example, Arnheim.

perspectives; the relative situating of objects in space; the locating of the self and a sense of 'mapping things out' (in relation to the fictional world, movements of characters or the literary work) – are likely to be as important as the tangible object of a map (which is occasionally given alongside the text).

An interest in mapping leads inevitably into the on-going theoretical debate over the distinction between place and space and the relative priority given to either. This is not the context in which to engage fully with a discussion that has been at the centre of the so-called 'spatial turn' in the humanities and social sciences.[2] Instead, my aim here is to consider the relevance of this 'real-world' debate for the understanding of *literary* place and space. The approaches that prove most productive for literary studies tend to be those which allow for active, performative interaction between the two concepts (Tuan; Lefevbre; Soja; Massey; De Certeau; Thrift) – albeit in a wide range of ways with different underlying political and ideological agendas. In such approaches, 'the focus lies in relating location to place through the experience of human beings as agents' (Agnew, 324). The binding together of place and space through objects and agents is essential in order to allow for the double dimension of representation and experience by characters, reader and writer and through the represented fictional, imaginative world as well as the material object of the book. It may be helpful to try and apply underlying concepts that inform the larger place/space debate to a particular literary example to understand the issues involved.

John Agnew identifies three fundamental dimensions to place and space that 'tend to reoccur across the various theoretical positions' (326) and that can help to ground discussion here. He defines these as: *location* or 'a site in space where an activity or object is located' (326); *locale* ('place as a series of locales or settings where everyday-life activities take place' [326]) and *sense of place*: 'identification with a place as a unique community, landscape and moral order' (327). Agnew also notes of this last that 'In this construction, every place is particular, and thus singular' (327). What happens, then, if we apply Agnew's three-part definition to *literary* place and space?

Thomas Hardy's *The Return of the Native* is strongly centred upon a particular site, Egdon Heath, with the author deliberately conforming to a unity of place. In a letter to his publisher, about the map for *The Return of the Native*, Hardy states: 'Unity of place is so seldom preserved in novels that a map of the scene of action is as a rule impracticable: but since the

2 For a succinct and highly informative, account of the distinction between the two terms (looking back to a fundamental distinction between Newtonian and Leibnizian accounts of space) see Agnew; a lengthier account is also given by Cresswell.

present story affords an opportunity of doing so I am of opinion that it would be a desirable novelty, likely to increase a reader's interest' (*Letters* 1: 61). Unsurprisingly, then, *location* is often given in detailed and specific ways:

> The distance to the boy's house was not more than three-eighths of a mile, his father's cottage, and one other a few yards further on, forming part of the hamlet of Mistover Knap: the third and only remaining house was that of Captain Vye and Eustacia, which stood quite away from the small cottages. (66)

These sorts of clear geographical description enable the reader to respond to the novel as if it describes a fixed objective coordinate upon the earth (corresponding to a traditional standard understanding of place as a specific location). However, the moment we attempt to *map* those fixed coordinates, problems arise. In the first edition of the novel (1878) Hardy provided a map of the Heath (Figure 6.1); but later, in the Wessex edition of 1895–96 and subsequent editions, this was removed and replaced with a smaller scale regional map given at the front of all the books (Figure 6.2).[3] Ostensibly, the verbal text provides locations that can be verified by means of the map at the front of the book. However, this is explicitly entitled 'sketch map' so that the kind of mapping it enables is less about 'accuracy' (impossible anyway since there is no real world space for it to be accurate to) than about relative positioning and what we might call a 'human geometry' created by the lives and movements of the characters. In fact, the 'accuracy' of the literary map – measurable in terms of visual/verbal correspondence – is also immediately brought into question, even for the quotation given above, by the absence of less important elements on the map (for example, the boy's cottage is not marked, only the house of Captain Vye). The outlines of the second map, given in the Wessex edition, correspond to the actual English coastline, and place-names on the map consist of a mixture of real and fictional towns and spots so that here imaginary, literary place is enfolded within actual place.[4] But none of the locations in the novel really 'exist', even though Hardy's blurring of the boundaries between fictional and real spaces encourages us to respond as if there *is* a real world correspondence.[5] What is *literary* location then?

3 This map itself was subject to numerous revisions. The map given here is that published in the 1912 Wessex edition based on Hardy's own sketch. For a helpful account see Plietzch.

4 See Pite's discussion of this and of Hardy's mapping of Wessex 169–77.

5 Egdon Heath is also reputedly drawn from heath land east of Hardy's birthplace at Higher Bockhampton, but in his Preface of 1895 Hardy states that in Egdon Heath 'are united or typified heaths of various real names, to the number of at least a dozen' (5).

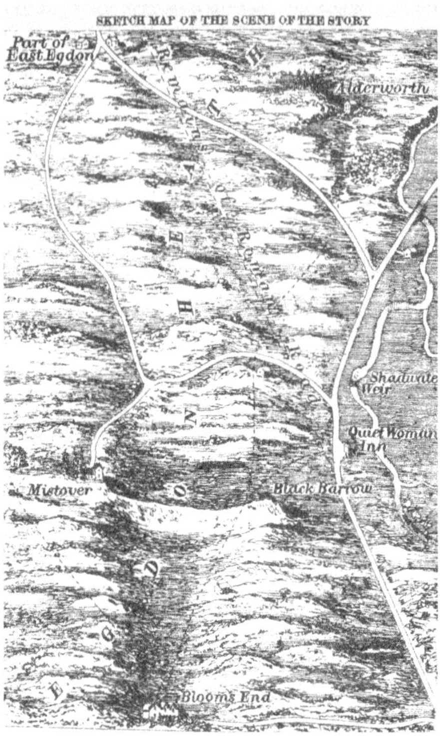

Figure 6.1 Map in first edition of *The Return of the Native* (1878).

Source: © The Author.

Figure 6.2 Wessex map in Macmillan edition of *The Return of the Native* (1912).

Source: © The Author.

Something that we are to experience *as if* it were real (since this is a realist novel)? Something that closely resembles actual mappable places that can be visited, but is always other by virtue of its existence in the form of literary representation?

The *locale* of the novel is Egdon Heath and those who dwell on and around it, but this is not merely a material setting: the heath is given its own distinct identity. It is presented from the start as a prior place, exceeding the human and of a different order:

> It could best be felt when it could not clearly be seen, its complete effect and explanation lying in this and the succeeding hours before the next dawn: then, and only then, did it tell its true tale. (8)

Cresswell states of locale that, 'Even imaginary places, like Hogwarts school in Harry Potter novels, have an imaginary materiality of rooms, staircases and tunnels that make the novel work' (7). Maybe so (although Hogwarts is actually unmappable since those rooms and staircases are also described as constantly moving in a way that actual interiors cannot), but *literary* locale is able to have an extraordinary richness and depth that far exceeds the account of it as background for social relations. The true nature of the Heath can only be described in terms that mark its absolute otherness from the human: it can only be fully experienced when there is no-one there to experience it.

According to Agnew's third category, *sense of place* is strongly depicted through individuals, who are defined by their unique feelings towards it. This functions for fictional characters in a directly comparable way to a lived sense of place in the world. However, within the form of the novel, it also allows for unifying themes and all kinds of spatial responses (for example, in terms of inside/outsideness; boundaries; paths; meeting points and so on) that explore the complex relationship between character and environment. So, the two main characters in *Return of the Native* are defined in and through a sense of place in relation to their immediate environment. Eustacia hates the Heath whilst Clem is the human embodiment of it: 'He might be said to be its product. His eyes had first opened thereon' (148). At the same time, there is a larger spatio-temporal opposition between the present 'here' of this remote provincial setting and the absent 'there' of fashionable Paris from which Clem has returned (and to which Eustacia seeks to escape). These basic acts of self-situating, through which identity is bound up with attitudes towards place, tell us from the outset that their relationship is doomed. Equally, though, a sense of place within the novel also extends outward to generate multiple readerly responses: to the fictional world and the characters within it; to the 'world' of Hardy; in relation to the reader's larger intertextual knowledge; to the reader's own life experiences and so on. Literary place, and the mapping of it, thus immediately presents particular problems both within the confines of a particular work and beyond.

We will return to these problems in the final part of the chapter. For now, though, it may be helpful to develop further the distinction between place and space by considering that made by Michel de Certeau in his work *The Practice of Everyday Life*. For de Certeau, 'place' corresponds to an imposed power structure associated with authority (in his terms a 'strategy') whereas 'space' corresponds to the effects and acts produced within place (a 'tactic') that challenge and undo the dominant order:

> A place (lieu) is the order (of whatever kind) in accord with which elements are distributed in relationships of coexistence. It thus excludes the possibility of two things being in the same location (place).
>
> A space exists when one takes into consideration vectors of direction, velocities, and time variables. Thus space is composed of intersections of mobile elements. It is in a sense actuated by the ensemble of movements deployed within it. (117)

In de Certeau's terms, then, things are fixed and specifically located in place, which is associated with the rigid imposition of order. In contrast, space is mobile, fluid, adaptable. He famously concludes that: '*space is a practiced place*. Thus the street geometrically defined by urban planning is transformed into a space by walkers' (117). The effect of this distinction is to make the concept of space less abstract (it is not about geometric structures so much as made patterns and connections) and to reverse the assumed priority of (empirical) place over (abstract) space by giving agency to the lived experience that creates 'space'.

De Certeau's value for literary mapping lies in the way he neatly connects his account of literal movement and pathways (space) through a city (place) to other 'everyday' systems (writing, reading, cooking). He describes ordinary reading as a kind of wandering through a system which can be resisted and undone by means of the plurality of meaning held within it. So, the text (and the assertion of authorial intention it holds) functions as the structure of power whilst the reader, making his or her own meaning through acts of reading, is creating resistant space by 'moving' in ways that the author may not have wanted. De Certeau states that, 'an act of reading is the space produced by the practice of a particular place: a written text, i.e. a place constituted by a system of signs' (117). In his terms, then, *the material text* is literary place, *the act of reading* brings into existence literary space. Literary place is associated with the grounded physical object of the work of art and encompasses the materiality of the book, meaning as held in the written marks upon the page and the intentions of the writer. Literary space concerns the activation of the imaginary world through the act of reading: indeed, space can *only* be created through this active engagement with place. If we relate this to literary mapping, then an actual map given within a work would be seen as a literary place (presenting a unified authorial sense of place that seeks to impose order and predetermine the world that has not yet been described) but the reader's *use* of that map, or negotiation of the fictional world when no map is provided, involves space-making. Crucially, for de Certeau, narratives are not seen as somehow secondary to lived spatiality but a vital element of it:

> The story does not express a practice. It does not limit itself to telling about a movement. It *makes* it. One understands it, then, if one enters into this movement oneself. (81)

If this is true then we might begin to consider that literature presents not mere representations of place but the possibility of a shared experience that is not at *one step removed* from how we experience place in the world but a more sophisticated articulation of that experience into which we enter and by which our own sense of place can be changed (even back out into the supposed realist point of reference in the world).

Distant Reading and Deep Mapping

The first part of this chapter has attempted to establish some key terms for literary mapping; I want now to turn to the *digital* exploration and mapping of literature. Key here is Franco Moretti and his two highly influential books: *Atlas of the European Novel* (1998) and *Graphs Maps Trees* (2005). The first point to make is that Moretti is not interested in working with authorial maps (such as Hardy's) but in generating maps from spatial information within the text. He is also not concerned directly with digital mapping, although his methods naturally extend in this direction (and have been highly influential here) so that it is essential to engage with his ideas.

Moretti's declared intention is to move beyond the use of maps as merely illustrative or decorative in relation to literature, and to use them instead as 'analytical tools: that dissect the text in an unusual way, bringing to light relations that would otherwise remain hidden' (*Atlas* 3). His techniques work to bring to the fore that which was previously treated as 'background' or 'setting' with the primary aim of 'mak[ing] the connection between geography and literature explicit' by means of a map, here defined as 'a connection made visible' (3). Moretti's succinct account of his own method makes clear the nature of his approach:

> in the end this is what literary geography is all about: you select a textual feature . . . find the data, put them on paper – and then you look

> at the map. In the hope that the visual construct will be more than the sum of its parts: that it will show a shape a pattern that may *add* something[.] (*Atlas* 13)

So, for example, he goes on to compare the city spaces of London and Paris in nineteenth-century fiction to argue that the spaces of London are socially demarcated, whereas in Balzac's Paris 'the magnetism of desire orients the city' (95) so that characters start in one kind of space but rapidly re-orientate themselves towards another. The series of maps generated by Moretti corroborates this hypothesis.

Although he does not directly articulate this himself, Moretti is effectively undertaking thematic mapping in relation to data derived from literary texts. A thematic map 'concentrates on the distribution of a single attribute or the relationship among several' (Robinson et al. 13). Its design is:

> concerned with portraying the overall form of a given geographical distribution. It is the structural relationship of each part to the whole that is important. Such a map is a kind of graphic essay dealing with the spatial variations and interrelationships of some geographical distribution. (Robinson et al. 317)

When they emerged in the nineteenth century, such maps were focused on the representation of data such as population numbers or the spread of disease (a famous example is Dr John Snow's use of the dot map to represent the spread of cholera in London in 1854). Thematic maps thus focus attention on a particular measurable issue represented in a powerful visual and spatial way. This is worth bearing in mind in terms of the intended purpose of Moretti's maps (and the maps of those who develop models based on his approach in the digital medium). The map may well function quite dramatically to enable us to see things anew, or to make a specific spatial point, but how does this bear in meaningful ways upon our understanding and interpretation of an entire text? Essentially what Moretti does is to apply spatial models to narrative structures and he does this extremely effectively. What he does *not* do is to convert the spatial model into anything *other* than a spatial model and this can make him strangely wasteful of the rich data he produces. In other words, he does not go on to show fully how a spatial revelation can fundamentally alter interpretation and, without this, his work has clear limits for Literary Studies.

As a result, Moretti's work tends to divide his audience. Within the context of Literary Studies it is subject to considerable critique by some, whilst adopted by others. In broader digital application, core ideas such as 'distant reading' or the generating of quantitative analysis from literary texts have been assimilated and adapted across a range of disciplines. Moretti explicitly defines his method as one 'in which the reality of the text undergoes a process of deliberate reduction and abstraction' (*Graphs* 1) and defines Distant Reading as a practice in which 'distance is however not an obstacle, but *a specific form of knowledge*: fewer elements, hence a sharper sense of their overall connection' (1). Such activities work well in terms of digital mapping, where information is typically drawn from across a database to generate a Geographic Information Systems representation of physical locations within literature or across a range of works in a particular genre or form.[6]

We can look briefly at how his ideas have been developed further, albeit with far greater awareness of cartography and cartographic practices, in the work of Barbara Piatti and others at ETH Zurich on the *Literary Atlas of Europe* project. Piatti takes Moretti's ideas forward into the digital environment by means of a more explicitly scientific and systematic project centred on 'literary geography' (a key term that Moretti adopts) understood as mapping fiction onto real-world locations. The project maps three distinct areas of Europe that have dense literary settings across a range of authors with the hope that:

> Creating maps based on the elements the author used to build up his fiction will not only better show where fiction takes place . . . it will also demonstrate new correlations between these two worlds. ('Mapping Literature' 2)

As in Moretti (who maps literary movements onto outline maps of real-world places), there is a degree of conflation between actual and literary geographies here. Different accounts of the project repeatedly make the claim that '[o]ne of its traditional starting points is precisely the assumption that a large part of fiction indeed refers to the physical/real world' ('Literary Geography' 4). This is problematic from a literary perspective since it assumes (or leads the user of such maps to assume) a straightforward correspondence between real world and fictional world. In other words, it fails to understand the complex nature of 'realism' in the novel or the fundamental point that the literary representation even of a supposedly 'real' place is always of a different order.

That said, there is something refreshing about the way Piatti and her colleagues focus upon the actual making of maps, as there is in their willingness to acknowledge and address the practical problems involved. In 'Cartographies of Fictional Worlds' they describe how fiction is mapped through five geographical components: setting; zone of action; projected space; route; marker (Piatti and Hurni, 220). Piatti states that 'Every topographic or geographic mention within the story belongs to one of these

6 For recent accounts of the potential of these ways of reading see Matthew L. Jockers, *Macroanalysis: Digital Methods and Literary History* and Franco Moretti, *Distant Reading*.

categories' (Mapping Literature' 2). These primary elements are then further broken down into attributes and categories. So, 'setting' exists in terms of a point, a line, a polygon, naming; and the system also allows for different uses of setting within literature as: simple scenery; thematic scenery; mythical/symbolic; physical protagonistical; protagonistical-psychic. Piatti also allows for indeterminate space and has developed graphic methods 'to represent uncertainty' ('Mapping Literature' 7) as well as acknowledging that 'some aspects of literary geography remain unmappable, for example completely imaginary spaces' ('Mapping Literature' 11).[7] The detailed level of spatial attributes fed into the database allows for the generation of a wide range of maps, to be followed by 'implementation of the second approach, using statistical methods for analysing literary elements from different texts' (10).

I want to consider the usefulness of thematic mapping by a crude application of it to the authorial map of *The Return of the Native.* The variable I have mapped here concerns *all the journeys on foot* recorded within the book.[8] I have chosen to map these onto the sketch map provided by Hardy (see Figure 6.3) simply by drawing a line for each journey made. This immediately reveals that almost all the journeys in the book *are* undertaken on foot: the space contains roads, but few people use forms of vehicular transport. In fact, only the reddleman regularly travels by cart along the roads and he is also the only principal character repeatedly to enter and leave the place of the Heath (within the space of the book).[9]

Within the confines of the book, the map is presented vertically with non-standard compass orientation but this presents a slightly misleading sense of the landscape that does not match Hardy's own verbal description of it. His account of the roads – 'The above-mentioned highway traversed the lower levels of the heath, from one horizon to another' (11) – clearly corresponds to the horizontal rather than vertical image which would also position North in the standard orientation at the top of the image. When we turn the map sideways (as in Figure 6.3) there is a far stronger sense of the long hill called Rainbarrow as 'the pole and axis of this heathery world' (15) and equally of Eustacia as 'standing so dead still as the pivot of this circle of heath-country' (49). The visualisation of human pathways also makes clear that the entangled relationship between the three main characters of Eustacia, Clym and Wildeve creates a strong spatial pattern of movement between their three houses (at Mistover, Bloom's End and the

7 See also Cooper and Gregory's account of four levels of mapping (base map; analytical map; exploratory map; interactive map) and attempts to map 'mood' in similar ways.

8 There are times where Clym or Wildeve or Eustacia wander around the Heath in unspecified ways but these are not represented.

9 For an alternate spatial reading of *The Return of the Native* centred on the Roman road, see Rode 19–58.

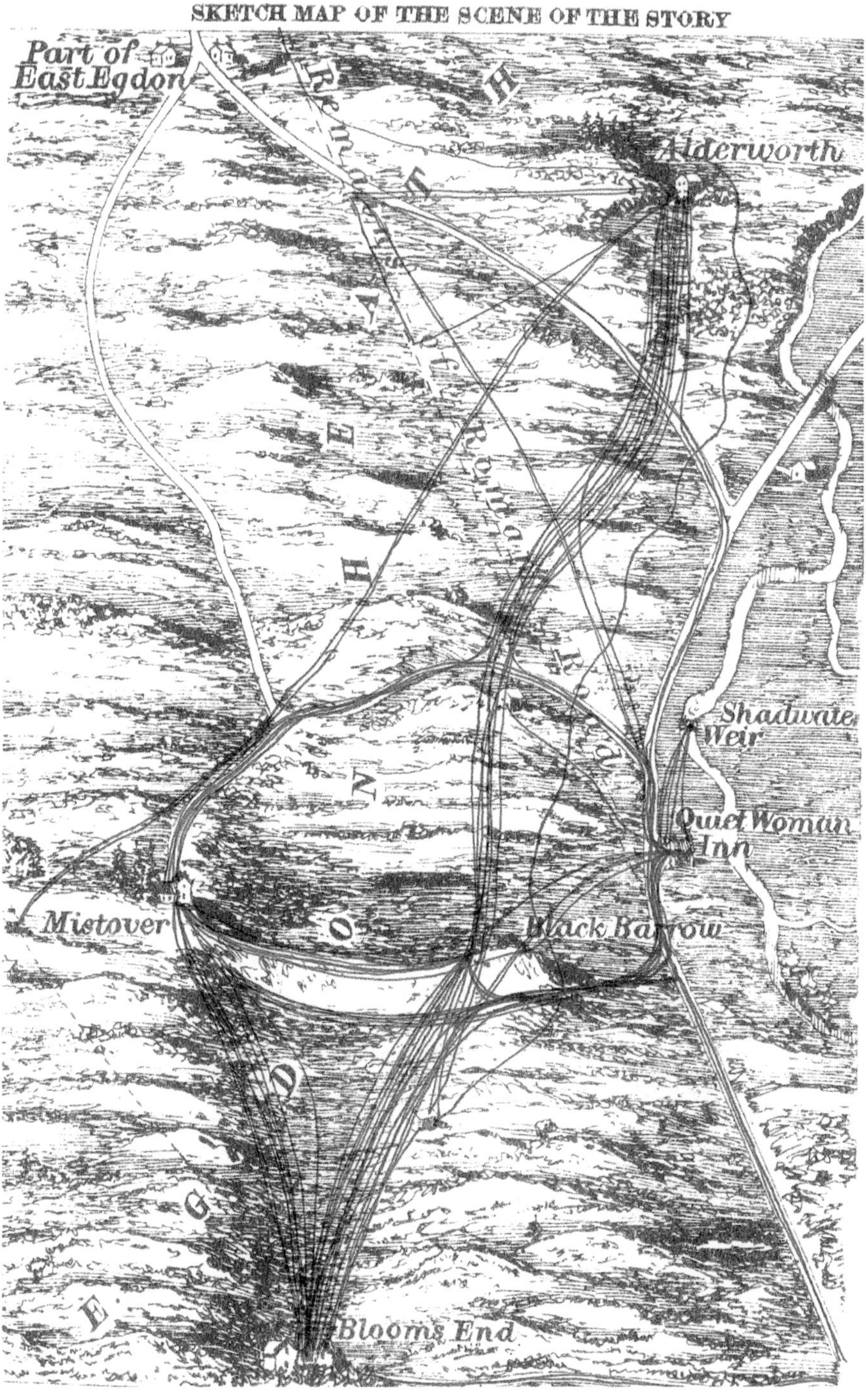

Figure 6.3 Static map depicting routes on foot in *The Return of the Native.*

Source: © The Author.

Quiet Woman inn) that dominates the entire novel. Mapping out the routes thus increases the sense of physical geography which, in turn, is absolutely bound up with a human geometry of feelings, relationships and meetings. A literal love-triangle of movement situates and establishes the emotional one. It also, perhaps more surprisingly, reveals the importance of the single location of Bloom's End (the home of Clym's mother). This site functions as a kind of funnel from which, or to which, many of the journeys begin or return. This makes clear the spatial centrality of Clym himself, but also the extent to which his identity is grounded in his mother as much as in place. In fact, when we view the heath horizontally the shape of the barrow and the roads connecting to it seems to take on the form of an eye, with the lens directed towards Bloom's End and the optic nerve leading back to Clym's new house at Alderworth.

This is strongly enhanced by the addition of routes onto the map which increase the sense of focus onto the location of Clym's mother and the relationship between them and shows the threads of connection stretched to breaking point. This reading of the landscape and movement across it may seem fanciful until we recall the description of the heath itself as 'full of a watchful intentness' (9) or bear in mind Clym's loss of sight that occurs when he leaves his birthplace, or the way in which sight/vision and being seen or not is an essential element in the tragedy of his mother's journey of reconciliation (travelling along that nerve) that goes horribly wrong. Hardy describes how Mrs Yeobright's 'eyes were fixed on the ground; within her two sights were graven – that of Clym's hook and brambles at the door, and that of a woman's face at a window' (238). Mapping onto the authorial map (in a way that Moretti and Piatti do not) also raises interesting questions about the nature of imaginative visualisation and between different forms of spatial representation. It partly releases one from the attempt to be absolutely 'accurate' (since the ground is not purporting to be in the real world) and uncovers inconsistencies between visual and verbal representation.

For the critic, such a map can clearly work to identify key areas of further spatial exploration within the text. Nevertheless, in terms of the full complexity of 'acts of mapping', which are central to our understanding and interpretation of literary spatiality, the static map only represents a first step. A literary critic would want to use the findings of this visualisation as outlined above to undertake a full textual analysis of a particular character's movements or a particular journey – such as the highly charged nature of Mrs Yeobright's last walk – and the powerful combination of external physical conditions, internal states, human actions and the misinterpretation of those actions, that bring about the tragedy. Distant reading clearly can uncover new areas of interpretation through the visual/spatial uncovering of core meanings held verbally within the text. The question is whether it must stand alone against its implied opposite (close reading) or whether it can only truly find fulfilment when reintegrated with it. Despite the attractions of these kinds of approach, then, it should be clear that I find them self-limiting, particularly for a discipline that engages in the interpretation of textual objects which themselves display a rich and complex understanding of the experience of place and space. I see no need for distant reading to set itself off against close reading or, for that matter, full socio-historical contextual reading. Only the uniting of multiple approaches and forms of analysis in the digital environment will adequately serve the needs of Literary Studies.

Deep Mapping and the Synchronic and Diachronic Axes

It should be clear by now that the digital mapping of literary place and space is problematic primarily because the represented place and space of the literary work of art (and many of the characteristics that *make* it literary) so significantly exceed in richness, complexity and depth the attempt to generate maps from them. Moreover, the digital mapping of literary space and place needs also to allow for *acts of mapping* by characters and readers not just the production of maps. How can such problems be overcome?

One possible solution is offered by 'deep mapping' as a newly emerging practice that begins to allow for qualitative as well as quantitative ways of mapping digitally. The origins of deep mapping are not absolutely clear, although the concept seems to emerge out of psychogeography and the work of early French Situationists such as Guy Debord in the 1960s. The term is first used by the American author William Least Heat-Moon in *PrairyErth: A Deep Map* (1991) to describe a composite work focused upon a particular circumscribed place, Chase County, Kansas. In an article on this writer, Gregory-Guider compares *PrairyErth* to an earlier work *Blue Highways* (a 'quintessential post-Kerouac road novel' [4]), to suggest that:

> *PrairyErth* exchanges breadth for depth concentrating entirely on a single relatively obscure county in Kansas. This shift from horizontal to vertical journeying is discernible in a subgenre of American travel-writing that seeks to discover the new directly beneath the feet[.] (4)

It is clear, then, that the first conception of 'deep mapping' is concerned with trying to access and create a different kind of engagement with place that draws upon the individual lives and histories that well up out of it.

If 'deep mapping' starts as a form of adapted psychogeography, it has since been adopted more broadly as a term to describe new ways of spatialising geography for the Humanities in the digital environment, whilst retaining some sense of being a voice for the ordinary lived experience of place. Cliff McLucas from the Stanford Media Lab retains a sense of deep mapping as 'genuinely multimedia, not as an aesthetic gesture or affectation, but as a practical necessity' and argues for non-authoritative forms of mapping: 'Deep maps will not seek the authority and objectivity of conventional cartography'. Taking such ideas forward, David Bodenhamer (also in this collection) and others at the Virtual Center for Spatial Humanities, Purdue University, Indianapolis, have started to develop the concept specifically in relation to spatial narratives. Bodenhamer gives a fuller definition of deep maps:

> They are meant to be visual, time-based, and structurally open. They are genuinely multi-media and multilayered. They do not seek authority or objectivity, but involve negotiation between insiders and outsiders, experts and contributors, over what is represented and how. Framed as a conversation and not a statement, deep maps are inherently unstable[.] (174)

A second, apparently independent, development of the same idea is put forward by Shelley Fishkin at Stanford who defines 'Digital Palimpsest Mapping Projects' as DPMPs and suggests 'that we pronounce the acronym DPMPs as 'Deep Maps'(2). In a recent paper Fishkin gives the following account:

> Deep maps are palimpsests in that they allow multiple versions of events, of texts, of phenomena (both primary and secondary) to be written over each other – with each version still visible under the layers. They involve mapping, since the form of display – the gateway . . . – would be a geographical map that links the text, artifact, phenomenon, or event to the location that produced it, that responded to it, or that is connected with it. (3)

Fishkin then goes on to describe a range of possible deep map projects focused on particular sites or events and responses to them over time (for example, Hiroshima; Dachau) as well as on transnational cultural forms that might bring together different responses to the same phenomena in different cultures. Deep maps certainly seem to offer an alternative to distant reading in terms of the kinds of literary maps produced and the nature of the engagement with place that they offer. But I want to pull back at this point and look more carefully at the underlying structure that may be informing *both* deep mapping and distant reading to help us understand how they can be brought together.

In his late essay 'An Ontological Consideration of Place' Heidegger describes 'horizontal' and 'vertical' modes of Being:

> Man is involved in 'place' in two dimensions, horizontal and vertical. The horizontal dimension is determined by his political relationship. Vertically, being is a dimension hiding the uniqueness of Being, but at the same time it is the place of Being. (19)

The individual is held in place vertically in relation to the self (ontologically) and horizontally in relation to others (existentially; politically); but the two dimensions are bound together and vitally inform each other: 'our horizontal "place" can never be totally eliminated without the consequent denial of our ontological "place". This is true because ontological place reveals itself through existential "place"' (25). In fact, Heidegger was himself drawing upon the earlier famous articulation of the distinction between synchronic ('all at one time') and diachronic ('over time') axes made by Ferdinand de Saussure in relation to language in his *Course in General Linguistics* (1916). As well as revolutionising his own discipline, Saussure's model has proved extremely influential across the humanities and social sciences. So, if we turn from Linguistics to Literary Studies, in a less dramatic way, the same two axes can also be understood to denote an implicit division between literary critics and textual critics/editors in relation to the textual stability of the literary work of art. The German textual critic, Hans Walter Gabler, makes this clear when he argues that 'works of literature are essentially based on this concept of the text as a synchronous and thereby . . . static structure' (306). However, the reality is that:

> the work may be said to comprise all its authorial textual states. By such definition, the work attains an axis and extension in time from earliest draft to final revision. Its total text presents itself as a diachronous structure correlating the discrete synchronous structures discernible, of which that conferred by publication is only one, and not necessarily a privileged one. (309)

So, literary critics generally seek to respond to the literary work through the superficially stable form of a published text in a single state in relation to the historical and philosophical context of a particular time (synchronic). By contrast, textual critics and editors seek to validate the multi-layered process of textual production over time and to allow for the evolution of texts across pre-publication and post-publication states (diachronic).

I want to suggest that the two concepts of 'distant reading' and 'deep mapping' considered above *could* be understood to correspond to the synchronic/diachronic distinction. 'Distant reading' primarily treats elements within texts as single fixed entities whose value is revealed only through comparison to other similar forms understood on a horizontal axis. In contrast, 'deep mapping' suggests a vertical model which contextualises and privileges an individual item within the database allowing its full history to emerge. In so doing, it also allows the possibility of readerly mapping and spatialisation. One question (emerging from Saussure's absolute distinction between the two axes) might be whether these two forms of digital mapping imply a major divide in terms of underlying principles and conceptions (as Saussure suggests) or can be brought together and even need each other in order to be fully understood (Heidegger). Can we also allow for a diachronic model in which each new response need not supplant the preceding one, but instead is cumulative and accretive in nature?

Such ideas are taken forward by Ridge, Lafreniere and Nesbit in their account of the relationship between deep maps and spatial narratives that allows that 'the spatial narrative may be developed as a route through a deep map' (178). Their account describes a pyramid structure for which the bottom of the pyramid is an ideal base consisting of a 'universe of potential data' with middle layers of a 'Deep map prime' and 'Deep map personal' (184) and the top of the pyramid as the spatial narrative. Deep maps provide the platform that enable personal routes through a site.

One other emerging approach is worth considering here. In a remarkably forward-thinking paper Christina Ljungberg poses the question: 'Are we, as some suggest, at the point of entering a new shift of mapping paradigm similar to the one that occurred in early modernity?' (37). Rather than starting with the *form* of map to be presented, she centres her consideration on the *user*, convincingly making the point that new technologies not only require new forms of mapping but ultimately have the potential to re-determine the subject:

> Although maps have to some extent always fulfilled these functions, what is different today are the technologies at our disposal which not only generate new dynamic spaces but also demand the development of new mapping strategies allowing for both improvisational and subjective positioning in constant negotiations for space. (38)

If in earlier periods, 'as maps were plotted a new self emerged which was partly defined by the relationship of the self to space' (39), then what new subject emerges from re-orientation through digital space? Ljungberg uses the example of the way in which contemporary artists interact with technology to suggest 'that we relinquish the subject-object framework for that of implicated agent and expansive field' in 'agential space' (39). Her model results in 'meshworks' (40) as 'diagrammatic thought of illimitable scope rather than closed systems of finite objects' (40). In the account given here, she provides a much more versatile and flexible form of synchronicity – of the horizontal axis – consisting of: 'a dialogic and communicative self immersed in incessant recontextualisation' (40). As she also points out, though, 'The Humanities have not yet taken full account of what this development implies' (40). In relation to Literary Studies, it would suggest a strong emphasis upon the reader/user but also a shift from 'the reader' to multiple creative readers; to readers as agents.

Spatialising the Literary Work of Art

The final part of this chapter begins to consider how such a dialogue might be developed for the place and space of literature. What we find in literary works (which Hardy's novels fully exemplify) is a deeply spatialised form in which place and space are fully bound up with the central meanings and structures of the work but, more than this, in which we are potentially enabled to experience *through* forms of representation a richer spatiality in relation to a non-existent place than we can ourselves experience out in the world. Should we use the term 'spatialise' rather than map? Spatialising texts *involves* mapping but extends far beyond this and allows for the possibility of mapping not just movement across and within place (geographical; physical) and space (projected; dreamed; imagined; interiorised response to place) but also for finding ways of exploring the significance of situated being through language – which would seem to be ultimately what the mapping of literature is all about.

In *Topographies*, literary critic and theorist J. Hillis Miller picks up where de Certeau and Moretti leave off by approaching the experience of literary landscape as a performed speech act. Place does not function as setting but as something that comes into being through the act of reading: 'The landscape "as such" is never given, only one or another of the ways to map it' (6). In a chapter on Heidegger and Hardy, Hillis Miller makes a number of points that are directly relevant to the experience of literary space and place. First, he allows the possibility that a novel can and must be mapped in the kinds of ways suggested by Moretti and Piatti's work:

> Every narrative, without exception . . . traces out in its course an arrangement of places, dwellings, and rooms joined by paths or roads. These arrangements could be mapped. They tend in fact to be mapped, at least implicitly, in the mind of the reader as he or she reads the novel. (10)

The realist novel – which presents a landscape 'as if' it corresponds to a place in the world that can be located – creates a relationship between literary place and geographical place that assumes the landscape already exists and that the literary version of it is an imitative form so that 'real country remains as a solid base giving a grounding in material reality to the act of transposition' (19). However, in Hillis Miller's account, there is an 'alogic of text and context, figure and ground, work and "hors d'oeuvre"' (20) with the result that '[i]f the landscape is not prior to the novel and outside it, then it cannot be an extratextual ground giving the novel referential reality' (21). The apparent realism of the realist novel is misleading because it creates a false relation between literary world and real world, *as if* the literary is referential when by its very nature it cannot be (by the nature of existing only in and through language).

If this is true, how is this landscape able to be so powerful and convincing? Here we need to return to de Certeau's account of space as 'a practiced place'. If human acts and movements in the real world make real places meaningful, this is also true in relation to place in the novel. Literary space is experiential not static. We might also want to note that the *creation* of space, making place meaningful, occurs twice over; for characters existing within that world and moving through it, and for the reader activating a work of literature through the act of reading. The novel itself functions as a network of relations 'a multitude of objects and persons distributed on a topographical surface connected by a reticulation of lines' (39). However, for the *reader* each reading is potentially *a new mapping* since there is no absolutely fixed and static place out of which space is generated or through which to locate ourselves. Hillis Miller makes just this point:

> the investigation of the spatial design of the action leads ultimately to the necessary hypothesis of the atopical or of the placeless. The true ground, the 'it', is everywhere and nowhere. It can be located on no map. (52)

The primary focus of deep mapping, in its current form, seems to be on bringing together geography and history rather than considering what it might mean for the literary work of art. Even those projects that have begun to develop in more literary ways are often concerned with networks *between* writers rather than the oeuvre of a single author or an individual work and, therefore, are as much historical as literary and primarily horizontal rather than vertical.[10] The point I seek to make is that there are layers and depths to real landscapes out in the world; but there are also layers and depths to the fictional landscape *and* to its manifestation in the form of a literary work which is *intrinsically layered* – not only in terms of its own history of composition and publication but also at each moment of reading (for writer,

10 See, for example: *Six Degrees of Francis Bacon* or Stanford University's *Mapping the Republic of Letters* project.

characters and reader) and in the experience of the represented fictional world (for characters 'living' there; for the narrator; for the reader; for the reader re-reading).

One way in which deep mapping for Literary Studies *could* work effectively is by a re-centring of the act of mapping *onto the work itself.* If we took the documentary model common in German editing, and enlarged it in ways that the digital space enables, then each version of a work in a particular form could be represented diachronically (to allow a total view of changes over time) whilst on the horizontal access, in theory, one could present the work in every manifestation (each republishing). In other words, one could be mapping the literary work across all forms and states and across all published forms, or alternatively select key stopping points along the diachronic axis. Fishkin outlines a good example of the kind of project I am proposing here in creating a deep map based upon the translation of a literary work (Mark Twain's *Huckleberry Finn*) into different languages (10). Using *The Return of the Native* again as an example, one could present this work in the forms of: surviving manuscript, first serial publication in the *Belgravia*; the first edition of 1878; the first collected edition of 1895–96; the Wessex edition of 1912. The text itself (in terms of content) is then also capable of a full range of historical and geographical contexts at each synchronic point along the diachronic axis through visual annotation, photographs, film and so on creating a network of contextual meanings.

In relation to the language and form of the text, more detailed acts of mapping could be undertaken. In *The Return of the Native* active visualisation of a mapped route could be presented visually alongside/behind the account of that route given in the text. One might compare the movement of different characters in a generated sequence, or focus on a single journey with each stage linked to the passage describing it. The text could either be presented cleanly or hold within it pockets of interpretation at key moments (stopping points) with the identification and analysis of key sentences (or of changes made to a key sentence over versions). This adapts the deep map/spatial narrative pyramid in ways that specifically address space and place in literature, allowing for a full range of response from macro to micro levels. Such a model works *inward* from distant reading that produces static maps of place illuminating particular concepts or movements, to spatialised criticism (integrating maps/acts of mapping and text) and acts of close-reading (detailed interpretation of complex ideas and images). The full oeuvre of certain British authors who communicate a strong sense of place throughout their work (such as Wordsworth or Hardy) offer perfect examples for exploring the deep mapping of literature and problematising the complex layers of actual, mapped, photographed, represented, fictionalised, dreamt, imagined and internally experienced place and space that constitute the experience of literary spatiality.

How can such literary deep mapping be opened up to the public? If real world deep mapping is about allowing a kind of open, shared, creative space for a region and those who inhabit or have inhabited it, then for Literary Studies it would have to be about readerly mapping of the narrative and the different ways in which readers respond to the world of the text. Readers could add their own annotations to particular sections or characters, or they could be directed to make particular contributions that might be geographically or historically subjective. To take a single example, one could ask readers to respond in terms of where they first read this book or where they are reading it *now* and build up a fascinating resource around the *place of reading* linked to a single text.[11]

Finally, (following Ljungberg) there also needs to be full consideration of how readers *actually* map either texts or digital sites. Crucial to the digital mapping of literary texts is the need for heightened self-conscious awareness about the way in which the user negotiates the multiple layers of space being created. An ideal project to explore such issues would use strategies of spatial reading and path-finding through digital space in a highly self-conscious way in relation to a particular body of literary texts that are *themselves* highly spatial in terms of the meaning held within them. At this point, spatial meaning becomes intertwined with materiality and the replication of the manuscript or printed text object in digital space, as well as with the means by which we negotiate and interpret that object, and that object's relationship to the content (meaning) of the text. If deep mapping is to achieve its potential for literature then it has to allow for the extraordinary spatial potential of literature itself.

Works Cited

Agnew, John. 'Space and Place'. *The SAGE Handbook of Geographical Knowledge.* Eds, John Agnew and David N. Livingstone. London: SAGE publications, 2011.

Arnheim, Rudolph. *Art and Visual Perception: A Psychology of the Creative Eye.* Berkeley and Los Angeles: University of California Press, 1954.

Bodenhamer, David. J., Trevor M. Harris and John Corrigan. 'Deep Mapping and the Spatial Humanities'. *International Journal of Humanities and Arts Computing* 7.1–2 (2013): 170–75.

Cooper, David, and Ian Gregory. 'Mapping the English Lake District: A Literary GIS'. *Transactions of the Institute of British Geographers* NS 36 (2010): 89–108.

Cresswell, Tim. *Place: A Short Introduction.* Oxford: Blackwell Publishing, 2004.

De Certeau, Michel. *The Practice of Everyday Life.* Trans. Steven Rendall. Berkelely and Los Angeles: University of California Press, 1984.

11 For a project that works in this kind of way see the *Digital Reading Network.*

De Saussure, Ferdinand. *Course in General Linguistics.* 1916. Trans. Wade Baskin. London: Peter Owen, 1959.

Digital Reading Network. 2014. Web. 20 March 2015. <http://www.digitalreadingnet work.com/>.

Fishkin, Shelley. '"Deep Maps": A Brief for Digital Palimpsest Mapping Projects (DPMS, or "Deep Maps")'. *Journal of Transnational American Studies* 3.2. (2011): 1–31.

Gabler, Hans Walter. 'The Synchrony and Diachrony of Texts: Practice and Theory of the Critical Edition of James Joyce's Ulysses'. *TEXT* 1 (1981): 305–26.

Gregory-Guider, Christopher C. 'Deep Maps: William Least Heat-Moon's Psychogeographic Cartographies'. *eSharp* 4 (2007): 1–17.

Hardy, Thomas. *The Return of the Native.* Ed. Phillip Mallett. 1969. New York: W.W. Norton & Co., 2006.

———. *The Collected Letters of Thomas Hardy.* Eds, Richard L. Purdy and Michael Millgate. Volume 1: 1840–1892. Oxford: Clarendon Press, 1978.

———. 'General Preface to the Wessex Edition of 1912'. *The Return of the Native.* The New Wessex Edition. Ed. Derwent May. London: Macmillan, 1974. 423–27.

Heidegger, Martin. 'An Ontological Consideration of Place'. *The Question of Being.* Trans. William Kluback and Jean T. Wilde. London: Vision Press, 1959. 18–26.

Jockers, Matthew L. *Macroanalysis: Digital Methods and Literary History.* Illinois: University of Illinois Press, 2013.

Lefebvre, Henri. *The Production of Space.* Trans. Donald Nicholson-Smith. Oxford: Blackwell, 1991.

Ljungberg, Christina. 'Mapping Practices for Different Geographies'. *Mapping Different Geographies.* Eds, K. Kriz, William Cartwright and Lorenz Hurni. Berlin: Springer, 2010. 37–55.

Mapping the Republic of Letters. 2013. Web. 20 September 2014. <http://republico fletters.stanford.edu/>.

Massey, Doreen. *Space, Place, and Gender.* Minneapolis: University of Minnesota Press, 1994.

———. 'A Global Sense of Place'. *Marxism Today* 38 (1991): 24–9.

McLucas, Cliff. 'Ten Things to Say about Deep Maps'. *Michael Shanks.org.* N.d. Web. 20 March 2015. <http://cliffordmclucas.info/deep-mapping.html>.

Miller, J. Hillis. *Topographies.* Stanford: Stanford University Press, 1995.

Moretti, Franco. *Atlas of the European Novel*: 1800–1900. London: Verso, 1998.

———. *Distant Reading.* London: Verso, 2013.

———. *Graphs, Maps, Trees: Abstract Models for a Literary History.* London: Verso, 2005.

Piatti, Barbara, and Lorenz Hurni. 'Editorial: Cartographies of Fictional Worlds'. *The Cartographic Journal* 48.4 (2011): 218–23.

Piatti, Barbara, Lorenz Hurni and Anne-Kathrin Reuschel. 'Literary Geography or How Cartographers Open up a New Dimension for Literary Studies'. *Proceedings of the 24th International Cartography Conference,* Santiago de Chile (2009): 1–12.

———. 'Mapping Literature: The Prototype of "A Literary Atlas of Europe"'. *A Literary Atlas of Europe.* 2015. Web. 20 March 2015. <http://www.literaturatlas.eu/en/>.

Pite, Ralph. *Hardy's Geography: Wessex and the Regional Novel.* Houndmills: Palgrave Macmillan, 2002.

Plietzch, Birgit. *The Novels of Thomas Hardy as a Product of Nineteenth Century Social, Economic, and Cultural Change.* Berlin: Tenea Verlag Ltd., 2004.

Ridge, Mia, Don Lafreniere and Scott Nesbit. 'Creating Deep Maps and Spatial Narratives Through Design'. *International Journal of Humanities and Arts Computing* 7.1–2 (2013): 176–89.

Robinson, Arthur H., Randall D. Sale, Joel L. Morrison and Phillip C. Muehrcke. *Elements of Cartography.* 5th Edition. New York: John Wiley & Sons, 1984.

Rode, Scott. *Reading and Mapping Hardy's Roads.* New York: Routledge, 2006.

Six Degrees of Francis Bacon. 2014. Web. 20 March 2015. <http://sixdegreesoffrancisbacon.com/>.

Soja, Edward W. *Thirdspace: Journeys to Los Angeles and Other Real-and-Imagined Places.* Oxford: Blackwell Publishers, 1996.

Thrift, Nigel. *Non-Representational Theory: Space, Politics, Affect.* London: Routledge, 2007.

Tuan, Yi-Fu. *Space and Place: The Perspective of Experience.* Minneapolis: University of Minnesota Press, 1977.

7 The Spatial Practices of Writing

Arnold Bennett and the Possibilities of Literary GIS

Angharad Saunders

Introduction

This chapter is interested in the cartographies that are immanent within and central to the process of literary creation.[1] It takes inspiration from Sally Bushell's concept of the 'text as process', which urges us to look not at the final published text, but at the compositional materials that come before the written word: the rough drafts, the revisions, and the manuscripts that go into its making (Bushell 1–8). Taking account of these materials allows us to see literary creation as a simultaneously, continuous and discontinuous process: one that is both tied to the material text and its meaningfulness; and one that is separate from it and meaningful in its own right. In this chapter, the emphasis is less on the textual practices of creation and more on using Bushell's work as a point of departure for exploring the spatialities of creation: the gamut of geographical experiences, both real and imagined, that are part of a writer's being-within-the-world. It begins by offering a broad overview of critical approaches to literary creation, before moving on to address its absence from and potential for literary GIS. It closes with a focus on the life and work of Arnold Bennett (1867–1931) – a novelist, essayist, and journalist long associated with the region known as the Staffordshire Potteries – and explores how his everyday spatialities were an intrinsic part of his literary practice. Central to this is a consideration of whether (and if so how) these spatialities can be mapped in a meaningful way.

Creative Practice and Literary GIS

In attending to that which comes before the text, this chapter is part of a resurgent interest in the idea and practice of literary creation. For much of the twentieth century the act of literary creation was a subject that

1 This chapter began life as a paper presented at the *Digital Texts and Geographical Technologies in the Digital Humanities* symposium held at Lancaster University in 2013, and I would like to thank the organisers for inviting me to participate in this symposium. My thanks also go to the Potteries Museum and Art Gallery for giving me access to their Arnold Bennett archive.

drew little attention from literary scholars. New Criticism, with its emphasis on the voice of the text not the author, and the associated assumption that creation was the work of a lone consciousness – and therefore, elusive, ephemeral, and unknowable – effectively stymied and discredited critical interest in creativity (Wimsatt and Beardsley 468–70; Brosseau 349). Meanwhile, work in textual studies retained some distance from these debates, allowing its interest in the materiality of the text to do much to acknowledge the role that editors and publishers play in textual creation (McGann 69–87). Much of this, though, came after the material fact of the text (an extant, coherent textual document); and it was not until the performance turn[2] that a more strenuous attempt was made to rescue the writer from ignominy and elucidate his or her relationship to the written word. Textual studies, for instance, have attended to both the aesthetics of creation (Attridge 10–16) and to the materiality of creative composition (Bushell 57–74), whereas influence studies have been more interested in the role that lived social relationships play in the creative process (Farrell 7–26; Glyer 1–26). The significance of space to creation has been implicit within both approaches, whether this is in terms of a writer's social world or as a source of inspiration. It has, however, fallen to geographers to flesh out the spatialities of the creative process, exploring the affective relationship between space and inspiration (Brace and Johns-Putra 400), examining the leakiness of place within the creative process (Rogers 665), and the manner in which creation is spun across space (Saunders 286). Although strongly qualitative in nature, what these works demonstrate is the lived social world that exists before the text. As yet, however, this is a world that has still to register within literary cartography or literary GIS.

Literary cartography is, as David Cooper and Ian Gregory observe, broadly divisible into two traditions: writerly and readerly mapping (91–2). Writerly mapping looks at how maps work as creative stimuli to, or as structural poetics within, the text, whereas readerly mapping explores the way readers rework the relationship between text and cartography through the reading process. The development of literary GIS has attempted to address both of these phenomena, exploring the role of maps and space within the narrative itself (Cooper and Priestnall 253–6; Reuschel and Hurni 293–7) and the new spatial relationships that reader-generated maps reveal (Cooper and Gregory 93–8; Gregory and Cooper 66–9). The former approach, however, could be pushed a little further. At present, it

2 The performance turn is interested in life as something lived, embodied and active. It 'turns' away from thinking about life in and through representation, which tends to regard texts as symbolic and to register life in quite static and reductive ways. Central to the performance turn in Geography is the work of Nigel Thrift and in Literary Studies that of John Austin.

is concerned more with the cartographic products not the cartographical practices of the writer. Thus, it is the spaces in the text and not the spaces of the writer that have appeared to motivate writerly-orientated GIS. This is not, as my own work suggests, an issue unique to literary GIS, as it pervades both literary studies and literary geography where the influence of post-structuralism and its emphasis on close reading has done much to obscure the lived reality of the author and his or her creative practice (Saunders 286–8). Yet, in the same way that studies of inspiration and creative practice are re-emerging within literary and geographic studies, there are complementary moves unfolding within cartographic studies. Here, there is an increasing awareness that cartography, far from being a secure form of knowledge, is instead, a practice or a way of knowing that is performative, relational and contingent to its time and place (Hewitt 157; Kitchin and Dodge 335; Walford Davies 14–19). In consequence, maps are never finished or complete, but are always being made through our lived practices. This post-representational cartography, as it has come to be known, (see Kitchin and Dodge 331–44; Perkins 381–91) has implications for literary GIS for it suggests that a writerly GIS is more than the material maps enfolded within the text, and is simultaneously, composed of what are here termed everyday cartographies: a writer's daily routines, journeys, incidental happenings, and social relationships.

Sometimes we are in luck and these everyday cartographies are recorded in visual form. David Cooper, for instance, explores Samuel Taylor Coleridge's map-making within the notebooks that record the poet's travels through the Lake District (Cooper 41–4). Cooper terms these 'literary cartographies', for they are the maps embedded within a writer's creative spaces, and bringing GIS technology to bear upon them begins to reveal the embodied, emotional, and productive relationship between writer and place. It enables us to understand creation as something lived and lively, multi-dimensional not planar, and something that is interwoven with and not inscribed in place. In other words, cartographically rendering the spaces of creation is one way of disclosing what Kathleen Stewart terms the ordinary affects of our being-in-the-world (Stewart 1–5). Few writers, however, are as assiduous as Coleridge in mapping their locations and opening up cartographically the place of their writing. Yet this does not mean that we cannot explore the relationship between everyday cartography and creativity in other ways and from other directions. Inspiration comes from Catherine Brace and Adeline Johns-Putra, who suggest that we contemplate creation as bound up with both the doing and being of authorship (404). It is a considered practice that happens in often material ways, as in the form of draft materials and notebooks, but it can simultaneously happen in unconscious or unplanned ways as we exist within the world. Thus, creation is something more than an inscriptive and located act (for instance, the putting of pen to paper), it encompasses the gamut of happenings and experiences that come before and can inspire the physical act of writing or recording. Consequently, another way to explore creation is to trace its spatial practice:[3] to consider it not as something that happens *in* place but, rather, as something that happens *through* place. It is, as I have elsewhere termed it, a process of *longue durée* (Saunders 287).[4]

Ideas of *longue durée*, however, do not necessarily fit with accepted practices of literary cartography (and literary GIS), which tend to centre, Rachel Hewitt argues, on Enlightenment notions of geometry and realism (Hewitt 161). The result is an almost indiscriminate mapping of a text's geographic references that displays little sensitivity to the spatial concerns of a text's happening; what Brian Massumi might term, the nagging problem of movement (Massumi 3). A more critical literary cartography arises, Hewitt observes, when we attend to the specific cartographic practices and imaginations of individual texts and authors (158; see also Walford Davies 14–19). Exploring the being and doing of authorship is one way of addressing this, for it takes us into the lived, social world of the writer and the everyday cartographies that give shape to this world. This is not a world of 'official' cartographic practices: of surveying or map-making; but rather, a world of everyday route-finding, way-making, and spatial knowledge. Literary GIS, by its nature, seeks to map this world, but unlike the maps that come after the text and work from the geographical and emotional knowledge encoded therein, mapping the world before the text is not so straightforward. It is a world that is slippery and messy, more subjective than objective, more mundane than exceptional, and more driven by movement than by stasis. This is not a world that lends itself to the methodological practices of literary GIS. It is, however, a world that provides opportunities for thinking through the practice of literary GIS in a way that is sympathetic to Hewitt's concern that we avoid a mass hermeneutic project in favour of a more culturally and spatially specific one.

The most pressing methodological issue that arises with a turn to the process of creation is how we recover these everyday cartographies. Studies of living writers often turn to interviews, or more generally methods of 'talking-with', to access, to some degree, the everydayness of an author's being and doing (Brace and Johns-Putra 405–411; Waldie and Campbell 228). Historical everyday cartographies are more problematic. Here, we are not able to 'talk-with' writers and so become reliant upon their textual traces: the letters, journals, draft materials, miscellaneous documents, and

3 Practice is often seen as non-textual, hence there is a tension in exploring spatial practices through textual sources. Writers such as John Austin and Tim Ingold, however, argue that texts teem with life and practices. Appreciating this viewpoint requires us to understand text as something more than a record or register, and to see it as something actively made and composed of social and spatial relationships. It is this approach that informs the discussion that follows.

4 *Longue durée* captures the sense of duration, over time and space, involved within the creative process.

material objects that are the archive of their lives. The apparent inertia of these traces is such that even those interested in post-representational cartography have tended to favour contemporary over historical studies (Perkins 383). Yet, if we follow Tim Ingold, and look beneath the veneer of stasis, textual traces and material artefacts are teeming with life (177–228). Whether it is in the loop and wobble of the script, the manner in which traces bear witness to life, or in the spatiality and transience of different forms, material objects are assemblages of living lines and lively social networks. This work locates life not in 'things' but in the relationships between things, and it is by exploring these relationships that we can begin to recover the historical cartographies of everyday practice.

A second methodological issue, closely related to the challenge of recovery, is that of completeness. Cartographies that come before a text are often very fuzzy and very partial. There is less likelihood, for instance, of discernible locational data, and where we are dealing with historical practices of creation in particular, we are confined to those draft materials that have been preserved and are accessible. In this way, the cartographies of creation may well be less comprehensive than the cartographies that start with the material text. However, historical geographers have begun to embrace this partiality, regarding it not as a problem *per se*, but as part of a creative and expressive mode of researching (DeSilvey 404; Lorimer 199; Patchett 20). Walter Benjamin's idea of historical constellations has been particularly influential to this work, motivating Caitlin DeSilvey to develop a historical approach in which redundant, partial, incomplete, and decaying objects are juxtaposed with one another in an effort to revivify past networks, associations, and affinities. This improvisational methodology that makes-do with what is at hand, rather than lamenting what is not present, inspires this chapter. It is an approach that has the capacity to reveal histories and relationships that more traditional methods may well obscure or overlook. As such, this chapter seeks to piece together rather than merely 'find' the everyday cartographies of creativity. It works through an inventive process of collecting, juxtaposing, and associating the material traces and fragments of Arnold Bennett's life, to disclose how his creative cartographies emerge between and not just in his texts.

A third and final methodological issue to bear in mind is that tracing the cartographies of creation may well be something of a clumsy and exhaustive endeavour. Mapping requires us to demarcate beginnings and endings, when in fact the creative process could feasibly encompass an author's entire lived experience. It is an issue that turns us to the broader question of what it means to map and what might be gained from mapping the elusive and discontinuous world of creativity. A useful place to start is with the difference between cartographic extension and intension. Whereas Franco Moretti advocates extension, mapping spatial relations on a flat plane (Moretti 5), Damian Walford Davies, echoing Hewitt, urges us to map in a way sensitive to the intension of place: to its inhabitable depth and lived quality (Walford Davies 14–15). Intension depends upon close reading, on attending to what Walford Davies terms, the cartographic sensibilities of a text; that is, its cartographic way of thinking, feeling, and knowing, or, put another way, its forms of triangulation that define the specificities of its conception, location, and language. This form of mapping is not resistant to visualisation within a GIS, but its visualisation is not a system of points on a flat plane; instead, it is a cultural meshwork in which lives, locations, routes, and readings are in dialogue with one another. It is a constellation that can, Walford Davies argues, 'prompt a critical and affective inhabitation of the cultural *dimensionality* of a literary work' (206). Such an approach has possibility in the world before the text, for it can accommodate the relational field of creation: the spatial, social, and temporal relationships that impel it forward. Within this context, demarcating beginnings and endings may well be less pressing, for attention focuses on the mingling of different life lines, routes and trajectories that comprise an author's life. It is a form of visualisation that enables us to discern both the affective web out of which creation flows and the places of creative co-production.

Arnold Bennett's Spatial Practices

This chapter probes these methodological issues through the writing practice of Arnold Bennett. It takes three moments in Bennett's early writing career and examines how, and to what end, everyday cartographies develop our understanding of the happenings that come before the text. Arnold Bennett was something of an amateur geographer and was explicit about the role that geographical knowledge and 'Experience' played in the creative process, observing that 'the makers of literature are those who have seen and felt the miraculous interestingness of the universe' (Bennett, *Literary Tastes* 19). Bennett's writing practice was driven very much by 'on-the-spot geography' (Bennett, 'Try Geography' 213). Before picking up a pen, or taking up his typewriter he would always insist on first undertaking a fact-finding mission. Quite often he would record and recount these fact-finding missions in his journal, offering a very detailed evocation of place. For example, on 10 September 1897, he records that:

> when I have been taking early morning walks . . . and when I have been traversing the district after dark, the grim and original beauty of certain aspects of the Potteries, to which I have referred in the introduction to 'Anna Tellwright', has fully revealed itself for the first time. Before breakfast, on the heights of Sneyd Green, where the air blows as fresh and pure (seemingly) as at the seaside, one gets glimpses of Burslem and of the lands between Burslem and Norton, which have the very strangest charm. (Bennett, *Journals* 46–7)

Many of the entries in Bennett's journals are replete with detailed, locational descriptions of this sort. Throughout the two volumes, for instance, there are numerous reflections on the role that being-in-the-world plays in the creative process. As he notes, for instance, on 8 February 1904, 'I walked yesterday morning into Italy, and got a few good ideas for the novel [*A Great Man*], and was rather impressed by all the pomp of the boundary business between France and Italy' (Bennett, *Journals* 151). If we followed Moretti we might map these locations, identifying them as sites of creation; but there are subtler geographies, or cartographic sensibilities, at play here.

The language that Bennett uses, particularly when recounting his journeys round Burslem, hints at a particular kind of movement: traversing. Traversing is a geographically-laden term. It is a tool of surveying and was, for a long time, D. Graham Burnett argues, a way of exploring, representing and making known *terra incognita* (66–80). It was a form of mobile mapping that took its cue from the route of an expedition and the natural landmarks encountered thereon. It was, therefore, a more improvisational and pragmatic form of surveying, one that was used where space was unknown and where landmarks had to be created not found. From this, it is possible to see how traversing, with a slight change of inflection, has come to convey a zigzagging, criss-crossing, meandering and thoughtful form of movement. There is no set destination or route, but rather, a desire to examine closely, to spend time and expend energy in place so that, as Bennett observes, things reveal themselves. This meant not just absent-minded meandering through the world, but rather multiple and considered goings-forth in order to experience its subtle textures and diurnal changes. There is, then, within the traces Bennett left behind, an evident cartographic sensibility; yet, other than the places Bennett mentions, we have no knowledge of the route of his traversing. Such spatial uncertainty, as Anne-Kathrin Reuschel and Lorenz Hurni observe, is not alien to literary mapping (299–304). It can be accommodated by degrees of transparency, fuzziness, colour saturation and resolution. Yet this does not capture the movement inherent in Bennett's creative process or the immersion within a landscape this movement suggests. Perhaps, therefore, what we need, if we are to map the spatial practices that come before the text, is to jettison the accepted vernacular of the map and play, instead, with Howard Horowitz's notion of the word map.

Horowitz offers a map of Manhattan not as a series of points and lines but as poetry (Horowitz 107). This cartographic poem features landmarks and historical events where they appear or occurred within the island. What is distinctive about word maps is that they register individual encounters with place. They are articulations of consciousness and this applies as equally to the reader as it does to the maker. Word maps suspend traditional ways of reading both poetry and maps: they are vehicles for the imagination, meaning that there are as many journeys through Manhattan as there are readers. Thus, whilst the place, the island of Manhattan, is stable, how it is understood, experienced and made known is under constant creation. Another way of visualising the landscapes Bennett traverses, therefore, is to impose text on place and traverse text-place as a reader, following the different lines of possibility the words suggest. The limit here, of course, is that such a text may appear unmoored: an expression or register of Bennett's life but yet with little sense of how this is part of, or related to, his creative practice. This returns us to the methodological issue of how we delimit the creative process. Tying text to place is too simplistic, it compromises the duration and depth of creativity. Instead, if we wish to visualise the spatial practices that happen before the text, we need to consider how to render process as something simultaneously temporal and spatial.

Another way into this is to turn to Bennett's correspondence, for like his journal, his letters disclose something of the contours of his craft. Bennett had many correspondents with whom he discussed his writing, but of particular interest here is his epistolary friendship with H.G. Wells that began in 1897. An early letter from Wells observed that the Potteries had made an immense impression upon him: 'I felt dimly then and rather less dimly today vast possibilities there' (Wilson 35). The 'then' he refers to was the summer of 1888 when, after a long illness, he spent several months recuperating with friends in Etruria, a town in the Staffordshire Potteries. Whilst there he recorded:

> I went out by myself to a little patch of surviving woodland amidst the industrial country, called "Trury Woods". There had been a great outbreak of wild hyacinths that year and I lay down among them to think. It was one of those sun-drenched afternoons that are turgid with vitality. (Wells 310)

Bennett was similarly brooding on these possibilities, writing to Wells that it is 'only during the last few years I have begun to see its possibilities . . . . It seems to me there are immense possibilities in the very romance of manufacture . . . in the tremendous altercation with nature that is continually going on . . . . Anyhow, I am trying to shove the notions into my next novel' (Wilson 36). It was only a month earlier that Bennett made his journal entry, noted above, about the 'grim and original beauty' (Bennett, *Journals* 46) of the Potteries and thus, what we begin to see is a shared textual making of place. Taking inspiration from histories of geography, which have explored the role that networks of correspondence played in the development of scientific knowledge in the eighteenth and nineteenth centuries (Withers 34–6; Ogborn 27–66), I have suggested that letters were similarly productive within the process of literary creation (Saunders 291–4). They were material spaces in which aesthetic ideas could be tested, defended and validated. Bennett, for instance, recognised that other writers had turned their hand to industrial settings, but as he wrote to Wells, it was a landscape that had not been captured in its full beauty:

> I am quite sure there is an aspect of these industrial districts which is really grandiose, full of dark splendours, and which has been absolutely

> missed by all novelists up to date. Tirebuck, in Miss Grace of All Souls was too much interested in his individual characters to note synthetically the general aspect, and Nevinson, in Valley of Tophet also let it escape him. (Wilson 36–7)

The letters that Bennett and Wells exchanged chart the emergence of the literary landscape of the Potteries. From the postmarks and addresses we can plot the spatial network of correspondence. We know, for instance, that Wells and Bennett were living in close proximity during the early years of their correspondence. Wells was in Worcester Park, Surrey and Bennett just ten miles away in Victoria Grove, Kensington, and we know that they were corresponding about another place 160 miles to the north. Such plottings, however, tell us little about the emergence of the literary place of the Potteries. Put another way, they conceal the shared creative process the content of the letters reveal. Trevor Harris, Susan Bergson and L. Jesse Rouse are acutely aware of the capacity of traditional GIS to denude life of its richness and suggest that geovisualisation and more immersive technologies may offer some means of reviving life as something more than a dot on a map (226–9). It is an approach that pivots on the digital reconstruction of past environments, wherein all kinds of textual and historical information can be layered in order to create a multisensory experience for the user. This kind of geovisualisation is very present-orientated; it is intended to deepen learning experiences within the now and, as such, it is not the kind of mapping of space that is sensitive to the time and place of Arnold Bennett. That said, it points to ways in which the living texture of Bennett's spatial practices, his traversing and his correspondence, may well be registered. Geovisualisation, for instance, could recreate the route of Bennett's correspondence, narrating content as one moves across the distance between Wells and Bennett. This requires, however, a little more thought and it is to ideas of social relationships that this chapter now turns, in order to suggest a more sensitive form of visualisation.

It is an example that once more underlines the importance of friendship to Bennett's creativity. In the extract above, we learnt of Bennett's familiarity with William Edwards Tirebuck's novel of working class poverty and the works of Henry Nevinson, a social reformer, who published short stories of and about the London slums and the Black Country. Both these writers became negative influences on Bennett's practice, for it was against their work and its limitations of landscape and setting that Bennett was writing. A more positive influence was Joseph Conrad, who Wells had recommended to Bennett towards the end of 1897:

> I owe you a good turn for pointing out Conrad to me . . . . I have just read his new book The Nigger of the Narcissus which has moved me to enthusiasm. Where did the man pick up that style, and that synthetic way of gathering up an impression and flinging it at you? (Wilson 38)

On finishing 'Joseph Conrad's superb book', Bennett records in his journal that 'I had in mind to go on at once with my Staffordshire novel, treating it in the Conrad manner, which after all is my own, on a grander scale' (Bennett, *Journals* 64). The correspondence between Bennett and Wells reveals a meshwork of influences, and although we could explain these away through recourse to intertextuality, there is something more at play here. A Polish national by birth, Conrad developed what Bennett variously called an impressionistic and psychological style of writing (256). It was a style of writing intended to be affective and one Bennett strove to copy: 'I have not quite got the trick of throwing them [impressions] into form instantly and of intensifying them to a degree sufficiently poetical' (265). Although Bennett first encountered Conrad as a textual reference in one of Wells's early letters, they later became better acquainted through their frequenting of the same literary salons and events.

Salons, whether held at homes or literary clubs, were often staging posts within the creative process at this time; here writers could share ideas, network with publishers and benefit from advice and even financial support (Vadillo 64–71). Bennett makes mention of many of these in his journal. We know, for instance, that through a network of London salons, Bennett developed working relationships with writers such as John Galsworthy, John Buchan, and Alfred Richard Orage. It was at Spade House, H.G. Wells's new home near Folkstone, that Bennett met Conrad for the first time. After this, Conrad was insistent on them meeting to talk about their craft:

> Nothing would give me greater pleasure than to have it out with you, the book there on the table, to be thumped and caressed . . . . I may have misunderstood your standpoint utterly. I want to hear what you have to say, if you think it worth while to say anything to me. Only let it be *viva voce.* Come when you can spare a day. (Hepburn 146)

Bennett's journal is peppered with many of these meetings. In the spring of 1910, for example, Bennett recounts a dinner at the Chelsea Arts Club, where he spent much time talking with Conrad. These face-to-face social relationships were important components in Bennett's writing practice. They were opportunities to discuss ideas, to catch-up on literary gossip and to foster professional networks. Attempts have been made to map such spaces of social interaction within fin-de-siècle London in order to reveal the clustering of creativity (McCracken 86–98), yet what is missing from these mappings is any sense of encounter: of what happened within these places and of the role these happenings played in the creative process. Once more, we only have fragments to guide us; but some form of historically sensitive geovisualisation may enable us to bring together fragments of place – the various recollections that exist of and about it – and create a record of its social happenings. In this way, a place like the Chelsea Arts

Club is not just a dot on a map, but rather, and to borrow from Ingold, a knot where various life lines come together (70).

One way of doing this is to turn to the forms of geovisualisation that Bennett was likely to be familiar with. Of these, the gazetteer stands out. This was, and is, a geographical directory that details the social, geological, physical, cartographic, and historical make-up of place. It is a form to which references are made throughout Bennett's fiction: *The Old Wives' Tale* (1908), for instance, makes reference to Dugdale's Gazetteer; *Clayhanger* (1910) names Lewis's Gazetteer; and we know that Bennett kept a historical gazetteer of place at hand to aid the writing process (AB 1994.5). In fact, Bennett's fiction, in its acuity to place, was itself likened to the form of a gazetteer (Howells 634). If we are to elucidate and attempt to visualise the creative world, or spatial practices, that existed before Bennett's material texts – the world of meanders, textual influences, and social relationships – in a manner sensitive to time and place, a digital gazetteer seems a most appropriate form. It is, perhaps, a temporally sensitive form of deep mapping,[5] for herein place as location can be overlain with place as encounter; alongside the factual information it could be possible to read about particular happenings. Thus, a virtual reconstruction of the 10 September 1897, a day on which Bennett records traversing the environs of the Potteries, could be crafted through hyperlinks that take us to accounts of local events, contemporaneous photographs of the area and records of weather and climate of the time, in order to build up a gazetteer not of place *per se*, but of the author-in-his-world. To illustrate this further, take the Chelsea Arts Club: we know that Bennett visited in the spring of 1910 and spoke with, amongst others, Conrad. If Conrad made a similar record of this encounter, or if the textual heritages of other writers make reference to Bennett in this place, a digital gazetteer may bring to light new influences on his craft. At the same time, we may come to understand the Chelsea Arts Club as a site of other encounters, a knot as Ingold terms it, where various life lines come together (70). Although this deforms the gazetteer somewhat, emphasising the relationality between, rather than the rational accounting of place, it remains true to its function as a compendium, albeit one impelled by the author.

Conclusions

In taking as its starting point the cartographies that go into, rather than those that come out of a literary text, this chapter suggests that

5 Deep mapping was first coined by the travel writer William Least Heat-Moon. It seeks to move away from 'thin maps' that record only factual information and singular perspectives to develop more complex, multi-layered maps. Deep maps layer different forms of information, photography, prose, image and audio about place. They can capture emotions and meaning in order to evoke the different patinas of place.

'text-as-process' – or put another way, that which comes before the materiality of the text – is far from straightforwardly mappable. Source material is always partial and often disparate, locational detail is more sporadic and the beginnings of the creative process are always difficult to pin down. To map this in conventional Cartesian ways not only underplays this discontinuity, but it denigrates the historical and personal specificity of everyday cartographies and the motility that inheres therein. Arnold Bennett created in place but he also created across space: he undertook field studies, he generated creative networks and he sought-out conversation. Where work within GIS has begun to grapple with these everyday flows of life, this needs to be pushed further and developed in ways that are both more historically sensitive and accommodative of life's relationality.

In doing this, however, we must question whether GIS is the appropriate vehicle for this kind of study. Urška Perenič has recently cautioned against the uncritical acceptance or imposition of maps and mapping practices within spatially orientated literary studies (Perenič 23–4). A map, the argument runs, can only take us so far and can often limit the ways in which we read or compose meaning. Everyday cartographies and creation are related, but their *longue durée* or procedural nature makes them, to borrow from Attridge, resistant to rational accounting; they are slippery and elusive (Attridge 3). A GIS that maps without a map, revealing instead linkages, networks, and relationships over time and space is, perhaps, more productive in the world that comes before the text. This is what a digital gazetteer can do. It can disclose spatial relationships and, if this gazetteer can bring together fragments from various authorial archives, it can potentially make known forgotten, lost, or previously unknown communities of influence.

Works Cited

AB 1994.5: Correspondence between Hugh Walpole and Arnold Bennett, 29 May 1913. *The Potteries Museum and Art Gallery Archive.*

Attridge, Derek. *The Singularity of Literature.* London: Routledge, 2004.

Austin, J.L. *How to Do Things with Words.* Oxford: Oxford University Press, 1971.

Benjamin, Walter. *The Arcades Project.* Ed. Rolf Tiedemann. Trans. Howard Eiland and Kevin McLaughlin. Cambridge MA: Harvard Bellknap, 1999.

Bennett, Arnold. *The Journals of Arnold Bennett, 1896–1910.* Ed. N. Flower. London: Cassell and Company Ltd., 1932.

———. 'Some personal memories of Conrad: "Cad" as a new word: his "Twilight"'. *Evening Standard* 3 November 1927: 7.

———. 'Try Geography for Example'. *The Bookman* (October 1918): 213–14.

———. *Literary Tastes.* London: Jonathan Cape, 1909.

Brosseau, Marc. 'Geography's Literature'. *Progress in Human Geography* 18 (1994): 333–53.

Burnett, D. Graham. *Masters of All They Surveyed: Exploration, Geography and a British El Dorado.* London: Chicago University Press, 2000.

Bushell, Sally. *Text as Process: Creative Composition in Wordsworth, Tennyson, and Dickinson.* London: University of Virginia Press, 2009.

Brace, Catherine, and Adeline Johns-Putra. 'Recovering Inspiration in the Spaces of Creative Writing'. *Transactions of the Institute of British Geographers,* 35 (2010): 399–413.

Cooper, David. 'Critical Literary Cartography: Texts, Maps and a Coleridge Notebook'. *Mapping Cultures: Place, Practice, Performance.* Ed. L. Roberts. Basingstoke: Palgrave Macmillan, 2012. 29–52.

Cooper, David and Ian Gregory. 'Mapping the English Lake District: A Literary GIS'. *Transactions of the Institute of British Geographers* 36 (2011): 89–109.

Cooper, David, and Gary Priestnall, 'The Processual Intertextuality of Literary Cartographies: Critical and Digital Practices'. *The Cartographic Journal* 48.4 (2011): 250–62.

DeSilvey, C. 'Salvage Memory: Constellating Material Histories on a Hardscrabble Homestead'. *Cultural Geographies* 14 (2007): 401–24

Farrell, Michael. *Collaborative Circles: Friendship Dynamics and Creative Work.* London: University of Chicago Press, 2001.

Glyer, Diana. *The Company They Keep: C.S. Lewis and J.R.R. Tolkien as Writers in Community.* Kent: Kent State University Press, 2007.

Gregory, Ian and David Cooper. 'Thomas Gray, Samuel Taylor Coleridge and Geographical Information Systems: A Literary GIS of Two Lake District Tours'. *International Journal of Humanities and Arts Computing* 3.1–2 (2009): 61–84.

Harris, Trevor M., Susan Bergeron and L. Jesse Rouse. 'Humanities GIS: Place, Spatial Storytelling, and Immersive Visualisation in the Humanities'. *Geohumanities: Art, History, Text at the Edge of Place.* Ed. M. Dear, J. Ketchum, S. Luria and D. Richardson. London: Routledge, 2011. 226–40.

Hewitt, Rachel. 'Mapping and Romanticism'. *Wordsworth Circle* 42 (2011): 157–65.

Hepburn, James. *Arnold Bennett: The Critical Heritage.* Abingdon: Routledge, 1971.

Horowitz, Howard. 'Wordmaps'. *Geohumanities: Art, History, Text at the Edge of Place.* Ed. M. Dear, J. Ketchum, S. Luria and D. Richardson. London: Routledge, 2011. 107–11.

Howells, W.D. 'Editor's Easy Chair'. *The Harper's Monthly* March 1911: 633–6.

Ingold, Tim. *Being Alive: Essays on Movement, Knowledge and Description.* Abingdon: Routledge, 2011.

Kitchin, Rob, and Martin Dodge. 'Rethinking Maps'. *Progress in Human Geography* 31 (2007): 331–44.

Least Heat-Moon, William. *Prairyerth.* New York: Houghton, 1991.

Lorimer, Hayden. 'Telling Small Stories: Spaces of Knowledge and the Practice of Geography'. *Transactions of the Institute of British Geographers* 28 (2003): 197–217.

McCracken, Scott. 'Voyages by Teashop: An Urban Geography of Modernism'. *Geographies of Modernism.* Eds, Peter Brooker and Andrew Thacker. Abingdon: Routledge, 2005. 86–98.

McGann, Jerome. *The Textual Condition.* Princeton: Princeton University Press, 1991.

Massumi, Brian. *Parables for the Virtual: Movement, Affect, Sensation.* Durham, NC: Duke University Press, 2002.

Moretti, Franco. *Atlas of the European Novel.* London: Verso, 1998.

Ogborn, Miles. *Indian Ink: Scrip and Print in the Making of the English East India Company.* London: University of Chicago Press, 2007.

Patchett, Merle. 'Tracking Tigers: Recovering the Embodied Practices of Taxidermy'. *Historical Geography* 36 (2008): 17–39.

Perkins, Chris. 'Cartography-cultures of Mapping: Power in Practice'. *Progress in Human Geography* 28 (2004): 381–91.

Perenič, Urška. 'An Overview of Literary Mapping Projects on Cities: Literary Spaces, Literary Maps and Sociological (re)Conceptualisations of Space'. *Neohelicon* 41 (2014): 13–25.

Rogers, Amanda. 'Butterfly Takes Flight: the Translocal Circulation of Artistic Practice'. *Social and Cultural Geography* 12 (2011): 663–83.

Reuschel, Anne-Kathrin and Lorenz Hurni. 'Mapping Literature: Visualisation of Spatial Uncertainty in Fiction'. *The Cartographic Journal* 48.4 (2011): 293–308.

Saunders, Angharad. 'The Spatial Event of Writing: John Galsworthy and the Creation of "Fraternity"'. *Cultural Geographies* 20 (2013): 285–98.

Stewart, Kathleen. *Ordinary Affects.* Durham NC: Duke University Press, 2007.

Thrift, Nigel. *Non-representational Theory: Space, Politics, Affect.* Abingdon: Routledge, 2008.

Vadillo, Ana Parejo. 'Aestheticism "at-home" in London: A. Mary, F. Robinson and the Aesthetic Sect'. *London Eyes: Reflections in text and image.* Eds, G. Cunningham and S. Barber. Oxford: Berghan Books, 2007. 59–78.

Waldie, D.J., and Neil Campbell. 'An assemblage of habits': D.J. Waldie and Neil Campbell – A suburban conversation'. *Western American Literature* 46 (2011): 228–49.

Walford Davies, Damian. *Cartographies of Culture: New Geographies of Welsh Writing in English.* Cardiff: University of Wales Press, 2012.

Wells, Herbert George. *Experiment in Autobiography: Discoveries and Conclusions of a Very Ordinary Brain.* 1934. Philadelphia and New York: J.B. Lippincott, 1967.

Wilson, Harris, ed. *Arnold Bennett and H.G. Wells: A Record of a Personal and a Literary Friendship.* Urbana: University of Illinois Press, 1960.

Wimsatt, William K., and Monroe C. Beardsley. 'The Intentional Fallacy'. *Sewanee Review,* 54 (1946): 464–88.

Withers, Charles. 'Writing in Geography's History: Caledonia, Networks of Correspondence, and Geographical Knowledge in the late Enlightenment'. *Scottish Geographical Journal* 120 (2008): 33–45.

8 Between 'Distant' and 'Deep' Digital Mapping

Walking the Plotlines of Cardiff's Literary Geographies

Jon Anderson

Introduction

As this collection demonstrates, digital literary mapping is an exciting, dynamic and growing field. From 'distant' to 'deep' cartographies, digital technologies offer a range of opportunities to recast written words and the literary worlds they project. This chapter introduces a new alternative with which we can conceptualise, understand and engage text and territory: the digital plotline. Plotlines are geographical and literary routes through the world. They are cartographic navigations that connect the page to the place and entangle them into an ongoing composition. This chapter outlines the notion of the plotline, and demonstrates its function through a digital mapping of one of Cardiff's literary geographies. Weaving together the writings of Peter Finch and John Williams and the cityscape of Cardiff Bay, the chapter demonstrates how the real gives shape to the imaginary, and how the texts of these authors and the territories of this place entangle in practice. In the process, it engages with the website *Cardiff Plotlines* to illustrate this entanglement. The chapter suggests that mapping plotlines digitally offers us fresh insight into the co-ingredient connections between reader, fiction and place.

Mapping Literature

> The map offers a fictional or figurative representation of the space in which we find ourselves, and the reassuring "You are here" arrow or dot or other marker provides the point of reference from which we can both imagine and navigate the space. (Tally 2)

Geography is essential to fiction (Reuschel and Hurni 2011 291). As Piatti et al. observe, it is 'impossible to [. . .] think of literature without any spatial context' (178). However, as this collection clearly illustrates, mapping literature is not a straightforward exercise. The mapping of literary works necessarily requires an engagement with the definition and practice of cartography. As David Harvey suggests, maps serve a wide range of social purposes. One of their clearest functions is to act as a store of knowledge about a territory, selectively portraying a 'patchwork representation of the world' (Tally 49). This representation reveals basic relationships between key elements in a place, and serves as a communication device to enable individuals to locate themselves and find their way (see Tally 2). With this in mind, maps can be seen to function in a similar way to fictional literature. As Robert T. Tally Jr. reminds us, the literary work can function 'as a form of mapping, offering its readers descriptions of places, situating them in a kind of imaginary space, and providing points of reference by which they can orient themselves and understand the world in which they live' (2). From this perspective, then, mapping literature is really a form of *re*-mapping, of mediating or translating one form of literary cartography into another. A literary map (in the sense of a drawn or digital image with pictorial signs and symbols) is therefore a kind of 'illiterature' (Manning, qtd. in Naramore Maher 23), in other words, the re-publication of a literary work (e.g. novel) in a different, often non-written, form (e.g. map). Two key questions arise when considering the production of such illiterature: What functions can this mediation fulfil? How can these functions be realised most effectively?

With reference to the first question, it is accepted that re-mediating literature can bring new understandings to literary cartography. By connecting representations together, literary maps 'will be more than the sum of their parts: they will possess "emerging" qualities, which were not visible' in the original form (Moretti, *Graphs, Maps, Trees* 53; original emphasis). Barbara Piatti and Lorenz Hurni suggest that this process 'creates knowledge'. As they explain, 'the mapping of fiction allows a better, deeper understanding of how fiction works – the mapping process supports the interpretation; it opens new horizons for literary scholarship, because some maps make aspects visible which have been invisible before' ('Cartographies of Fictional Worlds' 222). As a consequence, creating illiterature can 'change our understanding – not only of books, but of the world we live in' (222). Yet what new knowledges are sought through the digital mapping of literature, and how best can we capture them? Broadly speaking, two basic approaches can be identified to date: going 'distant' and going 'deep'.

Digital Mapping: Distant or Deep

> [D]evelopments in this field were primarily predicated upon the mapping and spatial analysis of quantitative sources using social science approaches. (Gregory and Cooper 266)

As Ian Gregory and David Cooper suggest, many attempts to map literature digitally have sought to create (social) scientific understandings of the relations between fiction and geography (Moretti, *Graphs, Maps, Trees*; Piatti and Hurni, 'Cartographies of Fictional Worlds'). In the positivist tradition scholars have abstracted themselves as readers and cartographers from the understandings being produced, 'extracting bits of information [from literature], transferring such bits onto a spatial diagram, and then interpreting the resulting diagram' (Tally 108). As Franco Moretti puts it, 'You choose a unit [from the fiction] – walks, lawsuits, luxury goods, whatever – find its occurrences, [and] place them in space . . . in other words: you *reduce* the text to a few elements, and *abstract* them from the narrative flow, and construct a new, *artificial* object' (*Graphs, Maps, Trees* 53). These 'distant' and 'artificial' readings (Gregory and Cooper 270) seek to minimise the 'uncertainty' and ambiguity of the literary work (Reuschel et al.), and enable 'data spaces', characterised by registered, counted and presented figures, to emerge (Benedikt 135). Through this process a literary cartography is produced in the form of 'a map [that] is . . . like a grid, . . . a geometric figure' (Tally 46). In the spirit of the abstraction that created it, this geometric cartography offers a view of the world 'read literally from above. Insofar as it represents a terrain, that terrain is seen from the heavens' (Idhe 67).

However, some literary cartographers seek to extract different understandings from their mapping exercises. Instead of going 'distant', they have attempted to go 'deep'. Deep mapping adopts a phenomenological rather than positivistic approach to understanding literature and geography. Drawing on the philosophies advanced by Martin Heidegger (1927) and Maurice Merleau-Ponty (1945), these approaches valorise 'being-in-the-world' rather than 'being-above-it', projecting worlds that seek to 'limn a matrix of communal and personal memory, competing histories and polyvocal narrative' (Naramore Maher xiv). Although these attempts are currently more drawn (or written) than digital, they nevertheless provide an alternative that offers maps best understood as 'place-hives'. These place-hives seek to create a multi-layered store of information, combining a matrix of personal and collective insight, local history and literary narrative. As Susan Naramore Maher puts it:

> Like a literary survey team, these [deep cartographers] map multiple measures. Inspired by geologists, [they] create cross-sectional narratives of natural history, illuminating the strata in which deep time and human time collide . . . . Following cultural geographers, they mark the shifts and migrations, booms and busts, erasures and additions, always keeping an eye open for the palimpsests of former worlds. (15)

These deep maps thus offer a micro view that contrasts distant mapping's 'macro' view. Between these extremes of distant and deep mapping, this chapter offers an alternative that seeks to foreground the experiential practice of reading fiction and it explores how this crucial dimension may be introduced to digital literary maps. Drawing on the relational turn in the social sciences, the chapter introduces the plotline as a way to enable the digital mapping of literature that prioritises and in turn produces an experiential encounter between text and territory.

Entangling Text and Territory

Rather than drawing directly on positivist or phenomenological perspectives, the plotline emerges from the relational turn within the social sciences (Haraway; Amin; Pierce et al.). Central to this turn is the valorisation of 'actor-centered' ways of seeing the world (Jones). In terms of the geography of fiction, this actor-centred approach returns us to the act of reading and how this can place us in the worlds of written fiction and material geography. This turn also emphasises how every inhabitant of a (real and imagined) place will have a different perspective on that world. This resonates strongly with Roland Barthes' seminal thesis in 'The Death of the Author' (1977). Here Barthes famously argues the reader should not be framed as a passive consumer of authorial intent. Rather they have the agency to produce their own reading of any story. As Hilary Mantel tells us, 'a novel [should be understood as] a co-operative effort, a joint venture between writer and reader. I [the writer] purvey my own version of events, but the facts change according to your [the reader's] own viewpoint' (134). By prioritising the 'co-operative effort' of writer and reader, it becomes crucial to consider how both parties perceive the relations between text and territory. In some cases, authors intersect their fiction directly with actual geographies. Narratives can be based in specific countries, regions and towns. This direct coincidence of fictional and geographical space can be seen in famous examples such as Thomas Hardy's Wessex, James Joyce's Dublin or Paul Auster's New York. Readers may intimately know these locations, and the laying down of fictional stories across these streets adds new matrices of meaning on their place palimpsest. In other fictions, real geographies are moulded by both the author and the reader, with distances reduced, streets folded and landmarks crumpled together. This process produces an array of 'counterfactual geographies' (complementing Piatti and Hurni, 'Mapping the Ontologically Unreal'), which 'select, distort and compress' material places into fictional cartographies that come to 'possess a truth of their own' (Harrison 119). These co-operative efforts of writing and reading emphasise the myriad ways that the 'physically comprehensible world' can become insinuated into the 'literary world', and vice versa. It is when we focus on elisions between 'geospace' and 'fictional space' (Piatti and Hurni, 'Cartographies of Fictional Worlds' 219) that the possibility for entanglements between text and territory begins to emerge.

As we have seen, by prioritising the relations between authors, readers, texts and territories, we become sensitive to the agency and influence that literature has on geography, and vice versa. As I have argued elsewhere (Anderson, *Page and Place)*, we can use the metaphor of 'entanglements'

(Foucault; Sharp et al.) to conceptualise these crossings between text and territory. This metaphor suggests how the literary plot of the story and the literal plot of a location are always implicated in, and constitutive of, one another. As Sharp et al. state, the:

> use of the term 'entanglements', suggesting an image of knotted threads, is intended to underline the deep 'spatiality' of this spinning together of [in this case, text and territory]. We talk about 'entanglements' to indicate that [text and territory are] *always* played out in [and] across [each other]. (Sharp et al. 1; emphasis added)

Framing the literary and the literal as entangled phenomena suggests, therefore, that text and territory cannot exist independently of each other. Taking this one step further, I argue that the literary relies on the literal for a grounded materiality, and concomitantly that the literal relies on the imagined for an alternative reality. In short, there is no complete separation between text and territory. From this entangled perspective, it makes more sense to suggest that we are not dealing here with text *and* territory, but text *in* territory (and territory *in* text). In other words, there is a 'constitutive coingredience' of text and territory (Casey 684). The page and the place are always connected. Storylines can take you to an actual site, whereas a real location is constantly shadowed by our literary emotions and imagination.

Capturing Text and Territory: From Plots to Plotlines

This idea of entanglement can be captured through expanding upon the concept of 'plot'. Plot can be understood in both a literary and a literal sense. Plot is at once a narrative or storyline, whilst also being a locatable, geographical territory. The entangled nature of these different, yet connected, aspects of the word plot is indicated by a third definition: the ability to intrigue, subvert, scheme or imagine. Just as we think we have nailed the plot as simply a storytelling device or location, it loosens its conceptual moorings and re-tangles itself with its other: our plots are therefore never singular, but always plural. From this basis, the connection between text and territory can be conceptualised as an ongoing composition of plot. Plots are composed. They have a position both in time and in place, and numerous actors perform them: the author of the book, its reader and others that contribute to the materialities and meanings of place. As a consequence, there is not simply one plot or narrative or geographical place. This conclusion presents us with a problem if we seek to map literature digitally in conventional ways. Due to the plurality of interpretations that compose every plot, distant or deep cartography cannot ever fully capture the geography of fiction. Yet if we no longer seek to map or locate a complete geography of fiction, it remains possible to direct one route, or 'trail', along which the geography of literature is 'lived' and experienced (Ingold 47).

Plotlines are particular routes chosen through the entanglement of text and territory. These plotlines are at once geographical and literal, referring directly to a position on the globe, but also inevitably entangled with literary plot and narrative. Knotted together with these components are the lives of readers, who bring their own outlook, opinion, culture and ideas to the plotline, making it actor-centred and personal, but also a potentially collective 'way' through the geography of fiction. Plotlines, in this way, can be compared with Aboriginal 'songlines' as famously described by Bruce Chatwin. As Chatwin explains, songlines are a form of indigenous oral cartography, forming maps through which Aboriginal tribes navigate the land, tell their histories and identify themselves and others. Although songlines are inherited through family and ancestry, plotlines are democratic and personal in nature. Any reader can 'call forth' a new world through their reading of a fiction. Plotlines therefore emphasise how reading is a creative act, bringing into being new routes of excitement, intrigue and experience to the geographies of fiction. These plotlines leave new 'traces' over this cartography (see Anderson, *Understanding Cultural Geography*), adding new trails of words and ideas that can themselves lead to new experiences.

If imagined and practised in this way, individual plotlines can function as 'ways' of communication (after Chatwin) between different people, ages and locations. Although not definitive or absolute, they can remain a partial and selective 'map' and 'history', functioning as both a 'library' and a 'museum' for a whole neighbourhood, city or region. In this way, plotlines can be mobilised for a whole country, and that whole country can be known via a plotline. However, plotlines are not definitive or objective, and their cartographical alignment with real places may be tentative and provisional. For some, their connection to 'real' places may be difficult to grasp scientifically, but in terms of personal affinity, poetic resonance or lateral memory, they offer fascinating new insights into the relations between people and place.

Digital Plotlines

This chapter illustrates the potential offered by digital plotlines by taking the reader on a journey through some of Cardiff's literary geographies by showcasing *Cardiff Plotlines* (see Figure 8.1). This website maps four plotlines through the streets of the city, entangling the real with the writings of Peter Finch, Tessa Hadley, Lloyd Robson and Andrew Craig Williams. The site offers virtual, fictional and actual routes through the city that set in motion literary cartographies that are both 'an analysis as well as an aesthetic provocation' (Smith 266).

In what follows, this chapter demonstrates the use of the website by taking one plotline through Cardiff Bay, a dockside neighbourhood of

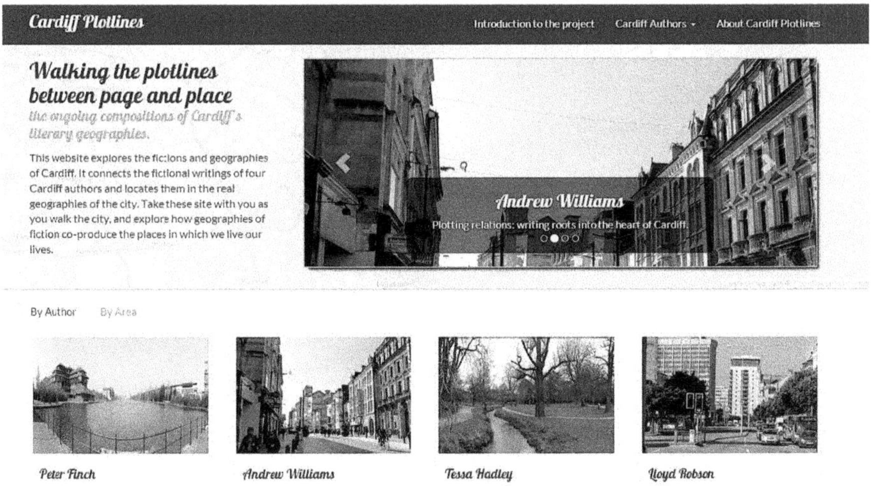

Figure 8.1 Cardiff Plotlines: Home.
Source: © The Author.

the city. This particular plotline brings together four key sources. Firstly, it draws on the writings of Cardiff authors Peter Finch and John Williams who set their works – *Real Cardiff* (2002–2009) and *The Cardiff Dead* (2000) – in the streets and alleys of Cardiff Bay. Secondly, it incorporates extracts from a personal interview with Peter Finch (March 2012) about the importance of place in grounding and defining his narratives.[1] Thirdly, this plotline includes a digital map that uses conventional street representations and landmark symbols to communicate the city to the reader. Those who know the area can bring their own experiences and reminiscences to the map, whereas those less familiar with Cardiff Bay can use the map's symbology to locate themselves (after Tally 2). Fourthly, this digital plotline waymarks its route with geo-referenced photos and short film clips that are embedded in the map interface. These multimedia

1 Importantly, this interview was undertaken whilst walking a route through Cardiff Bay. As I have argued elsewhere (Anderson, *Talking Whilst Walking*; *Understanding Cultural Geography*), the power of place can be used to give depth and texture to interviews. Emplaced interviews can realise an unstructured dialogue where all actors (interviewer, interviewee and location) can participate in a conversational and geographical pathway creation. As a consequence, the knowledge produced through this method is importantly different: atmospheres, emotions, reflections and beliefs can be accessed with regard to the plot, as well as intellects, rationales and ideologies. This approach also encourages the reader to enter into (as well as co-produce) an entangled landscape, engaging with or calling forth, particular plotlines through practice.

layers give depth to the distant cartographic view of the area and evoke some of the sounds of the Bay for the website user. With the ingredients entangled together, the reader or website user has a plotline to follow through the streets of Cardiff Bay. This can be followed via a computer or tablet from the comfort of an armchair, enabling readers to follow the footsteps of this plotline either virtually or by actively following the map and encountering the characters involved in this entanglement of text and territory.

Plotting Cardiff Bay

As John Williams tells us, Cardiff Bay was once known as Tiger Bay, and this neighbourhood, alongside the adjacent Butetown district, are distinct areas separate from the city of Cardiff itself. As Williams suggests, these places have fictional mythologies that define them:

> Tiger Bay was the original pirate town. The way I heard it as a child, the real old-time pirates, Captain Morgan and his crew, used this promontory off the then small town of Cardiff as a base: a little piece of Britain that was beyond the law. God knows whether it's true or whether my addled memory has simply cobbled together a new myth out of two or three old ones, but still, it's a legend that suits the wild side of the Welsh. ('Legend of Tiger Bay' 217)

At times a 'fortress', a 'ghetto', an 'island' and an 'isthmus' (Finch, *Real Cardiff* 222; Williams, 'Legend of Tiger Bay' 221), Cardiff Bay and Butetown are physically divided from Cardiff itself by both the river and the railway. The area has a rich and varied history which has often been ignored, silenced or destroyed by the development of the city and the Bay. As Williams states:

> So what that Tiger Bay is now remembered only as history – replaced by something new and shiny, and determinedly without history, called Cardiff Bay?
>
> Well so quite a lot – without memory and history we lack community and we lack soul. Community is something Cardiff has always been good at. But Tiger Bay gave [the city] its soul. So let us remember it for sure, but let us also try to preserve something of its anarchistic, tolerant, cosmopolitan soul in this new place that we have learned to call Cardiff Bay. ('Legend of Tiger Bay' 234)

Peter Finch seeks to rediscover some of this soul in his first book in the *Real Cardiff* series. Although the title of the series, *Real Cardiff*, suggests a preoccupation with strict ontological fact, the books present a 'reality' that merges fact with stories, myths and personal histories. In this sense,

Finch can be seen to be writing in the psychogeographic tradition (see Coverley), which, with its *avant-garde* pedigree, emphasises the necessity of accounting for both the physical and psychological dimensions of place (Pinder; Bonnet). As part of this tradition, *Real Cardiff* is more of an entangled mix of elements rather than a factual guidebook. For Finch this is an intentional, political statement. His 'general distrust' of sanitised, glossy guidebooks, which 'rarely give you all the information you want' (Finch et al.), moved him towards the psychogeographical tradition. As we walk out from the Millennium Centre, past the Water Tower and down to the waterside (see Figure 8.2), he reflects on this whilst explaining his intention for the series:

> I wanted to uncover that thin layer of city, the place which exists underneath the fabric that we see around us now . . . . So I walked around the city looking for bits that could make me think 'how did this use to be?' I wanted to find the stuff that was underneath, I wanted to find the myths, the ley lines, the sources of power, the places that actually felt like they were something magic, something ancient, something that were full of people that are no longer with us. (Finch, Personal interview)

Peter Finch's *Real Cardiff* series thus seeks to get under the skin of the city and root around for marginalised pasts, reawaken urban ghosts and better

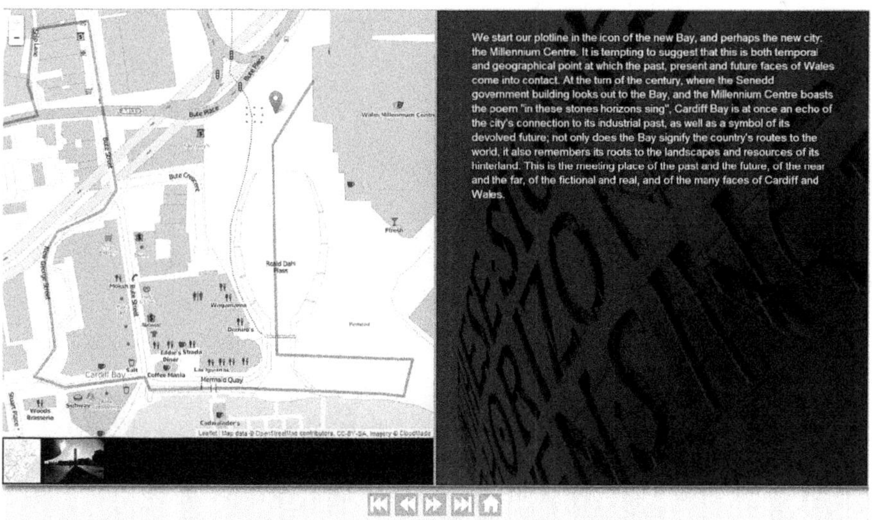

Figure 8.2 Peter Finch: Stalking the Soul of the City.

Source: © The Author.

understand – and retain – the 'real' soul of the city. We encounter some of this past as we continue our plotline away from the waterside and into some of the alleys that connect the streets of the Bay (Figure 8.3):

> This [Finch points down an alley in Cardiff Bay] is an old shipyard lane and I think this is very reminiscent of what Cardiff must have been like, the way it bends and the shape of those buildings there, the curved tops on the door, if you try and think of this without the cars and without

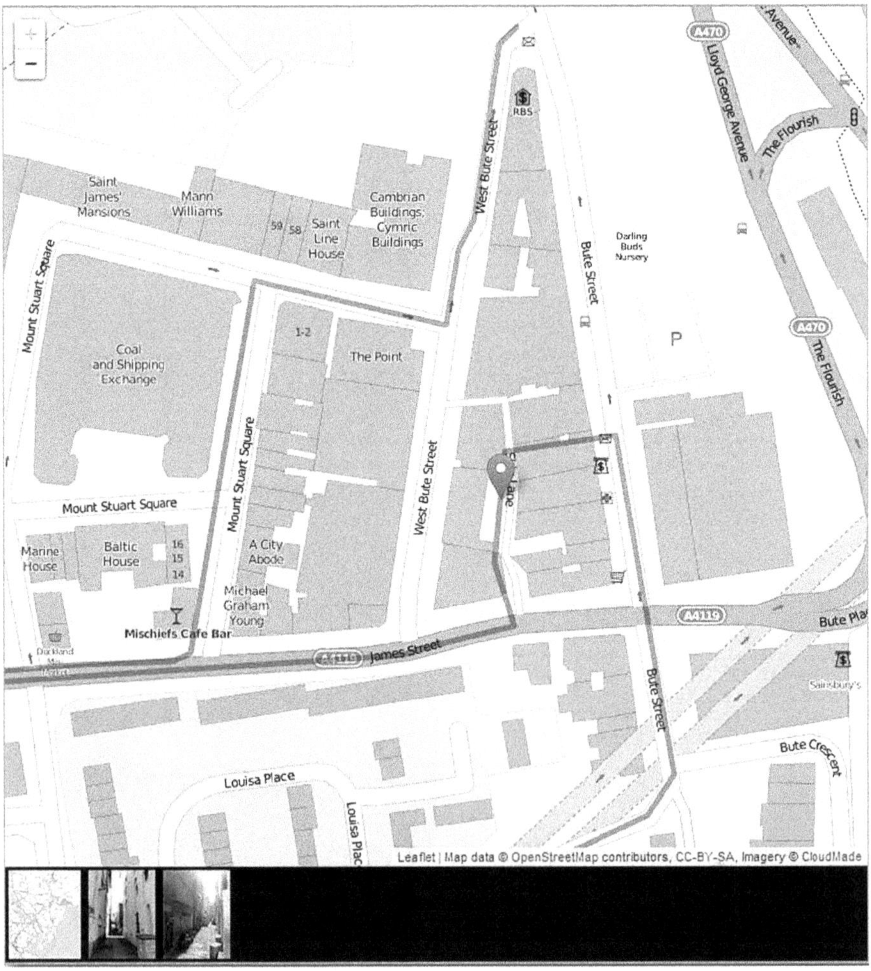

Figure 8.3 Butetown's Alleys.

Source: © The Author.

the dustbins and the ends of those air conditioning machines you can imagine it. So I did a lot of this, walking around the city looking for bits that could make me think 'how did this use to be?' and get me into that mind-set. (Finch, Personal interview)

Figure 8.3 Continued

We walk from the alley and arrive in Mount Stuart Square. The Square is an area close to the old dock community of Butetown, and has become part of the broader swathe of redevelopment occurring in the Bay. As we do so an excerpt from John Williams' *Cardiff Dead* comes to mind. John Williams,

Figure 8.4 Mount Stuart Square.

Source: © The Author.

Figure 8.4 Continued

a native of the city, has written many fictions about characters hustling in the communities of Butetown. *Cardiff Dead* is the second story in a trilogy about the area, which also includes *Five Pubs, Two Bars and a Nightclub* and *The Prince of Wales*. In *Cardiff Dead* the character Tyra lives on these streets:

> You came out of Tyra's place, cut through Mount Stuart Square and over James Street, and you were in a different world, and one getting more different every time she walked through it . . . . It was funny, the bay development – you'd heard about it for years; since the eighties they'd be knocking stuff down getting ready for it. But for ages that was all it had been, just knocking stuff down and leaving it there. Nice old pubs like the Mount Stuart and the Sea Lock all gone and nothing to replace it. Then gradually it started happening. They moved the Norwegian church right on to the bay. Looked nice, to be fair. They built the brand-new Techniquest building and the Harry Ramsden's next to it. Fair enough, kids loved Techniquest. The UCI, of course, suddenly landing like a giant space ship the far side of Bute Street – twelve screens, bowling alley and all. Kids loved that too, and it was packed from the start, all these people never came down the docks in their life piling in, driving there in their cars and straight out again . . . . Pace of it was getting frightening and she could see it starting to work. Another

> couple of years and the place would be swarming with tourists and that. (Williams, *Cardiff Trilogy* 325)

Calling forth Tyra's 'fictional' experience as I walk into Mount Stuart Square reminded me how locals in the area approach the redevelopment of their neighbourhood. Tyra exemplifies the instinctive scepticism about the imposed changes, but the willingness to concede that they may benefit some people (even if those do not currently live in the area). She also portrays pride in what is left of Tiger Bay, and an affective urge to value and see it protected from these changes. As she remembers her family's stories of Butetown and the docks, she is aware of how the phases of current redevelopment will wholly change the area in which she was brought up:

> [Tyra] had powerful nostalgia for the docks life her dad used to tell her about, a nostalgia all the more powerful for being on the very fringes of her memory . . . . She missed all these places she couldn't really remember . . . . Maybe they'd open up a Tiger Bay theme bar to remember [them] by: cute murals and cute prostitutes and sailors, a little whiff of airbrushed long-ago vice, give the visitors a thrill. Still, it has to be better than letting it rot, she supposed. (Williams, *Cardiff Trilogy* 364, 326)

Through my plotline, Tyra's 'fictional' experience of the redevelopment of the area merges with Finch's 'real' experience, with the latter exposing the tensions produced when the gentrified new threatens the working class-old in this part of the city:

> Today, in the Ship and Pilot [on James Street], Darren, or someone with a name like that, tells me he's not out to waste his useful time talking to muppets. He knows I'm not from round here. This is a small community. Everyone knows everyone else. I'm from Lisvane. No. Well, you're from th'apaartmunts, then, you should be drinkin in Bar38. Norere. This aint the Bay. It is not. I buy him a drink, no use falling out over nothing, dangerous in deep Butetown, whatever the night. He has a double Southern Comfort and Coke. I don't take my notepad out. The Bay is two communities existing in the same space-time continuum. They pass through each other as they move. They rarely interact. (Finch, 'Cardiff' 15)

As Finch and I leave Mount Stuart Square these readings and recollections echo through my head. This plotline is weaving together the embodied (the aches in my shoes, the streets resonating through my joints), with the real (my conversation with Peter Finch, and his words from his psychogeographical guidebooks) and the fictional (Williams' characters). This entanglement becomes so tangible that I forget whether Tyra or Darren is a fiction, or indeed whether the Finch I am talking to is the same Finch in the Ship and Pilot, or some amalgam of Finch and Williams and Darren

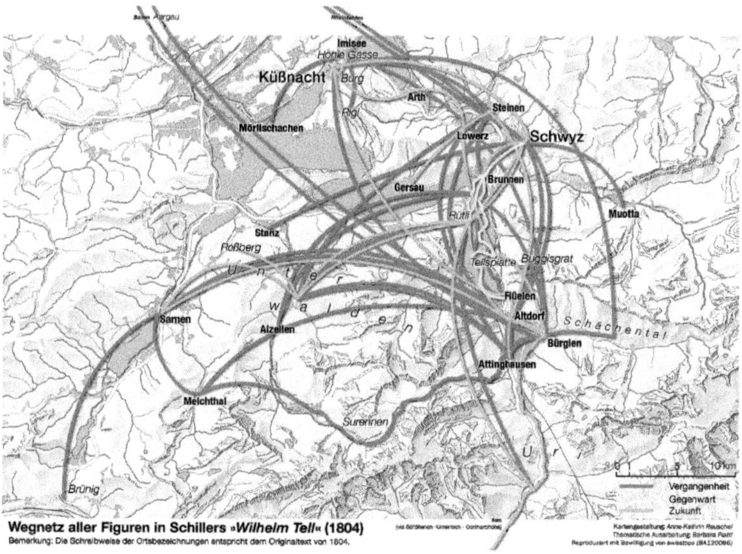

Plate 3 The movements of characters through the fictional space of Friedrich Schiller's *William Tell* (1804). Textual analysis: Barbara Piatti; Map design: Anne-Kathrin Reuschel; Background relief reproduced by permission of Swisstopo (BA 130086). Source: © Barbara Piatti/Anne-Kathrin Reuschel. *See page 94.*

Plate 4 GIS analysis of characters' journeys through model region North Frisia. Based on 57 novels, published between 1860 and 1890, the map shows the fictionalised North Frisia of Theodor Storm and contemporary writers such as Detlev von Liliencron and Klaus Groth. Textual analysis: Kathrin Winkler and Kim Seifert; Map Design: Anne-Kathrin Reuschel. Source: © ETH Zurich, Institute of Cartography and Geoinformation. *See page 94.*

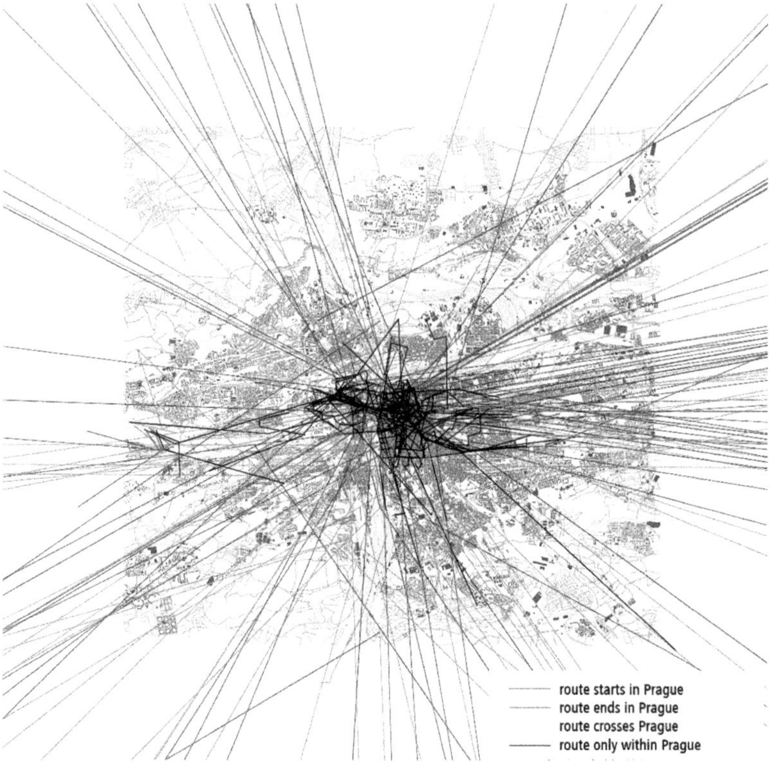

Plate 5 Movements of characters in and around model region Prague. The analysed text corpus contains 73 fictional accounts (prose) published between 1880 and 1918. Textual analysis: Marie Frolikova und Eva Markvartova; Map Design: Anne-Kathrin Reuschel. Source: © ETH Zurich, Institute of Cartography and Geoinformation. *See page 94.*

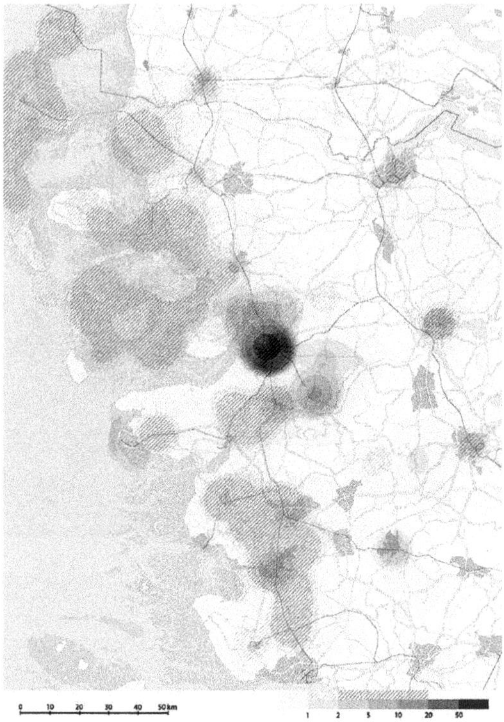

Plate 6 Bivariate Map of Theodor Storm's North Frisia (in green, with the centre point of Husum) and the Frisian settings of other writers (red). Textual analysis: Kathrin Winkler and Kim Seifert; Map Design: Hans Rudolf Bär. Source: © ETH Zurich, Institute of Cartography and Geoinformation. *See page 95.*

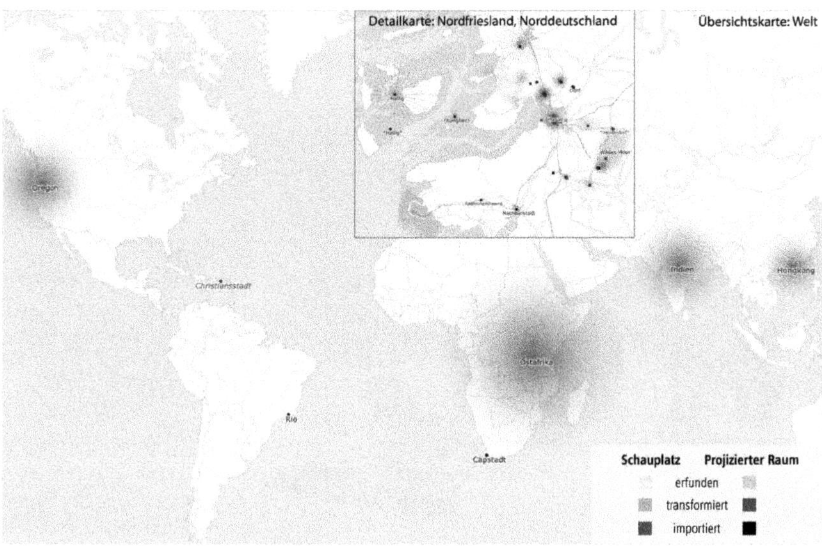

Plate 7 A World Map of fictional works written by Theodor Storm (most of the novellas have their main setting in and around the small town of Husum in Northern Frisia, but some include projected spaces in regions far away). Textual analysis: Kim Seifert and Kathrin Winkler; map design: Anne-Kathrin Reuschel. Source: © ETH Zurich, Institute of Cartography and Geoinformation. *See page 96.*

Plate 8 Rebecca Harding Davis's Wheeling neighbourhood generated in City Engine. Source: © The Authors. *See page 231.*

Plate 9 A street scene described by Davis and generated in Unity3D. Source: © The Authors. *See page 232.*

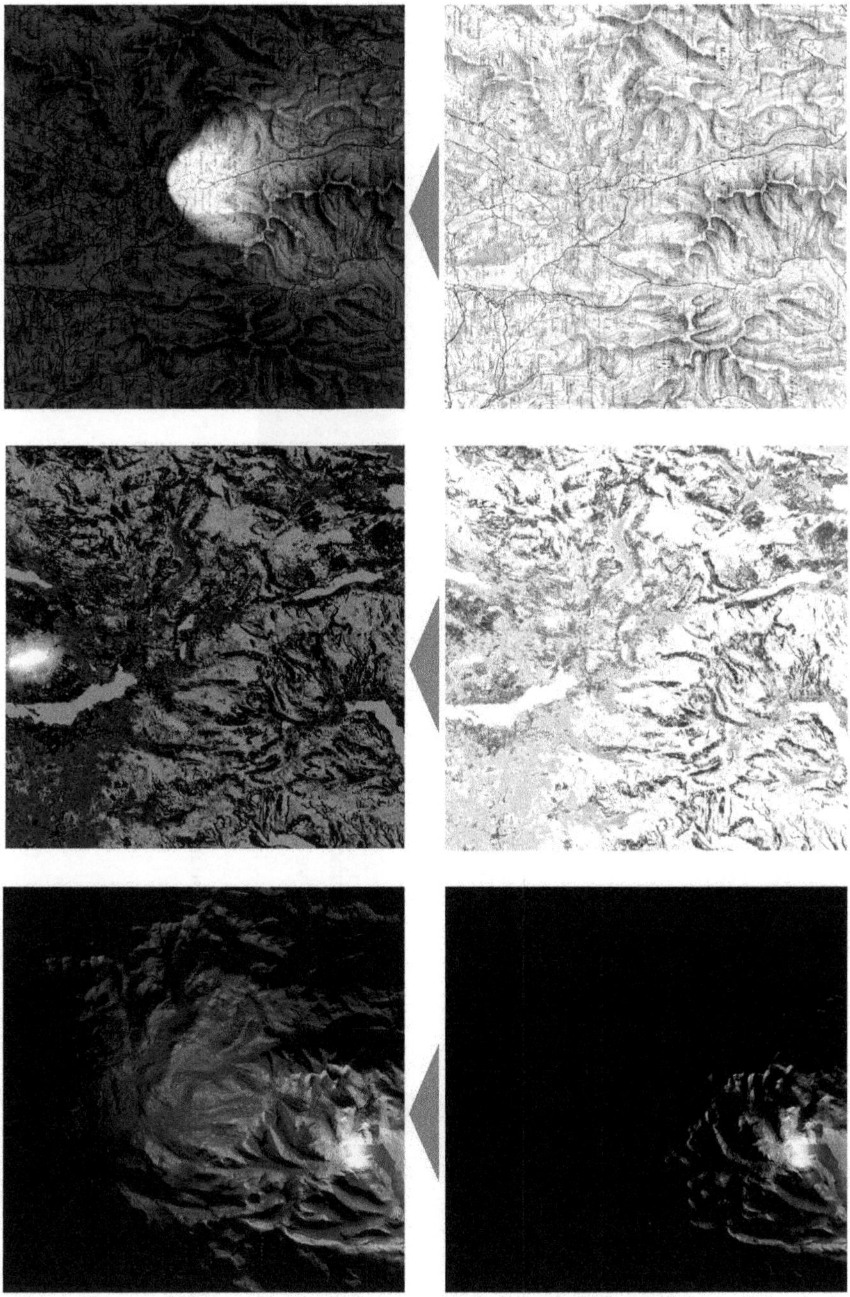

Plate 10 Authoring projection content related to extracts from the historical manuscripts. Source: © The Authors. *See page 248.*

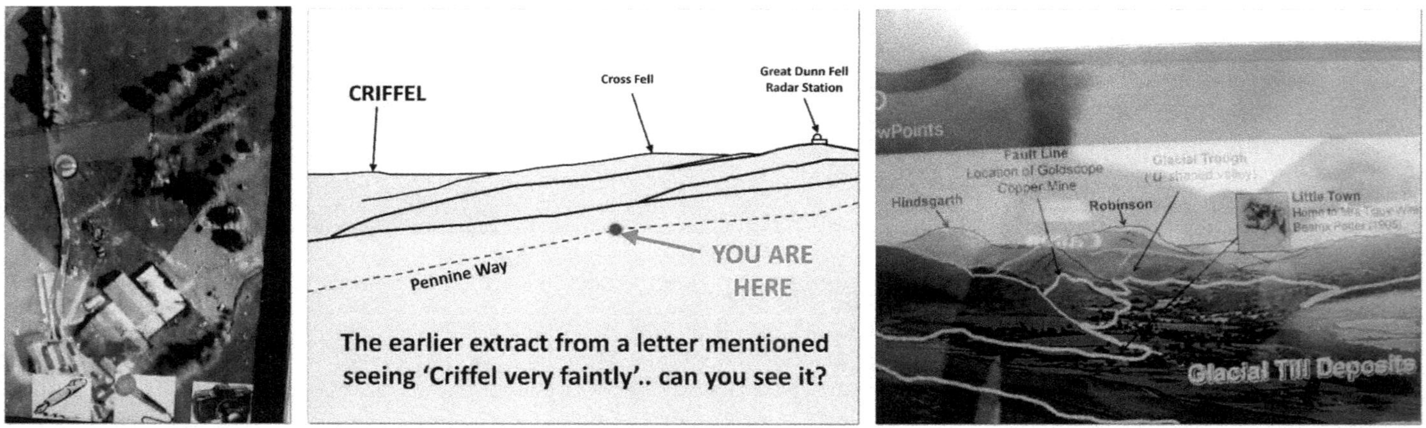

Plate 11 Locative media and the role of landscape models to help user orientation. Source: © The Authors. *See page 251.*

himself. To an extent, the definitions of 'real' and 'imagined' no longer hold sway. The plotline no longer distinguishes between 'fact' and 'fiction', from the active reader's perspective, the text is now the territory and vice versa.

As we walk further along James Street (the 'we' now including Finch, Tyra and I, as Darren remained in the bar with his Southern Comfort), the entanglement of text and territory takes on a material as well as metaphorical dimension. 'I'm bringing you here to show you this', Finch says: 'This is my poem' (Finch, Personal interview).

Outside the police station we stop at a piece of public art: a stone cone with a mirrored lighthouse emerges from the pavement and thrusts toward the sky. This piece forms a counterpoint to the Millennium Centre at the other end of James Street, and its stones which speak of horizons.[2] This stone cone has its own words that project the area's connections to the world. As Finch explains:

> The current text we are standing in front of represents a specific part of the history of Cardiff: this is what remains of the Ballast Bank. The Ballast Bank was made when early wooden ships came into Cardiff in order to take out iron and coal for the docks when all we had was the canal basin. They'd come in with the ballast in the hull of the ships, they'd take the ballast out and they'd load up with iron or coal and off they'd go .... So I said: why don't we put the ballast bank back? This is stylised [of course], but the poem is made by words derived from the races, languages and activities of the people that flowed in and out of Cardiff in the past 200 years. The Portuguese petroleum came in, the Poles came in, the potato trade came in; there were Indians, there were Irish, there were Italians .... There are also a couple of jokes in this. [... I wanted] to put the word Vulcan in there, which related to the vulcanising works [in Butetown, [... but] that word was censored from the piece. However, they missed this one. Klingon. (Finch, Personal interview)

On encountering the Bank the plotline takes on 'deeper' dimensions. Finch's guides in themselves not only scratch beneath the surface of the city to discover the history of Cardiff Bay, but also attempt to reinsert this history through memorials such as The Ballast Bank. Being introduced to the Bank by Finch, and its history emphasised in detail in *Real Cardiff*, has the effect of explicitly connecting the plotline to the international and cosmopolitan history of the city. It feels as if a host of fellow travellers have joined the plotline from the past, as well as some from a galaxy far, far away.

2 The Millennium Centre boasts the 'the best five word poem in the English language' (Griffiths 123). These words, written by Gwyneth Lewis, Wales's first national poet, are designed into the facia of the Centre's grand frontage, they state: 'in these stones horizons sing'.

Figure 8.5 Ballast Bank.

Source: © The Author.

Walking the plotline and calling forth forgotten pasts reminds us how a city absorbs and stores both personal and collective memories. As Italo Calvino tells us, 'the city soaks up [our lives] like a sponge and expands. A description of [a city] as it is today should contain all [of its] pasts' (Calvino 9). In

Figure 8.5 Continued

this reading, cities are affective, living, landscapes. As stones sing of horizons, buildings store memories and ballast becomes part of the city's material soul, a city's identity can be called forth by documenting lost pasts into our contemporary plotlines. In this way, Finch's guides seek to memorialise these transient pasts. As the past slips by, Finch seeks to create memorials to forgotten episodes, not only materially in terms of The Ballast Bank, but also in terms of his written works. Finch's *Real Cardiff* series and Williams' *Cardiff Dead* can thus be seen as entangled armatures that allow the past to be seen alongside the present, co-existing but not necessarily limiting the future.

We leave the plotline in Cardiff Bay, but from this geographical vantage point we can view the skyline of the city, including Millennium Stadium to the north. In this landmark it is possible to see the future of the city embodied. As John Williams narrates:

> Wales are about to play Western Samoa in the Millennium Stadium which has somehow opened on time and forever changed the way Tyra saw the city she'd spent her life in.
>
> The old stadium, the Arms Park, was almost invisible from the centre of town. You were only really aware of it when you looked at it from over the river. The new stadium dominated the centre and you could see it from almost anywhere in Cardiff. And, like pretty much everyone else,

> Tyra found it surprisingly inspiring. It suddenly made you aware that Cardiff was changing, that all the bollocks you heard people spit on the TV about being a European capital for the new century was really true. Even made you believe that all the bay development could come to something. (Williams, *Cardiff Trilogy* 443)

Conclusion

Digital plotlines offer an alternative to the distant and deep mapping of literature. They foreground the relations that readers and authors have to geographical places, and how these influence their production and consumption of a text. In tracing these plotlines, moreover, we come to recognise that 'as human beings, we are part of a world that we may also observe' (Weibel 344). As a consequence of this positioning, different qualities 'emerge' from the digital mapping of plotlines (Moretti, *Graphs, Maps, Trees*). Plotlines reveal the historical context of an area as well as cultural attitudes and local colour. They project meanings as well as facts onto our understanding of page and place. In Rosenbaum's terms, plotlines become 'a metaphor engine rather than an abstraction from the world' (57). Plotlines therefore become personalised and resonant accounts of how text engages with territory and, through the reader's and author's co-operative efforts, actually take and make this territory, changing its meaning and materiality. As a result, digital plotlines contribute significantly to our 'understanding – not only of books, but of the world we live in' (Piatti and Hurni, 'Cartographies of Fictional Worlds' 222), demonstrating how text and territory will inevitably entangle into an ongoing composition.

The digital mapping of plotlines is a crucial step. Digital mapping enables the evocation of sights, sounds and sensations occurring in a particular location through the capacity to embed georeferenced photos, film segments or audio clips. In the future, such mapping could be enhanced with the presentation of filmed interviews in a particular location, audio recordings of literature readings, or the addition of film clips which contribute to the definition and understanding of a place. (In Cardiff Bay, for example, one could embed clips from *Tiger Bay* (AcesHigh1916), documentaries such as Neil Sinclair's *Stroll through Tiger Bay* (Cardiffians), or even the BBC's *Torchwood*, that all thread into the Cardiff Bay plotline.) It would also be possible to enable readers and website users to add their own comments and tags to the site (through typed, recorded or filmed contributions), realising in practice the plural and ongoing traces of words and ideas that can lead to new experiences.

Mapping plotlines digitally offers the opportunity for the characteristics of deep mapping's place-hives to be connected to the charted co-ordinates of geospaces. In the spirit of the concept of plot, the new knowledges produced from this digital mapping are likely to differ from user to user.

Depending on an individual's familiarity with the chosen works of fiction, place and plotline, allied to the various ways the plotline is read (either virtually, physically or in combination), different histories will be recovered, different characters imagined, and different combinations of text and territory called forth in practice. It remains a critical question both to explore the different aspects of this new knowledge and to assess how digital plotlines can stimulate reading, local knowledge and a fuller understanding of the complex relations between page and place.

From my own perspective, digitally mapping this Cardiff Bay plotline has demonstrated how this area of the city is not simply an abstract location on a map, or the site of a new bowling alley to which we can drive in our cars and 'straight out again' (after Williams, *Cardiff Trilogy* 325), but a place with an insightful history and intriguing character. It is in this personal rediscovery and re-enchantment of places that the key contribution of plotlines, and perhaps mapping literature in general, lies. In contrast to Moretti's suggestion that 'one looks at a [distant] map, and thinks' (*Atlas of the European Novel* 7), when one engages with a digitally mapped plotline, one feels and imagines. Digital plotlines thus offer the opportunity to move between the extremes of distant and deep mapping, and find our own routes into the entangled worlds of literature and geography.

Works Cited

AcesHigh1916. *Tiger Bay*. 2012. Web. 20 March 2015. <https://www.youtube.com/user/AcesHigh1916>.

Amin, Ash. 'Regions Unbound: Towards a New Politics of Place'. *Geografiska Annaler: Series B* 86.1 (2004): 33–44.

Anderson, Jon. 'Talking Whilst Walking: A Geographical Archaeology of Knowledge'. *Area* 36.3 (2004): 254–61.

———. *Understanding Cultural Geography: Places and Traces*. London: Routledge. 2010.

———. *Cardiff Plotlines*. 2014. Web. 20 March 2015. <http://www.cardiffplotlines.com/>.

———. *Page and Place: Ongoing Compositions of Plot*. Amsterdam: Rodopi, 2014.

Auster, Paul. 'Book Club Archive'. *BBC Radio 4*. 2004. Web. 20 March 2015. <http://www.bbc.co.uk/programmes/p00fc3vt>.

Barthes, Roland. *Mythologies*. Trans. Annette Lavers. London: Paladin, 1973.

Benedikt, Micheal. *Cyperspace: First Steps*. Cambridge, MA: MIT Press, 1991.

Bonnett, Alastair. 'The Dilemmas of Radical Nostalgia in British Psychogeography'. *Theory, Culture, Society* 25.1 (2009): 45–70.

Calvino, Italo. *Invisible Cities*. Trans. William Weaver. London: Vintage, 1997.

Cardiffians. 'A Stroll through Tiger Bay'. *YouTube*. 2013. Web. 20 March 2015. <https://www.youtube.com/watch?v=tqu1b84WDis>.

Casey, Edward. 'Between Geography and Philosophy: What Does it Mean to be in the Place-world?' *Annals of the Association of American Geographers* 91 (2001): 683–93.

Chatwin, Bruce. *Songlines*. London: Vintage, 1987.

Coverley, Merlin. *Psychogeography*. Pocket Essentials: Harpenden, 2006.

Finch, Peter. *Real Cardiff.* Bridgend: Seren, 2002.
———. 'Cardiff – The City that Floats'. *Cardiff Central: Ten Writers Return to the Welsh Capital.* Eds, Dannie Abse and Francesca Rhydderch. Llandysul: Gomer, 2003. 13–26.
———. *Real Cardiff 2: The Greater City.* Bridgend: Seren, 2004.
———. *Real Cardiff 3: The Changing City.* Bridgend: Seren, 2009.
———. Personal interview. 16 March 2012.
Finch, Peter, Ian Davidson and Zoe Skoulding. '*Peter Finch Interview*'. *The Argotist Online,* 2006. 20 March 2015.
Foucault, Michel. *Power/Knowledge: Selected Interviews and Other Writings, 1972–1977.* Ed. Colin Gordon. New York: Pantheon, 1980.
Gregory, Ian and David Cooper. 'Geographical Technologies and the Interdisciplinary Study of Peoples and Cultures of the Past'. *Journal of Victorian Culture* 18.2 (2013): 265–72.
Griffiths, Niall. *Real Aberystwyth.* Bridgend: Seren, 2008.
Haraway, Donna. 'Situated Knowledges: The Science Question in Feminism and the Privilege of Partial Perspective'. *Feminist Studies* 14.3 (1988): 575–99.
Harrison, Frazer. *The Living Landscape.* London: Mandarin, 1986.
Harvey, David. *Science, Nature and the Geography of Difference.* Oxford: Blackwell, 1996.
Heidegger, Martin. *Being and Time.* 1927. Trans. Joan Stambaugh. Albany NY: State University of New York Press, 1996.
Idhe, Don. *Technology and the Lifeworld.* Bloomington: Indiana University Press, 1990.
Ingold, Tim. 'Up, Across and Along'. *5th International Space Syntax Symposium* 13–17 June 2005: 45–54.
Jones, Martin. 'Phase Space: Geography, Relational Thinking, and Beyond'. *Progress in Human Geography* 33.4 (2009): 1–20.
Mantel, Hilary. *A Place of Greater Safety.* London: Fourth Estate, 2006.
Merleau-Ponty, Maurice. *Phenomenology of Perception.* 1945. Trans. Colin Smith. London: Routledge & Kegan Paul, 1962.
Moretti, Franco. *Atlas of the European Novel 1800–1900.* London: Verso, 1998.
———. *Graphs, Maps, Trees: Abstract Models for a Literary History.* London: Verso, 2005.
Naramore Maher, Susan. *Deep Map Country: Literary Cartography of the Great Plains.* Lincoln & London: University of Nebraska Press, 2014.
Piatti, Barbara and Lorenz Hurni. 'Mapping the Ontologically Unreal – Counterfactual Spaces in Literature and Cartography'. *The Cartographic Journal* 46.4 (2009): 333–42.
———. 'Editorial: Cartographies of Fictional Worlds'. *The Cartographic Journal* 48.4 (2011): 218–23.
Piatti, Barbara, Hans Rudolf Bär, Anne-Kathrin Reuschel, Lorenz Hurni and William Cartwright. 'Mapping Literature: Towards a Geography of Fiction'. *Cartography and Art.* Eds, William Cartwright, Georg F. Gartner and Antje Lehn. Berlin: Springer, 2009. 179–94.
Pierce, Joseph, Deborah Martin and James Murphy. 'Relational Place-making: The Networked Politics of Place'. *Transactions of the Institute of British Geographers* 36 (2010): 54–70.
Pinder, David. 'Arts of Urban Exploration'. *Cultural Geographies* 12.4 (2005): 383–411.
Reuschel, Anne-Kathrin, and Lorenz Hurni. 'Mapping Literature: Visualisation of Spatial Uncertainty in Fiction'. *The Cartographic Journal* 48.4 (2001): 293–308.

Reuschel, Anne-Katrin, Barbara Piatti and Lorenz Hurni. 'Modelling Uncertain Geodata for the Literary Atlas of Europe'. Understanding Different Geographies. Eds, William Cartwright, Karel Kriz and Michaela Kinberger. Berlin: Springer, 2013. 135–57.

Rosenbaum, Sarah. 'The Map: Shifts, Overwrites, Overlays'. *Performance Research* 6.2 (2001): 54–9.

Sharp, Joanne P., Paul Routledge, Chris Philo and Ronan Paddison, (eds) *Entanglements of Power: Geographies of Domination/Resistance*. London: Routledge. 2000.

Smith, Phil. 'Mythogeography Works: Performing Multiplicity on Queen Street'. *The Journal of Applied Theatre and Performance* 16.2 (2011): 265–78.

Tally Jr., Robert T. *Spatiality*. Abingdon: Routledge, 2013.

Weibel, Peter. 'The World as Interface'. *Electronic Culture: Technology and Visual Representation*. Ed. Tim Druckery. New York: Aperture, 1996. 338–51.

Williams, John. 'The Legend of Tiger Bay'. *The Big Book of Cardiff*. Eds, Peter Finch and Grahame Davies. Bridgend: Seren, 2005. 217–34.

———. *The Cardiff Trilogy*. London: Bloomsbury, 2006.

9 *The Cestrian Book of the Dead*

A Necrogeographic Survey of the Dee Estuary

Les Roberts

Introduction

To paraphrase a line from the 1999 film *The Mummy*: the geography of death is only the beginning. What might follow is the subject of this chapter. Though remaining at best a niche area of geographical enquiry (and pretty much unchartered territory for literary geographers), *necrogeography*[1] is introduced here not as a central thread of discussion or as an attempt to consolidate scholarship in this area. Rather, the geography of death – or, more precisely, *a* geography of death by drowning – functions as a jumping off point to explore a wider set of questions which are in themselves not intrinsic to death or mortality. If anything, the opposite is the case. It is asking what necrogeography might tell us about the symbolic worlds and spatial stories that populate the everyday landscapes of the living. How might necrogeography inform the theory and practice of *mytho*geography or spatial anthropology? And, in the case of liminal landscapes such as wetlands and marshlands (the topographic focus of the present study), how might a rumination on death shed insights into the symbolic geography of landscapes that are by their very nature uncertain and ambiguous, poised 'betwixt and between' otherwise stable topographies of place and memory? In pursuing these and other questions this chapter follows a trajectory in which the liminality of both death and landscape underpins an analytical framework which, although centred on a specific geographical location – the Dee Estuary on the north west coast of Britain (Figure 9.1) – is no less engaged with the practicalities of 'doing' cultural and literary mapping in the digital age. To these ends, it is necessary to start by sketching some theoretical ideas relating to what may provisionally be termed the 'digital Aleph'.

1 The term 'necrogeography' was coined in 1967 by the geographer Fred Kniffen to describe the study of social and cultural landscapes associated with mortuary practices (Kniffen). My use of the term extends this to encompass sites and spatial practices linked to death more generally.

Envisioning the Digital Aleph

Re-reading Jorge Luis Borges's 1973 short story 'The Aleph', firstly by way of geographer Ed Soja's discussion of the text in his 1996 book *Thirdspace*, I am confronted with a number of salient themes. We have a protagonist (Borges as narrator) reflecting on the death of a lover (Beatriz), and who vicariously returns – via mediated memories (a series of photographs) – to places and times that bear the archival imprint of the departed. We have further, in the shape of Carlos Argentino Daneri (a cousin of Beatriz), a poet who, musing on the technological accoutrements of 'modern man', opines that, for such a man, actual travel is superfluous, and that modernity has thus 'inverted the story of Mohammed and the mountain; nowadays, the mountain came to the modern Mohammed' (Borges 13). We have anxieties and dilemmas posed by untrammelled modern development (Daneri's ancestral home is under threat of demolition) and the implications this has on the preservation of the past and the in/tangibility of memory. But also we have, in the conceptual conceit of the 'Aleph', a tropic device by which tentatively to map the heterotopic configurations of language, space, history and memory. Conceived of as 'one of the points in space that contains all other points' (Borges 17), Borges's heterotopia *in extremis* (see Foucault; Hetherington) transforms the prosaic environs of a writing desk in a dusty basement into a portal through which to access the infinitude of space and simultaneity of time past, present and future.

Taken as a template, or loose assemblage of ideas and reflections, these Borgesian extractions each speak to the wider arguments that I explore throughout this chapter. First and foremost the significance of death and passing ('The Aleph' ends with the narrator's fears of losing, over time, 'the face of Beatriz') provides the main thematic framework around which I build my discussion of the literary and historical geographies of the Dee Estuary. Secondly, the role of technology, and more specifically, locative media and digital technologies, is considered in light of the 'mountain coming to Mohammed': in this sense, the capacity for place-specific information to come to us, as mobile and situated subjects, rather than the more routine Web 2.0 practice of our 'visiting' websites or online spaces (McCullough 26). Thirdly, the immaterial architecture of digital spaces raises questions as to the sustainability and ontological status of cultural memory as a residual component otherwise 'located' in material geographies of place and dwelling. This in turn prompts questions as to the functional and tactical role of digital tools in the cultural production of what Lefebvre describes as 'differential spaces' (52): a co-extensive configuration of multiple and plural spatialities. Fourthly, and most intriguingly, reflection on the spatial 'thingness' and ontology of the Aleph as part of an overarching discourse on creativity, memory and radical simultaneity offers a productive framework by which to think through some of

Figure 9.1 Google Earth map showing the location of the Dee Estuary.

Source: 'Dee Estuary' 53°14'29.66"N and 3°03'57.62"W. *Google Earth*. 10 April 2013. Web. 3 September 2014.

the different ways that imagination and imaginative worlds are themselves 'located' spatially. Moreover, it informs theoretical discussion as to how developments in a putative 'digital spatial humanities' (Bodenhamer et al.; Gregory and Geddes; Roberts, 'Navigating the "Archive City" ') might inform critical interventions in this multi-sited and multi-spatial geography of literary praxis.

Returning to 'The Aleph', it is instructive to note the description the narrator provides upon his eventual realisation of the Aleph's mysterious powers:

> The Aleph's diameter was probably little more than an inch, but all space was there, actual and undiminished. Each thing (a mirror's face, let us say) was infinite things, since I distinctly saw it from every angle of the universe. I saw the teeming sea; I saw daybreak and nightfall; I saw the multitudes of America; I saw a silvery cobweb in the center of a black pyramid; I saw a splintered labyrinth (it was London) . . . I saw the Aleph from every point and angle, and in the Aleph I saw the earth and in the earth the Aleph and in the Aleph the earth; I saw my own face

> and my own bowels; I saw your face; and I felt dizzy and wept, for my eyes had seen that secret and conjectured object whose name is common to all men but which no man has looked upon – the unimaginable universe. (Borges 20–21)

From a digital standpoint, visualising the Aleph as 'little more than an inch across' conjures the image of some sort of hand-held device that exerts a mesmeric grip over its user. Envisaged in terms of a wide-eyed neophyte whose inexorable gaze falls upon an 'unimaginable universe' where all space exists and 'each thing is infinite', Borges's device is perhaps not quite so 'secret and conjectured' as it would have seemed in the pre-digital age. But no less compelling is the image it shapes of this unimaginable universe; 'unimaginable' here a synonym for 'unmappable', referring to worlds that paradoxically lie beyond the imagination but which are nevertheless a product of the very capacity to imagine what it is the Aleph *is* as an 'unimaginable' space. 'Unmappable', in turn, can thus be looked upon as both a gauntlet thrown down to those engaged with the critical implications and possibilities of literary mapping in the digital age and a recognition that understandings of what maps and mapping practices in fact are demand a recognition of what it is they *do* (Wood). By extension, what might seem 'unmappable' from a Euclidean or Cartesian perspective does not necessarily translate to literary or textual spaces (howsoever defined) or 'imaginary spaces' in a wider sense (Kerrigan). 'Cartography', in other words, cannot be assumed to have a uniformity of meaning or practice without at the same time taking into account the disciplinary and epistemological 'buffer zones' that define, undo, rework, challenge or dissolve what 'mapping' is and does in any given circumstance. The processual 'doing' of cartography is thus in part a reflexive remapping of the contingent ground upon which its constitutive performativities are rehearsed and precariously embedded (Roberts, 'Mapping Cultures').

'Seeing Beyond': Liminality and the Hypertelescopic Imagination

Moving laterally from our basement-ensconced 'surfer' of the Aleph, we note that Borges's citation or mapping of its constituent topographic features (the teeming sea, the multitudes of America, London's splintered labyrinth) bears some similarity to the 'hawk flight of imagination' that Olaf Stapledon ('Star Maker' 6) embarks upon in the opening passages of his 1937 science fiction novel *Star Maker*: a text that geographically places us both in the landscapes of the Dee Estuary and the infinite space of the Aleph. However, whereas Borges's imaginative hawk flight proceeds from the darkness and enclosure of a residential basement, Stapledon's is a product of the expansive, unbounded landscapes of the estuary. *Star Maker* opens with the narrator walking up on to the 'dark heather' above his house (Caldy Hill, near Stapledon's home in West Kirby overlooking the Dee Estuary and Liverpool Bay) following an argument with his wife. He observes the 'suburban lamps' below, a lighthouse beyond the foreshore, a star ('one tremulous arrow of light'); he notices how the 'shadowy hills or the guessed, featureless sea extended beyond sight'. Taking all this in, his imagination starting to drift to infinite spaces beyond (completing 'what mere sight could not achieve'), the narrator, having shot a brief glance of reassurance back to the glowing windows of his home, begins a journey across the 'boundless finitude' of the cosmos (Stapledon, 'Star Maker' 3–8). As he soars further and faster through space and the intensity of the experience starts to take hold, a second glance back reveals that his home, the hill, the suburbs, the sea had all now gone:

> Instead there lay far below me an insubstantial gloom. And I myself was seemingly disembodied, for I could neither see nor touch my own flesh . . . . When I realized fully the change that had come over me, I wondered if I had died, and was entering some wholly unexpected new existence . . . . Though by now I must have been hundreds of miles above the ground, I was not troubled by the absence of oxygen and atmospheric pressure. I experienced only an increasing exhilaration and a delightful effervescence of thought. The extraordinary brilliance of the stars excited me. (Stapledon, 'Star Maker' 9–10)

The radical intentionality of this new existence and the falling away of the mundane terrestrial world is *felt* corporeally insofar as the body, like the hills and suburbs, is phenomenologically left behind. The more the stars 'pierce' the heather-carpeted reality of Caldy Hill, the more a 'hypertelescopic imagination' (Stapledon, 'Star Maker' 8) infuses the narrator's sense of being-in-the-world. In terms of Stapledon the writer this decorporealised world goes hand-in-hand with the creative process itself: if nothing else, a 'hypertelescopic imagination' serves admirably as a metaphor for the literary act as a performative practice: a migratory or nomadic 'seeing beyond' of the parochially banal. But equally pertinent here is the more weighted and grounded sense of self that positions Stapledon bio-geographically (as a local Wirralian, for example), but also bodily and sensorially.

Stapledon's affective bonds with the landscape, and the Dee Estuary in particular, are well illustrated in his short memoir essay, *Fields Within Fields*:

> The Estuary! Twice daily the tide transforms it. An even beauty of broad water, with here and there a fishing boat or a distant coaster, alternates with a more varied beauty of mud-flats, sand-banks, and channels . . . .

> Across the estuary lies Wales, a long hill-side, mottled with fields, woods, villages, but the shore bristles with factory chimneys, like stiff, plumed grasses. On still days faint sounds reach us from that foreign country, the murmur of trains, and muffled, inexplicable eructations, perhaps from the furnaces . . . . On a clear day sheep may be detected. A man is a minute erect bacillus. (254)

The writer's deep attachment to the landscape around West Kirby is also evident from his expressed desire to have his ashes scattered from cliffs overlooking the Dee Estuary, a request duly honoured by his family following his death in 1950 (Crossley 398). Although we can only conjecture whether the interstellar and infinite 'hawk flight' of *Star Maker* struck a metaphysical chord with his funerary ambitions, the theological (although Stapledon himself was an agnostic) and ritual connotations of 'crossing over' the liminal spaces of the estuary – the font of so much of his creative and intellectual inspiration – are certainly suggestive. It is almost as if the molecular composition of an interplanetary diffusion of self had been re-imagined in the form of a final windblown journey of dust and ash.

In his 1944 novel *Sirius* – a Frankensteinian tale of a dog (the titular Sirius) scientifically engineered to possess humanlike intelligence – Stapledon recounts a journey undertaken by the canine hero through a bomb-ravaged Liverpool, across the Mersey to Birkenhead and the Wirral, and then onward, across the Dee Estuary, to his home at Trawsfynydd in North Wales. In Liverpool, Sirius's scientist owner/creator (and travelling companion) Thomas is injured by falling masonry from a bomb strike. As a consequence Sirius is forced to negotiate the Queensway (Mersey) tunnel crossing on his own, re-emerging into the glare of an incendiary Birkenhead fray as he makes his way across the Wirral to Thurstastone Common (situated near to Stapledon's beloved Caldy Hill). From here he walks along the coastline before crossing – at an unspecified location – 'over the salt marshes to Queensferry' (Stapledon, 'Sirius' 163). Ploughing on in a south-westerly direction, Sirius continues over the Clwydian mountains and the Vale of Clwyd before finally arriving back home at Trawsfynydd.

What is particularly noteworthy here is the scant detail afforded to the estuary crossing, as if the ease and inevitably of a straightforward (i.e. hazard-free) passage was somehow a given. For the journey's narrator such detail is little more than incidental: 'Sirius gave me a very full account of his long trek from Birkenhead to Trawsfynydd, but there is no need to report it here in detail' (Stapledon, 'Sirius' 162). Whereas for centuries crossing the estuary by foot was a commonplace (although notoriously treacherous) activity, since the canalisation and re-routing of the river in the eighteenth century and reclamation of large swathes of former marshland (Figures 9.2 and 9.3), the estuarine terrain has become all but impassable to the walker intent on securing an easy passage from one foreshore to the other. The geographical ellipses that characterise narrative journeys such as those detailed in *Sirius* become, therefore, all the more intriguing in that they overlay a topography of *absence* upon an historical, archaeological and anthropological space that simultaneously both affirms and negates the constitutive mythopoesis of the Dee. A topos of place is drawn not by harvesting the layered histories that define what the estuary *isn't* today (a social landscape) but by cartographically re-rendering the estuary in terms of its inherent liminality as a textual and performative space; one variously giving over to tropes of 'flight', 'escape', 'border-crossings' or 'hypertelescopic' travel.

The implications of this bifurcated (but no less imbricated) topography of the textual/imaginative and lived/anthropological is that the challenge of mapping these geographies becomes all the more compelling the more the former becomes acquainted with the latter. Put another way, the more the locative properties of the narrative become embedded in real landscapes (replete with real histories) the more the rationale and practical utility of digital locative media in literary mapping becomes evident.

Liminality as Necrogeography

Jumping back for a moment to the estuary crossing as a mythopoeic journey – one that overlays rather than *excavates* in the Benjaminian sense of an archaeology of memory (Benjamin) – it is useful to follow briefly *Sirius*'s shaggy dog adventure with that recounted in the Middle English poem *Sir Gawain and the Green Knight*. Having set off on his quest, Gawain, riding through 'England's realm', soon arrives in North Wales, the first geographic location cited in the poem: 'He wanders near to the north of Wales / with the Isles of Anglesey off to the left'. Referring to poet Simon Armitage's 2007 translation, we note that Sir Gawain then moves along the coast before (we assume) making his way down the Flintshire coastline from where he crosses back over into England: 'He keeps to the coast, fording each course / crossing at Holy Head and coming ashore / in the wilds of the Wirral, whose wayward people / both God and good men have quite given up on' (Armitage 37). Following closely on from the Anglesey reference, 'Holy Head' might at first glance be taken as 'Holyhead', however the geography clearly does not stack up. It might be that the author of *Sir Gawain* was referring instead to Holywell in Flintshire, an important site of Catholic pilgrimage which is situated near to a point on the estuary coastline from where a river crossing could conceivably have been made. Or alternatively it might be that the location *is* meant to be Holyhead in Anglesey and that, for literary purposes, its long-standing historical identity as a port town trumped the necessity for geographical

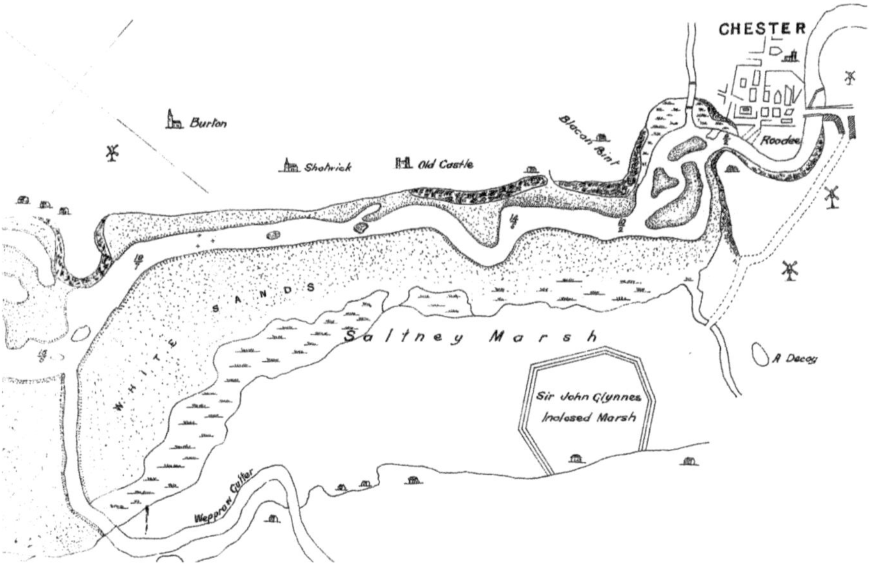

Figure 9.2 1732 map (by John MacKay) of River Dee and Saltney Marsh prior to canalisation.

Source: Courtesy of *Canalmaps Archive*.

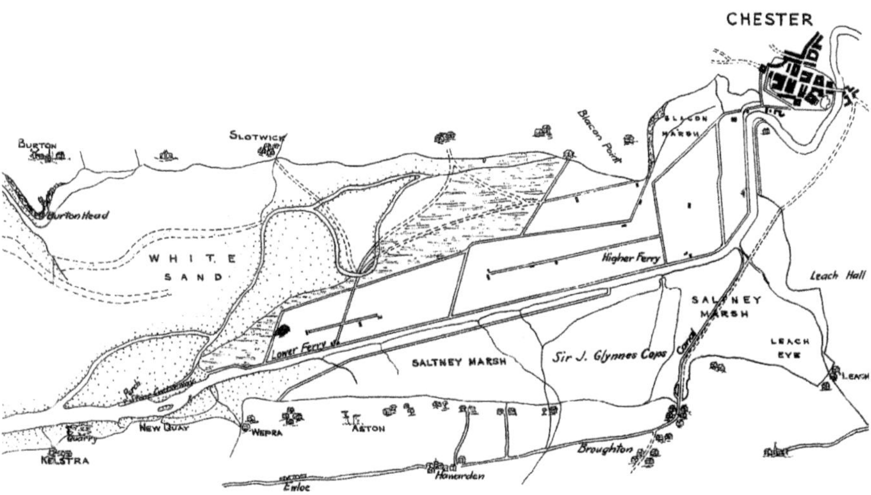

Figure 9.3 1771 map (by Thomas Boydell) of River Dee and Saltney Marsh showing the canalised stretch of river.

Source: Courtesy of *Canalmaps Archive*.

accuracy (a demand perhaps only placed on the text retrospectively by the literary geographer or cultural archaeologist). In a BBC television documentary produced to accompany the publication of his translation of *Sir Gawain and the Green Knight*, Armitage attempted to retrace, as far as was possible, the route of Gawain's Arthurian quest. When arriving at the Dee Estuary Armitage plumped for Connah's Quay as a crossing point (enlisting the services of a riverside scrap yard worker to ferry him across the Dee). Whether or not this (or indeed any other literary re-mapping of the crossing) is 'accurate' in any geographical sense is not really the point. What *is* the point is that the very process of taking the map to the territory – in other words, the text to the (or at least *a*) cognate topographic location – potentially feeds back into the mythopoesis of the originating text a spatial anthropology (Roberts, 'Mapping Cultures') that owes as much to contemporary affects of place and dwelling as it does to those palimpsestic layers that culturally anchor places in and across time. Differential space, once tapped and allowed room to breathe (the brief of the spatial anthropologist), is ineluctably – and thus politically – anchored in the present.

What is rarely far from view though – whether as a swirling undercurrent of narrative or in more explicit and imposing form – is the topoanalytic spectre of death: an existential, phantasmogoric and *necro*geographic grounding to a landscape that is forever slipping from both view and grasp. In *Sir Gawain* death is an omnipresent companion that Gawain carries with him as he journeys towards his inevitable confrontation with the Green Knight. Whereas in *Star Maker* it is almost as if Stapledon is contemplating his own mortality in the guise of an afterlife in which a sense of metaphysical belonging is made congruent with the cosmological 'fictions' of science rather than those bearing the hallmark of religions (Augé). Death here speaks less of a terminus than of a 'passing beyond': a consummation of a liminal experience in the terms adumbrated by Arnold Van Gennep and later developed by anthropologist Victor Turner (Thomassen, 'The Uses and Meanings of Liminality'; 'Revisiting Liminality') – a neophytic rite of passage where death is the prelude to new beginnings or the acquisition of a new identity or psychosocial selfhood: a *new chapter.*

But when more specifically applied to the Dee Estuary, the geographies of absence that are lain bare by the 'free-floating' textual spaces referred to above are no less shadowed by death. They map a sprawling landscape of ghosts: of phantom pathways, fords and forgotten journeys. Unlike the Chaucerian pilgrims who can be heard, some six hundred years out of synch, at the start of Powell and Pressburger's Second World War propaganda film *A Canterbury Tale* (1944), their song is mute. Theirs is no sacred journey: devotional footfall bound for a sainted site of martyrdom. The necrogeographic spaces that are mapped in *The Cestrian Book of the Dead,* which I discuss below, are more conspicuously mundane. The everyday journeys to which they play host are less easy to trace than those piously entrenched by the pilgrims. It is to an altogether different literary canon that we need to turn in order to tease out these other, more marginal cultures of mobility: official archives and public registers from which some semblance of a geo-historical topos might be evinced.

The Mythogeography of Drowning

I have elsewhere set out a provisional psychogeography of the Dee Estuary that took its lead from the idea that, as a liminal space, a topoanalytic attentiveness to affects of danger, precariousness and instability can offer a productive 'way in' for the cultural cartographer or spatial anthropologist (Roberts, 'The Sands of Dee'). When mapped on to the estuary in a more literal sense, 'danger' proved to be a trope that would quickly lead on to a preoccupation with an anthropology of *drowning*: who were the drowned? how did they drown? where did they drown? where were they going to or coming from when they drowned? Of these questions, 'where did they drown?' was the starting point that led to a number of other critical interventions, not least those which sought to populate what has remained remarkably fallow ground in terms of the 'mythogeographic' (Smith) provenance of the estuary. Firstly, it established a geographical rationale for the project of mapping the sites of drowning on and around the Dee. Secondly, and by extension, it was a means by which to map the activities and practices of the deceased on to site-specific historical locations. In other words it provided the rationale for a spatial anthropology of death by drowning: to shed light – however dimmed and fractured – on the estuarine lives of the living by taking into account the circumstances of their dying. Thirdly, having begun to flesh out the geography and historical anthropology of the drowning sites, this in turn informed the rationale for site-specific *practices* of mythogeographic engagement: the embodied 'layering', through the physical act of walking, of historico-textual topographies onto the industrial zones, edgelands, mudflats, retail parks, expressways, Ministry of Defence enclosures, power stations, recycling plants, sewage works, marshland and residential areas that variously comprise the post-reclamation landscape of the estuary today. Drowning is thus the topoanalytic prolegomena to a more expansive critical geography built around a loosely 'dissonant' framework of differential space.

The trope of drowning also connects the liminal landscapes of the Dee with other locations where the confluent themes of danger, ambiguity, precariousness and instability are strongly resonant. Morecambe Bay is one obvious example, not least on account of the legacy of the deaths of 23 Chinese migrant workers who were cut off and drowned by the incoming tide whilst cockle picking in 2004 (Roberts, 'The Sands of Dee' 108–9). Another is the Broomway, an ancient and notoriously hazardous tidal path situated off the Essex coast. For Robert Macfarlane, writing in *The Old Ways*, such places are elusive both topographically and typologically: 'We lack – we need – a term for those places where one experiences a "transition" from a known landscape . . . [to] somewhere we feel and think significantly differently. . . . Their traverse is generally unbiddable, and no reliable map exists of their routes and outlines' (78). Obvious geographical connections can also be drawn between the Dee Estuary and the Cambridgeshire and Lincolnshire Fens. Nathaniel Kinderley, whose River Dee Company was established to undertake the canalisation scheme that so dramatically and irrevocably transformed the estuarine landscape, was also, around the same time, an engineer on the scheme to drain the South and Middle Levels of the Fens (Young; Pritchard).

In both cases Kinderley's legacy has been profound. In *Edge of the Orison*, Iain Sinclair's retracing of poet John Clare's asylum flight from Epping Forest to his home in Peterborough, drowning, often conjoined with flying, serves as a recurrent motif: 'Drowning and flying were themes to be resolved in the Clare book that I was plotting, preparing to write. Drowned villages, flights across England' (190). Arriving, via poet Charles Olson, at the site of the drained Whittlesey Mere in Cambridgeshire, Sinclair invokes a poetic spirit of 'homecoming' – 'a "noble arc" of light and language' – to work 'against the void of the drained Mere. And our own drowned memories' (326). This idea of working against the void of memory and restoring bio-geographical detail to a landscape drained of its everyday histories strikes a particularly resonant chord with the Dee case study under discussion here. In both cases, topography functions as what Samuel referred to as a 'theatre of memory', where a performative tramping of the earth enacts a ritual invocation: the drowned and forgotten called up to bear witness.

These topographic connections were also traced by Celia Fiennes in the final few years of the seventeenth century, a pivotal moment for both landscapes as the convulsive impacts of industrialisation (personified in the figure of Kinderley) began to leave their mark. Commenting on Whittlesey Mere in her travelogue *Through England on a Side Saddle in the Time of William and Mary* (written in 1702, but published posthumously), Fiennes observes:

> The ground is all wett and Marshy but there are severall little Channells runs into it wch by boats people go up to this place. When you enter the mouth of ye Mer it lookes fformidable and its often very dangerous by reason of sudden winds that will rise Like Hurricanes in the Mer, but at other tymes people boate it round the Mer with pleasure. (Fiennes)

Passing through Cheshire and Flintshire a year later Fiennes's attention is drawn to the ford that connected Shotwick on the Wirral side of the Dee Estuary to Holywell and Flint on the Welsh side:

> I forded over ye Dee when ye tide was out all upon the sands at Least a mile, wch was as smooth as a Die being a few hours left of ye flood. Y^{e} sands are here soe Loose y^{t} the tydes does move them from one place to another at Every flood, y^{t} the same place one used to ffoard a month or two before is not to be pass'd now, for as it brings the sands in heaps to one place so it leaves others in deep holes wch are Cover'd wth water and Loose sand that would swallow up a horse or Carriages; so I had two Guides to Conduct me over. (Fiennes)

The first point to note here is that we come at Fiennes's travelogue (in this instance anyway) via the historical GIS / digital humanities resource *A Vision of Britain Through Time*: a website that collates historical maps and place-specific information, including census data, election results and travel narratives by writers such as Daniel Defoe, William Cobbett, George Borrow and others. Already, then, we are engaged with the 'doing' of literary mapping (or literary wayfinding) by virtue of the fact that the digital architectures of Web 2.0 facilities the otherwise elemental (and, from an analogue perspective, not especially new) task of stitching texts to places. However, given the somewhat desk-bound mode of literary 'navigation' that resources such as *A Vision of Britain Through Time* represent, there is less scope for inciting what could be described as a 'spatially intertextual collision' or 'mash up' (in a more purposively *detourned* sense) between cartographies of space, time and the body. Text and location are instrumentally bound together by no other apparent logic than that of rudimentary geographic elision. What such resources are less adept at is bringing into play the affective and emotional cartographies that allow 'spillage' between different spatialities (between performative and visual geographies of place, for example). This takes us to the second point to pick up on from Fiennes's travel writings: the ascription of danger – and the attendant fear of drowning – to landscapes that are essentially unstable and unpredictable, or 'unbiddable', to use Macfarlane's term.

As far as a poetics of the Dee Estuary go, the default literary port of call remains Charles Kingsley's 1849 poem 'The Sands of Dee', which recounts the tale of a young cattle girl, Mary, who ventures onto the sands of the estuary and is overtaken by the tides and drowned, her body later found caught amongst the fishing nets. To a certain extent the fame of the poem has outstripped any requisite local knowledge. On a website called *Poetry Atlas: Mapping the World in Poetry* the poem finds itself relocated to the River Dee in Galloway, Scotland. The poem was also afforded cinematic treatment by the early film pioneer D.W. Griffiths in his 1912 short *The Sands of Dee*, who transplanted the story to Newport Beach, California.[2]

Thomas De Quincey also has a minor role to play in the cultural geography of the Dee, an intervention which in part channels the same fears and uncertainties ascribed more generally to the river and its folkloric heritage. In his account of his time in Chester in *Confessions of an English Opium Eater* (1856), De Quincey describes a walk to an area of the city known as the Cop which is an open space overlooking the River Dee. Already in a state of some anxiety as a result of having to attend to a pressing financial matter, the writer is paralysed by 'a sudden uproar of tumultuous sounds rising clamorously ahead' (De Quincey 80). He later realises it was the tidal bore which in De Quincey's day would have come in close to the city centre. Clearly affected by the phenomenon – 'until that moment, I had never heard of such a nervous affection in rivers' (82) – De Quincey is moved to describe the river as 'hysterical', invoking lines from Euripedes's play *Medea*: 'Flow back to your sources, sacred rivers, And let the world's great order be reversed' (80; Roberts, 'Sands of Dee' 115–16).

It is hard not to read into De Quincey's invocation of Euripedes a certain revolutionary zeal and an implicit mythopoeic re-casting of the river in the guise of a provocateur. The radical liminality De Quincey alludes to – an inversion of the dominant political order; a convulsive landscape rising up and taking its browbeaten citizens along with it – conjures up a topographic imaginary in which the dangerous and uncertain physical terrain is interwoven with an altogether different sense of danger and instability, one tied to an insurgent political or revolutionary force that the river and the estuary are in some way seen to embody. A similar connection is made by Gary Snyder, who, in ' "The Great Clod" Project' writes of a tidal bore that engulfed the citizens of the Imperial City of Hang-chou in thirteenth-century China, drowning hundreds of people. For the people of Hang-chou, then capital of the Southern Song dynasty, the bore was an annual spectacle, with viewing platforms erected for the emperor and his family. In Snyder's hands the destructiveness of the aberrant tide visited upon the city that year becomes an augur for the impending Mongol invasion and the eventual collapse of the Song (310–12).[3]

2 It is worth noting that an aborted music video project by the celebrated British filmmaker Michael Powell, had it been completed, would have gone some way to geographically 'reclaiming' *The Sands of Dee* by shooting the video – storyboarded around Kingsley's poem – on location in the estuary – see Roberts ('Sands of Dee' 106–7).

3 The historical association of wetlands with acts of insurrection and revolution are highlighted by Purseglove, who cites the example of marshland villages in Essex that played a significant part in the Peasants' Revolt of 1381. He also notes that the origins of the English

Although the bore of the Dee plays no part in the narrative, a watery portent of upheaval, instability and political transformation flows through Arthur Behrend's novel *The House of the Spaniard* (1936). Set, amongst other locations, in and around the marshland near Burton on the Wirral coast, the plot centres on the adventures of a young shipping clerk, David Grey, who becomes unwittingly embroiled with the clandestine affairs of Liverpool-based shipping magnet and Spanish national Don Pedro. His suspicion that there is more to Don Pedro's business activities than meets the eye leads him to investigate activities at Don Pedro's home, Fen Hall (the house of the book's title), an isolated property perched on the edge of the expansive marshland that dominates this stretch of the estuary (Fen Hall is based on the real-life Denna Hall near Neston). Whilst out on the marsh Grey meets a game shooter from Birkenhead, who warns him against venturing too close to Fen Hall: 'there's a proper madman lives in it' (Behrend 25). On a second trip to Burton, Grey witnesses a body being carried off the marsh. He recognises it as that of the Birkenhead man from his previous visit. An inquest is held and the coroner records a verdict of death by drowning. Grey's well-founded suspicions that Don Pedro and his men were somehow involved in the death draws him inexorably into the intrigue of a plot which transports him to war-torn Spain where he is caught up with Republican forces fighting Franco's Fascists.

From a body in the marsh to the Spanish Civil War, Behrend's siting of the revolutionary's house on the estuary can be persuasively read as an attempt to harness the same liminal and heterotopic powers that writers such as Stapledon found so compelling. It is with these and other topographic allusions in mind that I turn to consider a digital necrogeographic survey of the estuary to which these literary geographies might, at first glance, seem little more than appendices. But it is more instructive to look upon them as but part of a wider mythogeography or spatial anthropology of the Dee; a project that takes us beyond the limitations of this chapter but which nevertheless productively feeds into debates surrounding the 'doing' of literary mapping and, more pointedly, the role of digital technologies in shaping an evolving literary spatial praxis.

Digital Necrogeography and Literary Spatial Praxis

It is fitting that Behrend's protagonist is prompted to throw himself into his researches off the back of a coroner's report of a death by drowning in the Dee Estuary. In what are admittedly very different circumstances this is exactly the same route I followed in pursuing what would

Revolution can in part be traced to resistance to fen drainage by so-called 'fen tigers' in seventeenth-century Cambridgeshire (34–5).

become, amongst other things, *The Cestrian Book of The Dead*: a digital map of geo-referenced sites of drownings (c.1500–1820) extracted from Chester Quarter Sessions Coroners' Inquest records. In the remainder of this chapter I will discuss the map in more detail and demonstrate the ways in which such resources can productively inform debates and practices of digital literary mapping. Moreover, working against the more routine extrapolation of literary geographies as extended to texts such as those discussed throughout this chapter, *The Cestrian Book of The Dead* also invites closer consideration of 'everyday' archival documents that are otherwise categorised as 'historical' (such as coroners' reports) as also having merit as 'literary' texts.[4] Accordingly, interspersed throughout the proceeding discussion are stanzas from what might best be described as a 'found poem' which has been assembled, with some minor modifications for literary effect, from the coroner's inquest records (these can be accessed in full via the National Archives online records – see 'Chester City Coroner's Records').

Alongside the map, the purpose of the poem is to populate the historical landscapes of the estuary by piecing together a series of everyday vignettes which, discounting the circumstances of their documentation, provide a window of sorts onto a real and imaginary world. The truncated nature of the stories, coupled with the iteration of toponyms that often prove no less elusive in pinning down a place as the narratives do in pinning down a person, have the curious effect of casting a relief map: a cartography of absence into which the imagination is left to roam at will and with abandon, should it so wish. The structure of the poem is thus intended to replicate the suggestiveness of what the coroner's report *doesn't* say but also to imbue it with a sense of geography, both in terms of the integrity and layout of the report itself (the map of the text), but also of the historical places to which it serves as index (the text of the map). Although this suggestiveness offers much that could be expanded on by zooming in on specific names or locations cited in the poem, the point here is weighted more towards consideration of the wider process and application of digital necrogeography and of a literary spatial *praxis* in which map and text are inextricably bound.

4 Although well-established in disciplines such as anthropology, in literary criticism the equal weighting afforded to literary and non-literary texts was a defining feature of New Historicist approaches that emerged in the 1980s. In the case of a coroner's report, Stephen Greenblatt's observation that 'I began with the desire to speak with the dead', which opens Harold Veeser's edited collection *The New Historicism* (ix), speaks directly to an approach in which non-literary documents are the starting point of a mythogeographic study of place and dwelling.

From a certain sloop near the Crane,[5]
John Blundell, in a stupor. At Denwall,
from a scaffold, Robert Mason, while caulking the stern.

Near Finchett's Gutter, cast up by the tide
a woman with no name, like the young male child
found at Brewers Hall, where the Roundheads took siege.

As noted earlier, the research proceeded from the question 'where did they drown?' This entailed the somewhat painstaking task of attempting to geo-reference those records where a location of death was clearly (or approximately) recorded and then compiling this information to create a GIS shapefile (a geospatial vector data file for use in ArcMap and other geospatial platforms). Many records do not have any location-specific information, meaning that the deaths could have occurred anywhere on the river from Chester up to Hoylake on the north-western tip of the Wirral peninsula. Of those that do have location information, many are not easily identifiable, some not at all (for example, 'Le Hollyne' or 'Stewbridge Gutter'). However, what can be determined reasonably accurately are key 'hot spots' where the highest concentration of deaths are located. Most of the drownings occurred near to three main locations: the Dee Bridge in Chester (a popular suicide spot), the port of Parkgate, and the Shotwick to Flint ford (Figure 9.4).

By the Shipgate at Le Posterne in 1586
Boniventrus Hanky, taken while bathing, whence,
six years on Lizzy Horton took her own.

Mary Turtle, while of unsound mind.
A yeoman at Burton who threw himself in.
William Cowpack, while gathering daisys in the spring.

Were the rationale for the mapping to end here – that is, the mere linkage of the coroners' records with points on a map – then projects such as this could quite justifiably be dismissed as little more than a slightly macabre exercise in abstract necrogeography. What lends added value is the fact that geography is a means to an end: the end being the cultivation of a spatial anthropology of the estuarine landscape whereby everyday historical narratives are both inscribed *in* space and extracted *from* space. Insofar as any literary mapping is concerned with the constitutive poetics of place and the ever-growing constellation of narratives by which places are imagined or (re)assembled, the imperative here is similarly the production of 'spatial

5 See Glossary below.

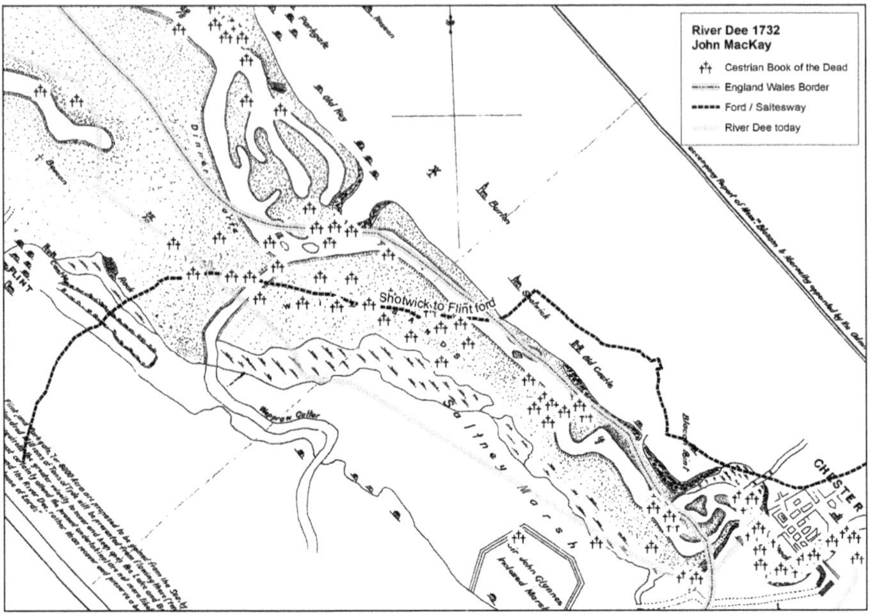

Figure 9.4 GIS map showing the approximate route of the Shotwick to Flint ford.
Source (base map): Courtesy of *Canalmaps Archive.*

stories'. This refers to narratives conceived of in terms of a temporalised notion of space as 'practiced place' (de Certeau 117): the everyday journeys, narrative trails and story-*lines* (Ingold) that humanise landscapes and inscribe them with meaning, historicity and sociality.

> William, son of James Eason, barber,
> while reaching for a Beesom's tail or wand
> floating on the necromantic Dee.
>
> On Ascension Day at Le Hollyne in 1552
> Agnes Coppocke and Elizabeth Symcoke went down.
> At the Cage while washing clothes, a spinster.

Accordingly, the hard-wired mechanics of digital cartographic tools are – or should be – similarly thought of only as a means towards an end if that end is the task of literary mapping (or, indeed, cultural mapping in a broader sense). If GIS tools cannot convincingly demonstrate what they are able to contribute that could not be done via other means then their constitutive role in this respect cannot and should not be taken as a given. In the case of *The Cestrian Book of the Dead* its utility value as a cartographic resource sits alongside what, for me, is arguably a more highly-prized goal: the cultivation of a literary spatial praxis whereby a more performative and 'non-representational' (Thrift) poetics of place is brought into play. This is not to say the map has limited value as a stand-alone resource, but is a recognition that such a resource, as with the GIS mappings conducted as part of the 'Mapping Medieval Chester' project (Clarke), can only ever be interventions as part of a much broader (and less demonstrably 'impactful') project: that of a politics and poetics of place more generally.

> Sailors by the fierce flow of water swept,
> Domyngo de Sasarendo and Farnando de Vitaer
> between Blaken Hed and Burton Hed.
>
> The children of John Justice, a labourer,
> while bathing near the fulling mill. Walter Blunkett,
> in a mill race at the walke milnes.
>
> Cut off by the tide while crossing the Salt Grass.
> A boat overturned near the Stewbridge Gutter.
> Plunge hole near the Sluices lately erected.
>
> Ann Harrison while crossing the Shotwick ford,
> one Tuesday in January 1753. Thomas, on Friday,
> went the way of his wife into the Book of the Dead.

If, like Borges's 'Aleph' or Stapledon's 'hypertelescopic imagination', we think of the *The Cestrian Book of the Dead* as a heterotopia, then it is not difficult to envisage how the relief map mentioned earlier – an imagined cartography of absence – might begin to be charted and plotted. It unfolds along pathways that connect to and from the documented points in time and space where Domyngo de Sasarendo or Farnando de Vitaer enter history. It spirals down the plunge hole of the 'Sluices lately erected' to re-emerge in a landscape reconstituted and drained of memory. It fans out across a meshwork of unknowable (and thus as yet unimagined) scenarios that determine the fate of a husband and wife, drowned whilst crossing the Shotwick to Flint ford, but several days apart. Such a map cannot be projected in any conventional cartographic sense, but it is possible to imagine what its features – its lines, contours, waypoints, edges, shadows, cartouches and scales – might look like. In this regard the geographical map – digital or otherwise – needs to be looked upon as the starting point. The literary work begins when the imagination starts to pull away and map its own spaces and itineraries.

Re-envisioning the Digital Aleph

In this chapter I have attempted to follow a deliberately oblique pathway, the objective of which has been: a) to examine more closely some of the fine-grained literary cartographies that have variously constellated around a specific geographical location: the Dee Estuary; b) to 'map' on to these spaces some of the heterotopic configurations of place, landscape and memory that open up the topoi of place to 'other' spatio-temporal and imaginative worlds; c) to populate a necrogeographic map of the Dee Estuary as a means to explore the everyday landscapes and affective attachments to place that are coalescent around tropes of danger, liminality, uncertainty and drowning; and d) to lay across all of this some reflections on the role and 'place' of digital tools in critical and creative practices of literary mapping and spatial anthropology.

On this latter point, without wishing to get sucked into the techno-determinist vortex that all too often bedevils discussion around so-called 'digital humanities', there is nevertheless a demonstrably practical utility that these technologies can offer and which justifies the necessary investment in time and resources that renders such utility sustainable. For example, the interoperability between proprietary GIS software tools (such as ArcGIS) and open-source and Web-GIS or Geo-Web resources means that project outputs that might otherwise have a limited reach in terms of users – such as *The Cestrian Book of the Dead* or *Mapping the City in Film*, both of which started out as ArcMap shapefiles – can be adapted for use in platforms such as Google Earth, accommodating both a larger constituency of users as well as making such resources adaptable for mobile and site-specific modes of cultural mapping practice. This basic recognition of the *practical* benefits of digital mapping tools goes hand-in-hand with the no less important point that it is *practices* as much as it is *texts* around which literary mapping is or should be principally oriented, and that what we might understand as textual practices need not necessarily preclude their also being conceived of as spatial practices.

In this respect, as a performative trope the idea of the Aleph introduced at the start of this chapter confers, albeit elliptically, an imaginary of space that the literary necrogeographer routinely intuits and puts into practice. As a means by which to consort with the dead the Aleph promises transcendence insofar as rational laws of space and time do not apply. But at the same time it anchors these flights of imagination in the prosaic geographies of worlds that are mappable and corporeally immanent. The dead are not so much 'visited' as walked alongside with in a relationship of co-presence (Ingold and Vergunst 7). Similarly, in the same way that the digital infinitude of the World Wide Web can be scaled down to a point of user interface (the situated body), Borges's Aleph, as a heterotopic portal to 'worlds beyond', is in the end only navigable via the earthly locale of a residential basement. The mountain that comes to the modern digitally-savvy Mohammed may not be one that can be scaled in any physical sense but that does not mean it cannot collide with or potentially transform those spaces where the capacity to imagine opens the way for new and reconstituted cartographies of mobility and memory. As a literary exemplar of differential space the Aleph, by stretching the geographical boundaries of imagination to impossible limits, offers the possibility for rethinking the scale and scope of what a literary spatial praxis might look like. By extension it is not too much of a flight of fancy to suggest that digital cultural mapping might conceivably offer the same, but only insofar as what it 'is' or 'does' is not merely reducible to the temporal fixity of the 'map'.

Glossary

Beesom's tail – *besom*: a broom made of twigs tied around a pole.
Cage, the – Salmon cage near the weir at Chester.
Crane, the – former boat yards located near the Watergate in Chester.
Denwall – probably Denhall, near Neston in Wirral.
Finchett's Gutter – drainage channel that feeds into the River Dee near Chester.
Le Hollyne – location not known.
Le Posterne – from Old French, meaning 'gate' or 'entrance'; the Shipgate or Bridgegate in Chester.
Stewbridge Gutter – location not known.
Walke milnes – fulling mills.

Works Cited

Armitage, Simon. *Sir Gawain and the Green Knight.* London: Faber & Faber, 2007.
Augé, Marc. *The War of Dreams: Studies in Ethno Fiction.* London: Pluto Press, 1999.
BBC. *Sir Gawain and the Green Knight.* BBC Four. 4 June 2009. Television.
Behrend, Arthur. *The House of the Spaniard.* New York: Country Life Press, 1936.
Benjamin, Walter. 'Excavation and Memory'. *Walter Benjamin: Selected Writings Volume 2, Part 2, 1931–1934.* Ed. Michael W. Jennings, Howard Eiland and Gary Smith. Cambridge, MA: Belknap Press, 1999. 576.
Bodenhamer, David J., John Corrigan and Trevor M. Harris, (eds) *The Spatial Humanities: GIS and the Future of Humanities Scholarship.* Bloomington: Indiana University Press, 2010.
Borges, Jorge Luis. 'The Aleph', in *The Aleph and Other Stories 1933–1969.* London: Picador, 1973.
Canal Maps Archive: Historical Maps and Plans of Inland Navigation. N.d. Web. 20 March 2015. <www.canalmaps.net>.
'Chester City Coroner's Records'. *The National Archives.* Web. 20 March 2015. <http://discovery.nationalarchives.gov.uk/browse/r/bf2894f1-c20c-4202-92a3-31501109cba7>.
Clarke, Catherine A.M., (ed.) *Mapping the Medieval City: Space, Place and Identity in Chester c. 1200–1600.* Cardiff: University of Wales, 2011.
Crossley, Robert. *Olaf Stapledon: Speaking for the Future.* New York: Syracuse University Press, 1994.
De Certeau, Michel. *The Practice of Everyday Life.* Trans. Steven Rendall. Berkeley and Los Angeles: University of California Press, 1984.

De Quincey, Thomas. *Confessions of an English Opium Eater*. London: Wordsworth Classics, 1994.

Fiennes, Celia. *Through England on a Side Saddle in the Time of William and Mary*. GB Historical GIS/University of Portsmouth. *A Vision of Britain through Time*. 2009. Web. 20 March 2015. <http://visionofbritain.org.uk/travellers/Fiennes/12>.

Foucault, Michel. 'Of Other Spaces'. *Diacritics* 16.1 (1986): 22–27.

Gregory, Ian, and Alastair Geddes, (eds) *Towards Spatial Humanities: Historical GIS and Spatial History*. Bloomington: Indiana University Press, 2014.

Hetherington, Kevin. *The Badlands of Modernity: Heterotopia and Social Ordering*. London: Routledge, 1997.

Ingold, Tim. *Lines: A Brief History*. London: Routledge, 2007.

Ingold, Tim, and Jo Lee Vergunst. 'Introduction'. *Ways of Walking: Ethnography and Practice on Foot*. Eds, Tim Ingold and Jo Lee Vergunst. Farnham: Ashgate, 2008. 1–20.

Kerrigan, John. 'The Country of the Mind'. *Times Literary Supplement* 11 September 1998: 3–4.

Kniffen, Fred. 'Geographical Record'. *Geographical Review* 57.3 (1967): 426–7.

Lefebvre, Henri. *The Production of Space*. Trans. Donald Nicholson-Smith. Oxford: Blackwell, 1991.

Macfarlane, Robert. *The Old Ways: A Journey on Foot*. London: Hamish Hamilton, 2012.

Mapping the City in Film. University of Liverpool. 2013. Web. 20 March 2015. <http://www.liv.ac.uk/communication-and-media/research/cityfilm/map/>.

McCullough, Malcolm. 'On the Urbanism of Locative Media'. *Places* 18.2 (2006): 26–29.

Poetry Atlas: Mapping the World in Poetry. 2015. Web. 20 March 2015. <http://www.poetryatlas.com/poetry/poem/1642/the-sands-of-dee>.

Pritchard, T.W. *A History of the Old Parish of Hawarden*. Wrexham: Bridge Books, 2002.

Purseglove, Jeremy. *Taming the Flood: A History and Natural History of Rivers and Wetlands*. Oxford: Oxford University Press, 1988.

Roberts, Les. *The Cestrian Book of the Dead*. *Liminoids.com*. 2013. Web. 20 March 2014. <http://www.liminoids.com/inthefield/sandsofdee/map.html>.

———. 'Mapping Cultures: A Spatial Anthropology'. *Mapping Cultures: Place, Practice, Performance*. Ed. Les Roberts. Basingstoke: Palgrave, 2012. 1–25.

———. 'The Sands of Dee: Estuarine Excursions in Liminal Space'. *Liminal Landscapes: Travel, Experience and Spaces In-between*. Eds, Hazel Andrews and Les Roberts. Abingdon: Routledge, 2012. 103–17.

———. 'Navigating the "Archive City": Digital Spatial Humanities and Archival Film Practice'. *Convergence* 21.1 (2015): 100–15.

Samuel, Raphael. *Theatres of Memory: Past and Present in Contemporary Culture*. London: Verso, 1994.

Sinclair, Iain. *Edge of the Orison: In the Traces of John Clare's 'Journey Out of Essex'*. London: Penguin, 2005.

Smith, Phil. *Mythogeography: A Guide to Walking Sideways*. Axminster: Triarchy Press, 2010.

Snyder, Gary. *The Gary Snyder Reader*. Berkeley, CA: Counterpoint, 1999.

Soja, Edward W. *Thirdspace: Journeys to Los Angeles and Other Real-and-Imagined Places*. Oxford: Blackwell Publishers, 1996.

Stapledon. Olaf. *Star Maker*. London: Methuen, 1937.
———. 'Fields Within Fields'. *An Olaf Stapledon Reader*. Ed. Robert Crossley. New York: Syracuse University Press, 1997. 250–62.
———. *Sirius*. London: Victor Gollancz, 2000.
Thomassen, Bjørn. 'The Uses and Meanings of Liminality'. *International Political Anthropology* 2.1 (2009): 5–27.
———. 'Revisiting Liminality: The Danger of Empty Spaces'. *Liminal Landscapes: Travel, Experience and Spaces In-between*. Eds, Hazel Andrews and Les Roberts. Abingdon: Routledge, 2012. 21–35.
Thrift, Nigel. *Non-Representational Theory: Space, Politics, Affect*. London: Routledge, 2007.
Veeser, Harold A. (ed.) *The New Historicism*. Abingdon: Routledge, 1989.
Wood, Denis. 'The Anthropology of Cartography'. *Mapping Cultures: Place, Practice, Performance*. Ed. Les Roberts. Basingstoke: Palgrave, 2012. 280–303.
Young, John, A. *The Physical and Historical Geography of the Dee Estuary*. MA thesis. University of Liverpool, 1926.

Part III

Mapping Futures

Collecting, Curating and Creating

10 Making the Invisible Visible

Place, Spatial Stories and Deep Maps

David J. Bodenhamer

Introduction

Humanists are acutely aware of the social and political construction of space and its unique expression as place. Space is not simply the setting for historical action. It is a significant product and determinant of change and the medium for the development of culture: 'space is not an empty dimension along which social groupings become structured', sociologist Anthony Giddens notes, 'but has to be considered in terms of its involvement in the constitution of systems of interaction' (364). All spaces contain embedded stories based on what has happened there. These stories are both individual and collective, and each of them link geography (space) and history (time). More importantly, they all reflect the values and cultural codes present in the various political and social arrangements that provide structure to society. In this sense, then, the meaning of space, especially as place or landscape, is always being constructed through the various contests that occur over power, or as Michel de Certeau reminds us, 'space occurs as the effect . . . of conflictual programs or contractual proximities' (117). There is nothing new in this argument – the earliest maps reveal the power arrangements of past societies – but humanities scholarship increasingly reflects what in fact may be the greatest legacy of postmodernism, the recognition that our understanding of the world itself is socially constructed.

Humanities disciplines now grapple routinely with space and its particular expression as place (Bodenhamer, Corrigan and Harris vii–xv). A cursory review of literary historiography reveals a wide array of approaches to problems of space and literature: the study of works and/or authors from a geographical perspective, including the spatial relationships or networks of influence it reveals; the parsing of spatial concepts within the work; analyses that privilege space over time; the spatial *imaginaire*; the influence of landscape and place on literature; and the ethno-poetics of place, by which literary texts construct and interpret the meanings of place (Cabo Aseguinolaza). The result is a sharper understanding of how places shape texts as well as how the language and forms of texts shape places (Alexander and Cooper 1). We have come to recognise, as historian Phil Ethington argues, that every past has a place made (and re-made) by human actions and human thought and that every text is by definition an interactive archaeology in which we read and write the varied meanings of place (463–5).

More recently, text mining has become a tool of choice for scholars who seek to understand literary representations of space and place. This approach offers significant potential for humanists who seek patterns in a large corpus to shape or to enhance insights gained from traditional methods. Tagged distant reading by Franco Moretti (Moretti *Graphs, Maps, Trees*; *Distant Reading*), it promises to reveal contextual information that cannot be gained by the close reading of one text or small groups of texts. References to space and place are prominent markers of context, and a number of projects in history and literature are testing the potential of text mining to reveal what Edward Said termed the 'geographical articulations' of literature (61). At Stanford University, a team of historians and computer scientists depicted the connections, volume and flow of letters that passed amongst 6,400 correspondents in the 55,000-record Electronic Enlightenment Database, revealing an even denser intellectual network than scholars of the Republic of Letters had imagined previously (Chang et al.). In the *ChartEx* project, an interdisciplinary team is mining medieval charters for rich descriptions of people and places, whilst another project based at Virginia Tech is mining newspapers to form an '*Epidemiology of information*' about the influenza epidemic of 1918.

But mining texts for their place markers, although perhaps essential for large-scale analysis, is insufficient to provide the rich bed of data required for a postmodern scholarship that understands the world as endlessly complicated, fluid, interlinked and flush with simultaneous, competing truth-claims representing the unique position of the observer. To achieve entry into this world, we must find a way to marry the results of text mining with their full spatial context – geographical, conceptual, metaphorical and imaginary – and then to make them visible. This work already is underway. A team of geographic information scientists (GIScientists), computational linguists and literary scholars at Lancaster University is mining literary and visual representations of the English Lake District to understand the way we have constructed and canonised this iconic landscape (Cooper and Gregory). Projects in Zurich and Brisbane are seeking to map the real and imagined geographies of works of fiction (*Literary Atlas of Europe*; *Cultural Atlas of Australia*). Joycean Dublin has long served as a test bed for the interplay of text and geography, and scholars now are extending its lessons in new directions (Travis). Similarly, classicists are exploring spatialised texts and images, as illustrated in the *HESTIA* project (Barker). In addition, a number of geographers are developing methods to extract spatial language from texts for use in a variety of applications from military surveillance to scholarship (Yuan).

These initiatives hold great promise for a more complex understanding of past space-time, yet even successes here are not taking full advantage of a computing revolution that is offering new ways to reimagine heritage and culture. As we explore new methods for integrating text and space, we will need to be clear about how we conceive our task and to re-conceptualise how to use what we discover.

Space, Place and Spatial Technologies

What may not be apparent at first glance is that the terms we use are themselves highly contested: space and place are everywhere, and their names and definitions have been legion. In modern practice space and place have become ideological battlegrounds, divorced from the geography and history that embraces them both. The separation first began at the end of the seventeenth century when, with the emergence of modern science, place became transformed into location or, simply, position. In Newtonian physics, space is abstract until it can be fixed, a characteristic that allows it to be measured and verified, thus giving it value within scientific method. It thus could be evaluated with other points for patterns that revealed a universal law. Places offered variations that were not interesting except as local cases; the pattern amongst them was what was useful.

This Newtonian model has obvious limitations for understanding the discursive and imaginative geographies that inform human society and behaviour. Nineteenth-century historians and antiquarians found the scientific construction of space and place especially unappealing. Augustin Thierry, George Bancroft and Sir Walter Scott, for instance, defined place as a fusion of culture and history that was manifested in the local and particular (Rigny). This fundamentally humanistic conception of space and place, grounded as it is in the desire to explore the human condition, has gained significant traction since the beginning of the twentieth century, spurred by the insistence of thinkers, such as Martin Heidegger, who famously argued that 'the "surrounding world" does not arrange itself in a previously given space . . . . The actual world discovers the spatiality of space belonging to it' (97).

Heidegger's insight reflects the natural human response of using the body as a reference point to organise space. Our language contains numerous words – up, down, left, right, back, forward, etc. – that keep us positioned to the environment we are in, whether actual or created in text. Indeed, an increasingly large body of thought known as embodied cognition suggests that what we know cannot be separated from what our bodies perceive as we move through space (Tversky; Tversky and Hard). But we also orient ourselves through a symbolic geography governed by beliefs, ontologies or otherwise created relationships: sacred and profane, colonising and colonised or town and country, for instance. Narrative arises when a character or event crosses these symbolically charged spaces (Buchholz and Jahn).

From this perspective, the narratives we construct, regardless of their form, are inherently spatial. So what sorts of spatial stories are we telling?

Deconstructionists have argued that narrative is so basic to our cultural beliefs that our stories bear little resemblance to reality. 'The past is not an untold story', Louis Mink once noted; we are so trapped within our narrative discourse that we have no way to reach a past that exists outside of our cultural assumptions (22). Heidegger offered another view, one more attractive to humanists who are uncomfortable with the post-modern divorce between stories and an inherent reality. In this view, we cannot experience reality other than through our narratives (Elden). As a result, historian William Cronon has noted, 'we inhabit an endlessly storied world' (1368).

Our stories, however, are not about abstract space but about place, or space made meaningful by human actions and memory. Edward Casey argues that the experience of place actually precedes knowledge of space. Resurrecting an ancient insight that we perceive the world through our experience of place, he concludes that humans are 'ineluctably place-bound'; we are, in his playful phrase, 'more even than earthlings, we are placelings' (19). Far from being particular, place is 'something general, perhaps even universal' (19). Phil Ethington echoes this idea when he claims that 'the past cannot exist *in* time; only *in* space. Histories representing the past represent the places (topoi) of human action' (465). In brief, place has meaning because human actions make the abstract concepts of time and space comprehensible.

Place-making is not an event to be measured; rather, it is a process of being in a 'configurative complex of things' (Casey 25). We are continually making place by our acts of living in space. Geographer Doreen Massey has even argued that place is an open and hybrid concept, a product of interconnecting flows and not something that is rooted or fixed (316). We invest these places with meaning, but they do not exist as isolated and independent spaces. Instead, they are both general and particular. We all experience place, but the places we experience are all different. The world is diverse and complex, and we can understand it only through an appreciation of the uniqueness of places and the events and cultures that they hold. Or, as Heidegger suggested, 'spatiality has its own unity by virtue of the world like totality of relevance of what is spatially at hand' (97).

This sense of spatiality mirrors the humanist's view of history and culture as extremely complex, with endless connections amongst events and actors and multiple causes for effects that exert continuing influence on the world of thought and behaviour. This web-like interrelatedness plays itself out within two dimensions: space and time. Although the past is always bound by these two elements, humanists often treat them as artificial, malleable constructs. We move freely across these spatial and temporal grids, ignoring issues of scale, as we compare and contrast one place or one time with another in an effort to recapture a sense of the whole, to illuminate differences, to discover patterns (Gaddis). For the humanist, space alone is an abstraction; it is occupied space, or place, that draws our attention. And like time, place exists not simply in a material world but also in memory, imagination and experience.

Trying to comprehend space, place and time in concert has always proven difficult, even in the most expert narratives. Historian Hugh Trevor-Roper noted the problem decades ago: 'How can one both move and carry along with one the fermenting depths which are also, at every point, influenced by the pressure of events around them? And how can one possibly do this so that the result is readable?' (Thomas 56). Or as digital humanities pioneer Edward Ayers has asked more recently, 'how might we combine the obvious strengths of geographic understanding with the focus on the ineffable, the irreducible, and the particular . . . ? How might we integrate structure, process, and event? In sum, how might we combine space, time, and place?' (4)

Archaeologists and historians have used geographic information systems (GIS) for at least two decades, and increasingly other humanists, including literary studies scholars, are turning to GIS and other spatial technologies to aid their search for a more sophisticated linkage between context and content. Within a GIS, users can discern relationships that make a complex world more immediately understandable by visually detecting spatial patterns that remain hidden in texts and tables. Maps have served this function for a long time, but GIS brings impressive computing power to this task. Its core strength is an ability to integrate, analyse and make visual a vast array of data from different formats, all by virtue of their shared geography. This capability has attracted considerable interest from historians, archaeologists, linguists, students of material culture and others who are interested in place, the dense coil of memory, artefact and experience that exists in a particular space, as well as in the coincidence and movements of people, goods and ideas that have occurred across time in spaces large and small.

The past two decades have witnessed a wide-ranging, if still limited, application of GIS to historical and cultural questions: What influence did the rapidly changing cityscape of London have on literature in Elizabethan England? What does the geographical extent of correspondence in the Enlightenment reveal about channels of communication and influence in the so-called Republic of Letters? What was the relationship amongst rulers, territory and culture in the chequered political landscape of state formation in nineteenth-century Germany? How did spatial networks influence the administrative geography and worldview of medieval China? Increasingly, scholars have turned to GIS to provide new perspective on these and other topics that previously have been studied outside of an explicitly spatial framework (Knowles; *IJHAC*).

Despite this flurry of interest and activity in GIS, the technology has limits when applied to the questions raised by humanists. The problems we pursue present epistemological and ontological issues that challenge the technology in a number of ways, from the imprecision and uncertainty of humanities data to humanists' reliance on time (and time linked to space) as an organising principle. Essentially, GIS and its related technologies currently allow users to determine a geometry of space, which requires precise data. Fuzzy evidence, conceptual space and relative time pose often insurmountable obstacles for these tools, although scholars are making progress in this area, as witnessed by Torsten Hägerstrand's concept of 'time geography' in archaeology (Mlekuz). In the context of the humanities, it will be necessary to replace this more limited quantitative representation of space with a view that emphasises the intangible and socially constructed world and not simply the world that can be measured. It also will be essential to match technologies with the traditions of argument and narrative employed by humanists (Bodenhamer, 'Beyond GIS'). It is here where the deep map, a new form for managing and visualising information, becomes important, perhaps essential.

Deep Maps

A deep map is a finely detailed, multimedia depiction of a place and the people, animals and objects that exist within it and are thus inseparable from the contours and rhythms of everyday life. Its intellectual roots spread in various directions: early maps with their elaborate cartouches, imaginative renderings and embedded texts; eighteenth-century gazetteers and antiquarian local histories, which were a pastiche of history, folklore, hearsay, statistics, natural history and anything else that could be discovered about a place; the *Annales* school of historiography that insisted on a thick description of the past, to include the physical and material culture in which actions occurred; the insistence of the 1950s Situationists that life could be understood only as a contextual stew; Certeau's spatial stories that captured the practices of everyday life; geographer Yi Fu Tuan's emphasis on the connection between emotion and landscape seen in the love of place (topophilia); the theatre/archaeology practices of Michael Pearson and Michael Shanks, who viewed their work as re-articulating fragments of the past as real-time events; and the exemplar work by William Least Heat Moon, *PrairyErth (a deep map)*, which explores time and space in a single Kansas county, much as William Faulkner did fictionally in his Yoknapatawpha novels.

This lineage suggests that deep maps are not confined to the tangible or material, but include the discursive and ideological dimensions of place (the dreams, hopes, imaginations and fears of residents); they are, in short, positioned between matter and meaning. They are also topological and relational, revealing the ties that places have with each other and tracing their embeddedness in networks that span scales and range from the local to the global. The spatial considerations remain the same: geographic location, boundary and landscape remain crucial. What is added is a reflexivity that acknowledges how engaged human agents build spatially framed identities and aspirations out of imagination and memory, and how these multiple perspectives constitute a spatial narrative that complements the prose narrative traditionally employed by humanists (Bodenhamer, 'Narrating Space').

A deep map is a platform, a process and a product. As platform, it is an environment embedded with tools to bring data into an explicit and direct relationship with space and time. As a process, it is a way to engage evidence within its spatiotemporal context and to trace paths of discovery that lead to a spatial narrative and ultimately a spatial argument. As product, it is the way we make visual the results of our enquiry and share the spatially-contingent argument enabled by the deep map. Within this environment, we can develop the event streams that permit us to see the confluence of actions and evidence; we can use path markers or version trackers to allow us (and others) to trace our explorations; and we can contribute new information that strengthens or subverts our argument, which is the goal of any exploration. It is, in short, a new creative space that is visual, structurally open, genuinely multi-media and multilayered. Deep maps do not explicitly seek authority or objectivity but provoke negotiation between insiders and outsiders, experts and contributors, over what is represented and how. Framed as a conversation and not a statement, they are inherently unstable, continually unfolding and changing in response to new data, new perspectives and new insights (McLucas).

What would a deep map look like, and how would it function? Ideally, the deep map exists as an immersive environment in which the user experiences the spatial and temporal context of the evidence in both written and visual forms. Consider the following example: A military battle is a discrete event bounded by space and time but with clear antecedents and discernible after-effects. But it exists not simply as a physical event but also as memory, in imagination, through artefacts and records, in images and often as myth. We can understand the battle in all of these ways, but only as discrete elements. How do we understand it as all things at once, and how do we experience it? How can we recognise the contingencies facing participants and with this knowledge gauge the counterfactual? How can we move fluidly across spatial and temporal scales when considering the decisions made or not made, because surely that is how the participants judged their choices. As humans we cannot understand the battle in any general way other than through analogy, metaphor, emotion and cause and consequence, all the ways that we have invented over time to aid our understanding and our ability to communicate but which separate us from the experience of the event itself, even though we acknowledge that individual and collective experiences shape our understanding of it. But if we could recreate the battle, populating it with all we can discover about it, and if we could view this evidence simultaneously and from its own unique perspective, as well as from the perspectives we bring to it – if we could *see* what they reported within its geographical, temporal and cultural context – would not the battle appear more chaotic, more contingent, more fluid, more uncertain, more ambiguous, more immediate – in other words, more fully human?

Deep maps can be useful in literary studies by bringing into play not only the work but also the author's world and the reader's experience. Two examples will illustrate. James Joyce wrote *Ulysses* whilst in self-imposed exile from Dublin even though he could not escape it emotionally or in his imagination. He recast Homer's *Odyssey* as a comment on modern life, but the journey through Edwardian Dublin that he created for Stephen Dedalus and Leopold Bloom also drew heavily on Dante's *Inferno*. To aid him in constructing the novel, he relied upon maps of Dublin, his own charts of Homer's and Dante's work, and numerous other aids (Groden). With a deep map, we can keep all of these items in tension, along with available evidence about where Joyce lived, his network of correspondents, what else he read, how he revised his work and what was happening in the world around him. It also can help us analyse the work's reception by revealing how, when and where people responded to Joyce's radical modernism. The deep map, in sum, can create a complex picture that allows us to understand more completely the larger context of his masterpiece by integrating the data dynamically and by visualising it in multiple ways.

We also can gain a different perspective on English Lake District writing by tracing the connections across space and time amongst texts, images, terrain and socio-economic change, as the *Spatial Humanities* project at Lancaster University is doing. In a deep map, users could trace the complex spatio-temporal relationships amongst developments that were occurring simultaneously with the same space or across different spatial scales. What was the temporal and spatial correspondence amongst Romantic writings, railroad development, population change, mass media and economic change on the development of the Lake District as a tourist destination, for instance? Space rendered in 3D could provide an experiential perspective, allowing researchers to 'see' the actual landscapes portrayed figuratively in the poetry of William Wordsworth and Samuel Taylor Coleridge. In this virtual environment we easily could compare the real and imagined geography to enrich our understanding of their symbolic language.

But how? A promising approach stemming from an Advanced Institute on Spatial Narratives and Deep Maps, sponsored by the National Endowment for the Humanities in 2012, uses a series of dynamically linked panes in which changes in one pane are reflected in the others. The interface employs faceted browsing rather than complicated search filters to provide an immediate sense of the richness and depth of data rather than rely on complicated search forms to construct queries and lead audiences to the data. The intent is to navigate through the content sources and their multiple relations to time, place and other sources. Timeline and place sliders are always present for easy and intuitive navigation, with an event stream or histogram that shows the density of different types of documents and resources over time to help users get a sense of the scope of the content and help alleviate the patchiness and messiness of humanities data. A zooming function also includes 'spatial bookmarks' that not only expose the administrative boundaries relevant to that location but also help the user jump quickly between scales or locate themselves precisely within space ('Spatial Narratives and Deep Maps'; Bodenhamer et al., *Deep Maps and Spatial Narratives*).

A more ambitious approach marries existing technologies and methods (for example, GIS, text mining, image capture) within a serious gaming engine that allows us to move through a 3D world filled with what we know so we can test more effectively what we assume or what we advance as argument, as well as how others have constructed their own meaning from events and actions. The Experimental Technologies Center at UCLA is working on such an environment, at least in part, with its Rome Lab, which employs GIS and procedural modelling to create a virtual Roman Forum, created first as the *Digital Roman Forum Project* in 2003, using Unity, a gaming engine. Classicists have adopted it to recreate and study Roman funeral orations, which were important state rituals. By linking historical evidence and modern acoustical physics within a virtual reality platform, they are developing an argument about how space was used to reinforce the power and class arrangements of Roman society. Most importantly, they have visualised the setting and the evidence about the occasion, thus inviting the user to experience the environment, albeit incompletely, and the actions within it to advance understanding of another place in time, another culture (*UCLA RomeLab*). At West Virginia University, Trevor Harris – see Chapter 11 – has moved GIS into the CAVE (Computer Augmented Virtual Reality) and recreated nineteenth-century Morgantown, the location of the university. Here, more senses are available for understanding place. Once in the CAVE, users experience not only the physical sensation of walking through historical Morgantown (created in Google Sketch-Up), but also something of the town's atmosphere. To this latter end, Harris and his students have introduced visual effects, such as smoke-generated haze, and the smells of the coal-burning furnaces and abattoirs. These effects are not mere tricks to delight the user. Rather they help the user grasp more fully what Rebecca Harding Davis, a nineteenth-century American novelist, intended in her evocative *Life in the Iron Mills* (1861), when she asked, 'do you know what it is like in a town of iron works? . . . [C]ome right done with me, here, into the thickest of fog and mud and foul effluvia' (Harris, Rouse and Bergeron 229). The recent introduction of Google Glass and the virtual reality head-mounted display Oculus Rift suggests that it will not be long before we all experience total immersion within an information-filled, richly sensual environment.

The experience of modern gamers suggests the rich potential of deep mapping. *EVE Online* is a futurist game in which users create avatars from one of four races, choosing hair, scars and clothes to distinguish individuals. The game revolves around three basic tasks: mining asteroids for minerals, using these minerals to build spaceships, deploying these vessels in battle. It is a dystopian world made up of thousands of interlocking solar systems, with independent agents to monitor safe zones but with vast spaces in which anything goes. It also has an economy but no legal system; everything depends upon trust and relationships. Now, many games provide similar experiences but *EVE Online*, an unbounded (no rules) massively multiplayer gaming environment, has been studied extensively by scholars. The findings are interesting – and suggestive: half a million players inhabit this universe at the same time and they continually make both the rules and interpret the meaning of their experiences. Users internalise the history of this fictive universe; they keep records of events, interview famous players and write their own narratives of what has happened. They are immersed in *EVE* and make sense of it through story-telling, with the multiple narratives allowing us to see the world from multiple perspectives. There is no master argument, no voice of authority (Suzor and Woodward; Glushko). Instead, like the deep map, its multivocality leads to negotiated meanings that reveal the contingent nature of this fictive yet all-too-real world.

Regardless of the ultimate platform, deep maps will depend upon scholarly traditions developed over generations by the various humanities disciplines. They must reflect the ways we discover knowledge even as they contribute new approaches to research questions. A deep map will not replace the close reading of a text, for example; rather, deep mapping exists alongside other methods of investigation, whether text mining or literary cartography or the text as artifact, as we seek to gain a more complete purchase on our inquiry. It also welcomes knowledge developed in non-traditional ways, such as reflected in neogeography, the non-expert or volunteered spatial information produced in everyday life (Warf), which may not follow the vetting practices or publishing formats normally associated with our disciplines. Scholarship may take new, more fluid forms but the aims of the monograph and essay will not be replaced or subsumed as much as they will be re-imagined and re-expressed.

What the deep map allows – no, encourages – is what John Corrigan has termed a 'genealogy of emplacement'. That is, it frames investigations of space and place as 'a kind of genealogy . . . into the historically contingent layering of structures, strategies, tactics, discipline, anti-discipline, environment, everyday practice' (Corrigan 63). This genealogy de-emphasises the predictably rational by focusing instead on locally specific and unique processes, as well as in interstices between everyday practice and structure. Although it embraces some of William James's sense that 'reality, life, experience, concreteness, immediacy . . . exceeds our logic, overflows, and surrounds it' (212), its aim is to understand how place is made through everyday practice and how it influences (and is influenced by) structure. In its observations of practice, emplacement also locates the networks and associations that constitute narratives of place. The deep map, in this sense, narrates through curation, constructing complex collages that reveal change over time as well as the inherently varied meanings we ascribe to place. Its means of representation is the spatial story that emerges from this act of curation.

A deep map, in brief, promises a different way to understand society and culture, past and present, fictive or real. Historian William Sewell, for instance, has argued that 'social life may be conceptualised as being composed of countless happenings or encounters in which persons and groups of persons engage in social action. Their actions are constrained and enabled by the constitutive structures of their societies' (100). As a result, ' "societies" or "social formations" or "social systems" are continually shaped and reshaped by the creativity and stubbornness of their human creators' (111). The deep map embodies these human engagements and the structures that accommodate (or flow from) them. It also mirrors what literary theorist Mikhail Bakhtin termed the chronotope, the intrinsic interconnectedness of time and space in literature:

> Spatial and temporal indicators are fused into one carefully thought-out, concrete whole. Time, as it were, thickens, takes on flesh, and becomes artistically visible; likewise, space becomes charged and responsive to the movements of time, plot and history . . . . Time becomes, in effect, palpable and visible; the chronotope makes narrative events concrete, makes them take on flesh, and causes blood to flow in their veins. (Bakhtin 250)

Deep maps have the potential to help us see the interactions of agents and structures in the narratives we construct, all bound in the space-time that defines place.

If place is made by stories – and if our stories (and memories) make place – then we define its myriad forms by fashioning open-ended narratives about place that reflect our own situatedness. Not only will we gain valuable perspective, but doing so will help us avoid what French philosopher Henri Bergson categorised as an illusion of retrospective determinism (Bergson 10–12). It will be difficult to trace inevitability between cause and effect when faced with multiple voices and multiple paths, and when reminded of where and how we confront the many forms of the story we seek to understand.

But will deep mapping enhance scholarship? It is a legitimate question. Making the process experiential and open-ended may lead to greater understanding; it also may provide so many narratives, so many voices, that it fragments our understanding and results in intellectual chaos. It may add nothing more than our essays and books provide collectively, except perhaps for whatever utility is gained through interactive, scalable visualisations of a wide array of evidence. It is also true that texts change when filtered digitally, just as maps change when rendered by computational technology rather than the cartographer's hand. But this risk of misreading is far outweighed by what the deep map offers: it opens discovery and invites multiple expert and naïve voices into the conversation about meaning; it reinforces the role of emotion in the construction of place and event, not to the exclusion of rational argument but as an alternate interpretation; it allows us to trace complex spatial narratives, with the paths dependent upon what we can see within a recreated landscape populated with all we have discovered about the events or actions under study; it moves us from the linear frame of prose into a world of simultaneity that more closely approximates what we know experientially and logically to be true; and it allows us to uncover the emergent realities and deep contingencies of the past that arise from the intentional and unintentional actions of people as they make their way through life and create the stories that define us. Here, then, is the value of the deep map: it makes the invisible visible. In the process, it enables a unique postmodern scholarship that is rarely possible, if at all, through traditional disciplinary forms.

Works Cited

Alexander, Neal, and David Cooper. 'Introduction: Poetry & Geography'. *Poetry & Geography: Space & Place in Post-War Poetry.* Eds, Neal Alexander and David Cooper. Liverpool: Liverpool University Press, 2013, 1–18.

Ayers, Edward L. 'Turning toward Place, Space, and Time'. *The Spatial Humanities: GIS and the Future of Humanities Scholarship.* Eds, David J. Bodenhamer, John Corrigan and Trevor M. Harris. Bloomington: Indiana University Press, 2010. 1–13.

Bakhtin, Mikhail. *The Dialogic Imagination: Four Essays,* Ed. Michael Holquist. Austin TX: University of Texas Press, 1981.

Barker, Elton. 'Mapping an Ancient Historian in a Digital Age: The Herodotus Encoded Space-Text-Image Archive (HESTIA)'. *Leeds International Classical Studies* 9:1 (2010): Web. 20 March 2015. <http://www.leeds.ac.uk/classics/lics/>.

Bergson, Henri. *The Creative Mind: An Introduction to Metaphysics.* New York: The Philosophical Library, 1946.

Bodenhamer, David J., John Corrigan and Trevor M. Harris, (eds) *Deep Maps and Spatial Narratives.* Bloomington: Indiana University Press, 2015.

———. 'Beyond GIS: Geospatial Technologies and the Future of History'. *History and GIS: Epistemologies, Considerations and Reflections.* Eds, Alexander von Lünen and Charles Travis. New York: Springer, 2012. 1–13.

———. 'Narrating Space and Place'. *Deep Maps and Spatial Narratives.* Eds, David J. Bodenhamer, John Corrigan and Trevor M. Harris. Bloomington: Indiana University Press, 2014. 7–27.

Buchholz, Sabine, and Manfred Jahn. 'Space in Narrative'. *Routledge Encyclopedia of Narrative Theory.* Eds, David Herman, Manfred Jahn and Marie-Laure Ryan. New York: Routledge, 2005. 550–55.

Cabo Aseguinolaza, Fernando. 'The Spatial Turn in Literary Historiography'. *CLCWeb: Comparative Literature and Culture* 13.5 (2011): Web. 20 March 2015. <http://dx.doi.org/10.7771/1481-4374.1903>.

Casey, Edward S. 'How to Get from Space to Place in a Fairly Short Stretch of Time: Phenomenological Prolegomena'. Eds, Steven Feld and Kenneth H. Basso. *Senses of Place.* Santa Fe, NM: School of American Research Press, 1996. 13–52.

Chang, Daniel, et al. 'Visualizing the Republic of Letters'. 2009. *Mapping the Republic of Letters.* Web. 20 March 2015. <http://www.stanford.edu/group/toolingup/rplviz/papers/Vis_RofL_2009>.

ChartEx Project. N.d. Web. 30 March 2015. <http://www.chartex.org>.

Cooper, David, and Ian Gregory. 'Mapping the English Lakes District: A Literary GIS'. *Transactions of the Institute of British Geographers* 36.1 (2011). 89–108.

Corrigan, John. 'Genealogies of Emplacement'. *Deep Maps and Spatial Narratives.* Eds, David J. Bodenhamer, John Corrigan and Trevor M. Harris. Bloomington: Indiana University Press, 2014. 54–71.

Cronon, William. 'A Place for Stories: Nature, History, and Narrative'. *Journal of American History* 78.4 (1992): 1347–76.

Cultural Atlas of Australia. 2012. Web. 20 March 2015. <http://australian-cultural-atlas.info/CAA/>.

De Certeau, Michel. *The Practice of Everyday Life.* Trans. Steven Rendall. Berkeley and Los Angeles: University of California Press, 1984.

Elden, Stuart. *Mapping the Present: Heidegger, Foucault, and the Project of Spatial History.* London: Continuum, 2001. 93–111.

Epidemiology of Information Project. 2012. Web. 20 March 2015. <http://www.flu1918.lib.vt.edu>.

Ethington, Philip J. 'Placing the Past: "Groundwork" for a Spatial Theory of History'. *Rethinking History* 11:4 (Dec. 2007): 463–530.

Gaddis, John Lewis. *The Landscape of History: How Historians Map the Past.* Oxford: Oxford University Press, 2002.

Giddens, Anthony. *The Constitution of Society: Outline of the Theory of Structuation.* Oxford: Oxford University Press, 1984.

Glushko, Bobby. 'Tales of the (Virtual) City: Governing Property Disputes in Virtual Worlds'. *Berkeley Tech Law Journal* 22.1 (2007): 507–32.

Groden, Michael. *Ulysses in Progress.* Princeton NJ: Princeton University Press, 1987.

Harris, Trevor M., L.J. Rouse and Sue Bergeron. 'Humanities GIS: Adding Place, Spatial Storytelling and Immersive Visualization in the Humanities'. *Geohumanities: Art, History, Text at the Edge of Place.* Eds, Michael Dear et al. New York: Routledge 2011. 226–40.

Heidegger, Martin. *Being and Time.* 1927. Trans. Joan Stambaugh. Albany NY: State University of New York Press, 1996.

IJHAC (*International Journal of Humanities and Arts Computing*) 3.1–2 (2009).

James, William. *A Pluralistic Universe.* New York: Longman, Green & Co., 1920.

Knowles, Anne Kelly, (ed.) *Placing History: How Maps, Spatial Data, and GIS Are Changing Historical Scholarship.* Redlands, CA: ESRI Press, 2008.

Literary Atlas of Europe. 2015. Web. 20 March 2015. <http://www.literaturatlas.eu/en/>.

Massey, Doreen. 'A Global Sense of Place'. *Reading Human Geography.* Eds, Trevor Barnes and Derek Gregory. London: Hodder Arnold. 1997. 315–23.

McLucas, Cliff. 'Ten Things to Say about Deep Maps'. *Michael Shanks.org.* N.d. Web. 20 March 2015. <http://cliffordmclucas.info/deep-mapping.html>.

Mink, Louis O. 'Narrative Form as a Cognitive Instrument'. *The History and Narrative Reader.* Ed. Geoffrey Roberts. New York: Routledge, 2001. 211–20.

Mlekuz, Dimitrij. 'Time geography, GIS and archaeology'. Proceedings of the 38th Conference on Computer Applications and Quantitative Methods in Archaeology. 2010. Web. 20 March 2015. <https://www.academia.edu/376168/Time_geography_GIS_and_archaeology>.

Moretti, Franco. *Distant Reading*. London: Verso, 2013.
———. *Graphs, Maps, Trees: Abstract Models for Literary History*. London: Verso, 2005.
Rigny, Ann. *Imperfect Histories: The Elusive Past and the Legacy of Romantic Historicism*. Ithaca, NY: Cornell University Press, 2001.
Said, Edward. *Culture and Imperialism*. London: Vintage. 1994.
Sewell, William H. Jr. *Logics of History: Social Theory and Social Transformation*. Chicago: University of Chicago Press, 2005.
'Spatial Narratives and Deep Maps: A Special Report'. *International Journal of Humanities and Arts Computing* 7:1–2 (2013): 170–227.
Suzor, Nicolas and Darryl Woodford, 'Evaluating Consent and Legitimacy Amongst Shifting Community Norms: An EVE Online Case Study'. *Journal of Virtual Worlds Research* 6.3 (2013): 1–11.
Thomas, Keith. 'A Highly Paradoxical Historian'. *New York Review of Books*. April 12, 2007: 53–57.
Travis, Charles. 'Abstract Machine – Geographical Information Systems (GIS) for Literary and Cultural Studies: "Mapping Kavanagh"'. *International Journal of Humanities and Arts Computing* 4:1–2 (2010): 17–38.
Tversky, Barbara. 'Spatial Cognition: Embodied and Situated'. *Cambridge Handbook of Situated Cognition*. Eds, Philip Robbins and Murat Aydede. Cambridge: Cambridge University Press, 2009. 201–17.
Tversky, Barbara, and Bridgette M. Hard, 'Embodied and Disembodied Cognition: Spatial Perspective-taking'. *Cognition* 110.1 (2009): 124–9.
UCLA RomeLab. 2014. Web. 20 March 2015. <http://romelab.etc.ucla.edu/>.
Warf, Barney. 'Deep Mapping and Neogeography'. *Deep Maps and Spatial Narratives*. Eds, David J. Bodenhamer, John Corrigan and Trevor M. Harris. Bloomington: Indiana University Press, 2014. 134–49.
Yuan, May. 'Mapping Text'. *The Spatial Humanities: GIS and the Future of Humanities Scholarship*. Eds, David J. Bodenhamer, John Corrigan and Trevor M. Harris. Bloomington: Indiana University Press, 2010. 109–23.

11 From Mapping Text in Space to Experiencing Text in Place

Exploring Literary Virtual Geographies

Trevor M. Harris, H. Frank LaFone and Dan Bonenberger

Introduction

From the perspective of geographers, the linking of geography and literature is not new (Pocock; Noble and Dhussa; Brosseau; Caviedes; Lando; Sharp; Hones). Although not the first to explore the geography embedded in literature, H.C. Darby, in 1948, was amongst the first geographers to examine a canonical literary landscape – Thomas Hardy's Wessex – and to emphasise the role of locality and place in literature. Darby claimed his work was merely an intellectual exercise. Yet, today, it can be seen as foundational, since the geographical exploration of literature with its integrated triad of person, plot and place, is now an essential subfield of human geography (Thorpe; Pocock; Brosseau). Whereas the emergence of structuralism in the 1950s and 1960s emphasised the role of literature as a resource for the social sciences (Brosseau), these early forays into literary geography were limited or unidirectional both because of the perception by geographers that literary works were highly subjective and because of the lack of a nuanced appreciation of spatiality amongst literary scholars. Increasingly, however, both geographers and literary scholars have come to recognise the ways in which geography and literature inform one another and how geographical setting and spatial behaviour contextualises and shapes our understanding and interpretation of authors and their works (Alves and Queiroz; Brosseau; Lucchesi; Moretti; Saunders; Yuan).

This renewed attention to the examination of geographical space in literature has been partly due to developments in the geographic information systems (GIS) based analysis of digital text corpora and in text mining techniques (Gregory and Cooper). These initiatives to identify, extract and analyse the latent geographical elements embedded in texts have several evolving and interlocking themes including advances in the bulk digitisation of texts, the burgeoning corpus of digital-born text, the conceptualisation of the spatial humanities and the development of the geospatial semantic web. The enhancement of automated geo-location identification and geo-encoding through the assignment of geographic coordinates to unstructured texts through geoparsers and spatial gazetteers such as GeoNames and GNIS have facilitated rapid place name lookup and spatial coordinate generation. In the process of emphasising mapped space, however, these developments have broadly led to the earlier focus within literary geography on place and locale being relegated to a secondary role.

This chapter seeks to extend the emergent interest in the geographical analysis of literary works by focusing not on the GIS-enabled spatialisation and mapping of unstructured literary place names (important though that is), but on the ability of geographic information science (GISc) including geovisualisation to drill deeper into the experiential, emotional and local associations that connect text and place. To this end, we explore serious game engines, 3D modelling and virtual reality within the immersive virtual environment of the CAVE (Cave Automated Virtual Environment) to create immersive geographies that link the experiential, the emotional and the symbolic elements of literary works to the nuanced, dimensional richness of places as inspired by authors and their works. The clinical dismembering of literary works through concordances (albeit with their important spatial context), often disassociates spatio-textual analysis from the powerful and evocative prose and content of the original text. Going beyond the mere cataloguing of space and incorporating evocative descriptions that capture an author's appreciation and characterisation of a region or locale has long been indicative of the geographer's art (Hart; Pocock). Aspects of place become pivotal to literary geography in converting events, locale and setting into experiences. This chapter proposes complementary approaches to the GIS-based mapping of literary works to explore both the reductionist spatial mapping of texts and the affective and symbolic characteristics of literature by adding reflexive and multi-sensory geovisual mappings through virtual immersive geographies. The powerful narrative and autobiographical writings of Rebecca Harding Davis, focused as they are on an industrialising, nineteenth-century American frontier town, are used as a platform for experimentation and demonstration.

Text, Space and the Geo-literary Tourist

D.C. Douglas Pocock suggests that literature is both a source and a tool for geographical exploration in that it can be used for data seeking and for examining the geographies of the lives and works of authors (Pocock 96). The somewhat confusing terms used to describe the linking of spatial technologies and texts, and the lack of a standard nomenclature, is problematic. Many of the terms that might be used to denote the particular aspects of a GIS-based analysis of literary works, such as text mapping, word mapping, sentence mapping, story mapping or even mapping texts, are not characterised by a focus on the cartographic norm but have attained specific educational or other meanings where the term 'map' is used metaphorically in the sense of actively planning, drawing up or mapping out a scheme. In preference, the term spatial text is used here to denote a text that has an associated spatial description to which a coordinate might be assigned and where the focus is on the geographical analysis of that text. This process of textual spatialisation has become increasingly important as the subfield of literary geography has evolved from comparative literary descriptions between real places and reconstructed fictitious geographies (such as in Darby and Andrew Radford), to one where text and literature represent more than just a location in GIS but become sources of geographical knowledge. Despite the tendency of GIS technology to record and explore landscapes, past and present, solely in terms of the visible fabric of physical features, infrastructure or demographic aggregates, the spatial analysis of text seeks to develop insight and understanding 'in terms of behaviour, sensations, ideas, feelings, hopes, and faith' (Lando 10). Sheila Hones, for example, suggests that the multiple material existences that can be developed within written texts allows literary geography to destabilise existing geographies because of these multiple perspectives. In a reciprocal fashion, texts are not imagined, formed, created, written or even read in a spatial vacuum but are embedded in cultural, social and geographical contexts (Sharp).

It is interesting that in the past geographers have tended to take greater interest in fictional literature and to neglect equally evocative and informative non-fiction texts such as works of travel writing (Pocock; Brosseau). Although the distinction between factual sources and fictional sources, and between realism and the imaginary are often blurred (as our case study of Rebecca Harding Davis's nineteenth-century 'Wheeling' will illustrate), recent investigations have tended to correct this imbalance by opening up massive data repositories in the form of newspapers, travelogues and other archival materials and publications. Certainly, the degree of a writer's fidelity to geographical reality is a constant challenge to interpret, and perhaps may even be part of the appeal of literary geography. Literary geography might, at one level, be seen as an analysis of the literal rendition of landscape based on an author's knowledge or perception of a place. In contrast, Marc Brosseau suggests that readers often assign the subjective element within fiction to the physical elements of a place whilst assuming that the interpretation of the relationships between places and people is reliable: 'novelists are seen as some kind of spokesperson . . . . They immerse us in the various attitudes, values and conflicts shared by the people of a particular region in relation to their environment' (Brosseau 336). Darby, for example, suggests that Hardy manipulated some topographical details in his novels to facilitate his storyline and yet remained true to the underlying human-environment relations. Assessing the presumed geographical realism within texts raises questions about accuracy, verisimilitude, representativeness and reliability; and corroborative evidence of a place must bear witness to the accuracy or falsity of the writer (Brosseau). Yet, arguably, it is the insight that the author brings to human-environment relations rather than the accuracy of spatial description that is most revealing, and especially so in historical contexts where corroboration of these social-cultural-economic-environment relationships are so difficult to establish. Identifying and interpreting the symbolism portrayed in literary works may say more about the geography of a place than an unsubstantiated reliance on geographical accuracy (Porteous).

The emergence of literary geography at the interface between the two disciplines of geography and literary studies has brought about a shift in emphases over time as well as its own challenges. From the early focus on comparing fictitious and 'real' landscapes in fiction; to an increased interest in the analytic methods of literary criticism; to a focus on the text itself in terms of structure, composition, narrative modes, variety of languages and style: the coupling of literature and geography continues to evolve and influence our understanding of literary places. Yet the implicit role of space in a given literary text may require a geographical perspective that represents a mode of regard that differs from that of the literary critic (Bordessa; Brosseau; Hones). Whereas both literary critics and geographers are 'bent to the same task of understanding the world . . . and benefit from being informed by the progress of the other' (Hones 1304), geographers seek to extract meaning from texts for the sake of the geographical insights they can afford. This is an important distinction that becomes especially apparent in the spatial analysis of literary texts.

Our understanding of these geographical aspects of texts continues to evolve (Hones). The early focus on the geographical perceptions of authors and their descriptions of local places and regions, for example, has evolved into the re-imagining of space and place meanings. Critical literary geography and the evaluation of space and texts as culturally produced creations are fascinating ventures, yet they represent real challenges to a positivist GIS-based spatial exposition of texts. Interdisciplinary space-text engagements that embrace the work of spatial theorists, critical literary traditions and geospatial technologies are yet to emerge largely due to the duality of 'the textuality of space' and 'the spatiality of texts' (Stainer 103). For the geographer, literature allows insights into places and manifestations of place identity based on the writer being simultaneously a witness, interpreter and shaper of social worlds real and imagined (Stainer 103–4). Literary geography, however, to Andrew Thacker, is rooted in literary and cultural criticism and is a 'process of reading and interpreting literary texts by reference to geographical concepts' (60). The main point to make here is that any engagement between literary works and space constitutes multiple audiences and perspectives, from literary critics interested in spatial theory to geographers interested in the consumption of text and what it can reflect about places that are relational, unfinished and dynamic (Schweickart and Flynn 1). Text in this latter case is articulated in explicitly spatial terms and, as Massey would suggest, represents 'the coming together of the previously unrelated, a constellation of processes rather than a thing' (Massey 141). Spatialising text is central to these endeavours and opens up significant opportunities to include the rich content of textual media and meaning into spatial analysis but so too are engagements between geography and literature that transcend the spatial and engage social space and place.

The Writings of Rebecca Harding Davis and the Industrialising American Frontier Town

By way of illustration we propose to explore the writings of Rebecca Harding Davis (1831–1910), an early pioneer of literary realism in the United States. Born Rebecca Blaine Harding in Washington, Pennsylvania, she was the eldest of five children and her family relocated to Wheeling (then in Virginia and now in West Virginia) in 1836. She received her early schooling at home, and later attended a female seminary from which she graduated in 1848. According to her autobiography, *Bits of Gossip*, she was well read and knew 'enough math to do accounts, enough astronomy to point out constellations, a little music and drawing, and French, history, literature at discretion' (Davis 101). She worked for the local Wheeling *Intelligencer* newspaper in the 1850s writing reviews, poems, stories and editorials. Much of her early work focused on those people marginalised within industrial society, including immigrants, the working poor, blacks, women, Native Americans and the disabled. Davis witnessed dramatic changes as Wheeling grew from a small rural village to become a major industrial town. This had a profound influence on her writings, which remained largely unknown until Tillie Olsen's republication of her novella *Life in the Iron Mills*, which is now recognised as a landmark work of nineteenth-century American literature (Baym et al.).

Despite Davis's descriptions of the rustic Wheeling of her childhood, there are also indications in her writings of the town's emerging identity as an industrial centre and as a major gateway for westward migration. Due to its location on the Ohio River, the National Road and its later intersection with the Baltimore and Ohio Railroad, Wheeling became a transportation hub linking the eastern seaboard with the South and the interior West. Even in her childhood Davis observed stagecoaches from the East unloading passengers for the steamboats heading south on the Ohio River to St Louis and New Orleans. She also witnessed the passing of huge vans laden with merchandise for the plantations, or bales of cotton destined for the northern mills, and she saw Conestoga wagons bearing immigrant families to the West (Davis 3–4). These scenes, which were portrayed in contemporary landscape paintings (Figure 11.1), changed rapidly during Wheeling's industrialisation and they stand in stark contrast to the social-industrial scenes that Davis went on to depict in her writings:

> A cloudy day: do you know what that is in a town of iron-works? . . . . The air is thick, clammy with the breath of crowded human beings. It stifles me. I open the window, and, looking out, can scarcely see

Figure 11.1 Wheeling as depicted in 1850.

Source: © The Brown Photo Collection, The Ohio County Public Library, Wheeling, West Virginia, USA.

> through the rain the grocers shop opposite, where a crowd of drunken Irishmen are puffing Lynchburg tobacco in their pipes. I can detect the scent through all the foul smells ranging loose in the air. The idiosyncrasy of this town is smoke. It rolls sullenly in slow folds from the great chimneys of the iron foundries, and settles down in black, slimy pools on the muddy streets. Smoke on the wharves, smoke on the dinghy boats, on the yellow river, clinging in a coating of greasy soot to the house-front, the two faded poplars, the faces of the passers-by . . . Smoke everywhere! . . . I look on the slow stream of human life creeping past, night and morning, to the great mills. Masses of men, with dull, besotted faces bent to the ground, sharpened here and there by pain or cunning; skin and muscle and flesh begrimed with smoke and ashes; stooping all night over boiling caldrons of metal, laired by day in dens of drunkenness and infamy; breathing from infancy to death an air saturated with fog and grease and soot, vileness for soul and body. (Davis 11–12)

With this *mise en scène* Davis introduces the industrialising antebellum town of Wheeling and its people. Having provided an initial portrait of the town, moreover, she proceeds to invite her reader to explore the world of Wheeling through her writing, anticipating, as it were, the interest that historians and literary geographers alike take in her work today:

> Stop a moment. I am going to be honest. This is what I want you to do. I want you to hide your disgust, take no heed to your clean clothes, and come right down with me, – here, into the thickest of the fog and mud and foul effluvia. I want you to hear this story. There is a secret down here, in this nightmare fog, that has lain dumb for centuries: I want to make it a real thing for you. (Davis 13)

In order to make it a 'real thing' for the reader Davis writes in powerful evocative prose to create a literary experience of travelling through mid-nineteenth century Wheeling. If spatial text analysis is the reconstitution and articulation of experiences that are latent in texts, then Davis's call should resonate well with literary geographers for the events in Wheeling have a geographical context that grounds and shapes those experiences and informs our own spatial stories. There is also good reason why Davis is regarded amongst the earliest socially conscious realist writers to explore industrialisation in America. Her prose, through its frank descriptions, exposes the reader to the consequences of the pollution and oppression that characterised industrial centres such as Wheeling (Baym et al.).

Davis's writings contain references to locations and landmarks in and around Wheeling (including the Ohio River and specific streets and buildings in the town), and she describes them and explains their functions.

Looking at a mid-nineteenth-century map of Wheeling, however, only takes the reader so far in understanding the empirical aspects of the town as narrated by Davis. The geography one can draw from Davis's writings is more than a mere map of locations or of connections to locations elsewhere. It is also more than one can adequately convey through the use of GIS. Davis's writing is unquestionably about the direct, human experience of place, and this is something that exceeds the ability of GIS to capture and to convey, in spite of its powerful mapping and spatial analytic powers. Davis's world is evocative, multi-dimensional and sensuous. Her writings are awash with qualitative references to smell, texture, light, touch and sound, all of which stem from the creative investment in place-making to which she gestures in her aforementioned address to the reader: 'I want to make this real for you'. Analysing the experiential world of Davis's Wheeling through quantitative GIS-based spatial analysis would, therefore, seem to be insufficient. Yet, as we demonstrate below, augmenting the reading of Davis's writings through geovisualisation and immersive, 3D GIS mapping can help us to understand better the literary world woven together in her works.

Immersive Geography, 3D Worlds and Sensuous Cues

Translating conceptions of place from rich literary descriptions into GIS is challenging. A phenomenological approach to exploring, interpreting and experiencing text-based place in GIS requires new forms for understanding and representing the unique characteristics of place: it requires us to explore the familiar using innovative technologies. And so we step beyond the 2D mapped world of GIS into the immersive world of serious game engines, virtual reality, 3D GIS and the CAVE to open up interpretative possibilities not readily available to previous generations of literary geographers. This extension into virtual environments is premised on the goal of extracting more from the geography contained within text than the spatial associations embedded within it. To this end, we seek not only to generate visual imagery that portrays an interpretation of an author's writings about place, but also to do so within an immersive world where users can experience the imaged text.

One defining characteristic of place is the notion of meaning gained through experiencing space and this involves moving beyond the locational, cartographic vector and raster representations of GIS. In contrast to the traditional 2D world of maps, the CAVE enables users to become immersed in a virtual rendered world with extensive interactive and navigational capabilities. The CAVE is a specially designed structure comprising (in this study) three walls and a floor made up of 10 foot screens (Figure 11.2). Onto these walls stereo images are projected from the rear, and from above for the floor, and are computationally made into one seamless image projected across the CAVE walls (Cruz-Neira et al.; Buchroithner).

This sense of immersion is attained through stereo depth perception, peripheral vision capture, head tracking devices and interactive hand-held controllers. The effect is to make images appear holographic and for the user to manoeuvre through and around the image as if it were a real environment. A user standing in the CAVE wears 3D glasses and experiences a feeling of immersion within a lifelike sensory environment.

Immersion is a powerful psychophysical experience that situates the user as being present in a seemingly real but virtually rendered environment and seemingly disengaged from the reality of the four projected walls (Harris and Baker). This immersive experience is powerful and draws on the creative power of the mind to move seamlessly between the physical, virtual and imaginary worlds whereby the user experiences a sense of 'being there'. Users can interact virtually with the displayed scene in real time and perform fly-through navigation. The effect is a powerful one in transporting the user into a multidimensional and multisensory environment embedded within GIS and yet capable of supporting investigation of place as well as space. The visual-cognitive model within immersive environments provides a highly dynamic and interactive display environment and transforms the user from a passive observer to an active participant capable of accessing and experiencing complex information through the display medium (Harris and Hodza; Harris, Bergeron and Rouse).

Through a fusion of recent advances in immersive technologies, computer graphics and computing power, as well as automated 3D rendering software, serious gaming engines and geospatial technologies, it is possible to display complex 3D images in CAVEs. Together, these technologies and software enable impressive virtual representations to be created that render images in a manner that moves far beyond static 2D maps to dynamic, multidimensional and highly interactive representations that more closely emulate how we view the world around us (Fisher and Unwin). The CAVE environment enables phenomenological approaches to be linked to geospatial technologies generating experiential aspects of place and a visual immersive sense of being in the world portrayed in the text.

Immersive environments offer differing levels of immersion and experiences and have been used in the 'edutainment' industry for some time in the form of sophisticated video games. Everyday users can experience landscapes and virtual worlds through the medium of powerful graphics. Indeed, the graphical quality and power of these systems has often far exceeded what is available to many university scholars within classroom and even research settings. Serious gaming engines utilise concepts, technology and software drawn from the video gaming industry to develop applications for a range of training, simulation and education applications (Lewis and Jacobson; Dovey and Kennedy). The term 'immersive learning simulation' has been proposed as a way of distinguishing these 'serious' applications from 'game-based' virtual technologies. The additional speed and response time, the high quality graphics, the ability to perform collision detection,

Figure 11.2 The WVU immersive four-wall CAVE.

Source: © The Authors.

and the availability of particle physics that provide dynamic movement to objects in a scene such as wind blowing through the leaves on a tree or water flowing, often elevate serious gaming engines as superior 3D visualisation systems over geospatial engines such as City Engine. The corollary, however, is that the analytical functionality of GIS is not available in serious gaming engines. Using 3D modelling software such as Trimble SketchUp and ESRI City Engine, the CAVE is able to elevate the immersive geospatial experience of users to a different level. SketchUp provides an excellent platform for generating 3D buildings for incorporation into a virtual scene and supports an extensive repository of 3D buildings that are available for free from its 3D Warehouse. The use of GIS is also advancing into the 3D world and has begun to develop links to serious game engines. The recent release of City Engine provides a 3D modelling, rendering and visualisation platform that integrates powerful geoprocessing functionality with complex scene-building capabilities. A further advantage of City Engine is that it draws heavily on procedural rules that through Python-like scripting enables the mass generation and rendering of building structures within a virtual scene to be automated. Although detailed custom building models can be generated individually this can also be both time consuming and tedious. City Engine also provides a powerful platform that connects to ESRI's other GIS products to produce 3D GIS platforms that are ideal for display in a CAVE and for 3D geoprocessing. In addition to LumenRT, which adds scene generation and lighting conditions to 3D GIS scenes, this software provides a suite of programmes capable of supporting many of the virtual scene generators discussed in this chapter.

Rebecca Harding Davis's Virtual Wheeling

To situate Davis's house within its surrounding Wheeling neighbourhood, 3D models were created using ESRI City Engine (Plate 8). The building footprints were obtained from the 2D paper Sanborn fire insurance maps of 1884, which provide a valuable insight into the geography of Wheeling at that time. Significantly, the map also provides information not only about the location of buildings around Wheeling, but also about their dimensions and the materials from which they were constructed. Historical photographs and surviving buildings were used to generate architectural styles and building appearance typical of the era. The intent here is not to be photorealistic but to capture a scene representative of the geography of Davis's neighbourhood. Custom building 3D models using software such as Trimble SketchUp can generate very accurate architectural reproductions, but it can also be time consuming. In this representation (Plate 8) the buildings were generated using the rule-based City Engine that extrudes buildings from their footprints to determined heights, allocates building facades based on a number of prepared render images, reproduces several styles of roofs and roof textures and adds windows and doors according to the orientation of the building. The terrain and surface imagery is ingested into the scene from a GIS spatial database. Attempts to use the Sanborn map as a surface cover for the scene or contemporary high resolution aerial imagery were unsuccessful because of the paucity of land cover in the Sanborn maps and the significant changes that had occurred to the urban morphology of Wheeling since the mid-nineteenth century. City Engine is capable of generating high quality urban streetscapes in a relatively short time and being tightly coupled to ESRI's ArcGIS, ArcScene and 3D Analyst, it provides 3D geoprocessing capabilities such as identifying line-of-sight, viewshed generation, shadow estimation and light capture.

Useful as this image and the software platform are for moving beyond 2D mapping, it still lacks the sense of place engendered by Davis's writings. To that end Unity3D was used to model the neighbourhood surrounding Davis's residence and from which she wrote so evocatively about the area (Figure 11.3 and Plate 9). Unity3D is a game development platform with powerful interactive 3D graphical rendering, animation and illumination capabilities. A number of similar systems, such as Unreal4 and Crytek are also available, and many of them are licensed gratis to educational users. In addition, a number of powerful 3D modelling and animation applications such as 3ds Max, Maya and Blender are also available to support scene and object development. The street scenes generated within Unity are built on the GIS spatial database used within City Engine and provide powerful rendering, animation and atmospheric effects to replicate the smoke laden, dark and somewhat forbidding street scenes described by Davis. The intent is not to seek photorealistic images but to provide a sense of place based on the reading of Davis's texts. Subjectivity is clearly evident here and an issue when developing virtual environments that requires evidential substantiation or justification, in much the same way as when creating a map.

In the images reproduced here (Figure 11.3 and Plate 9) the scenes appear static because of the format of this collection, whereas in reality there is a high degree of interaction and navigation exercised with the scene. To enhance the sense of 'being there' in the scene, the virtual world of Wheeling is ingested into the stereo-enabled geovirtual CAVE. In this immersive environment the user has a profound sense of being in the virtual world of Davis's Wheeling as if it were real. The scenes become three dimensional, holographic and seemingly real enough to reach out and touch the buildings or objects. Such an experience is difficult to describe and yet with the full navigation provided, the user becomes not just an observer of a scene but a part of the virtual world displayed. The experiential sensation of immersion in the virtual world is profound. Furthermore, not only does the virtual environment provide a visually stunning and immersive insight into Wheeling as a place and as described in Davis's writings, but there are also further ways in which to engage the senses to reinforce immersion and the sense of place.

Davis describes in her many autobiographical writings the areas and streets within her neighbourhood where she felt particularly uncomfortable

Figure 11.3 A street scene described by Davis and generated in Unity3D.

Source: © The Authors.

and somewhat fearful to enter because of concern for her personal safety. Based upon these writings we have constructed a map using the Sanborn map of Wheeling at the time of her writing to represent an interpretation of her geography of fear (Figure 11.4). Although these spaces could be displayed in the form of a map overlaying the virtual scene, instead the geography and emotion of contested place is portrayed through the use of ambient sound through the CAVE's 7.1 sound system. Here, the GIS database is used to track the user's movement through the virtual world, employing the geography of Davis's map of fear to inform the detection and collision parameters that buffer Davis's geography. As a user navigates through the virtual world scene, the sound of a beating heart is heard which maintains a steady beat until the user enters into an area defined by Davis as being of personal concern or somewhat dangerous. At this point the sound of the beating heart increases in pace as it emulates Davis's reality and replicates her emotional state. Similarly, as the user leaves these areas and enters places where Davis felt more at ease so the heart beat returns to a normal pulse rate. In this way not only can we draw upon the visual sense to reinforce a sense of immersion in Wheeling as a place but we can also draw upon the sense of sound and smell to reinforce these feelings. Davis describes a street in her neighbourhood where a tannery and a brewery were in close proximity and the resulting mix of smells permeated the area. Smell has a powerful effect in triggering and evoking memories and the sense of places. Attempts to reproduce smell in the CAVE using smell diffusers such as ScentAir have been relatively successful and further research may provide an interesting component of smellscapes to augment soundscapes and visualscapes. This ability and desire to draw upon other senses to augment the sense of immersion through holophonic sound and smell moves GIS into sensuous areas where phenomenological and experiential frameworks combine with the positivist world of GIS.

Other ways of embedding spatial narrative and story-telling into the virtual literary environment are being explored such as interaction with avatars that provide guidance and information about a place and which can be interrogated in a variety of ways, including conversational interaction. A user, in this case, is provided with structure in the telling of a spatial story through the use of virtual signposts and pathways. All of these interactive forms potentially enable the 'reader' of the virtual literary world to enrich his or her understanding and appreciation of the geography embedded in the original literary text.

Conclusion

In this chapter we have sought to go beyond the traditional text-map metaphor and to propose ways in which both space and place as embedded in literary works might be explored. Recreating the landscapes and structures

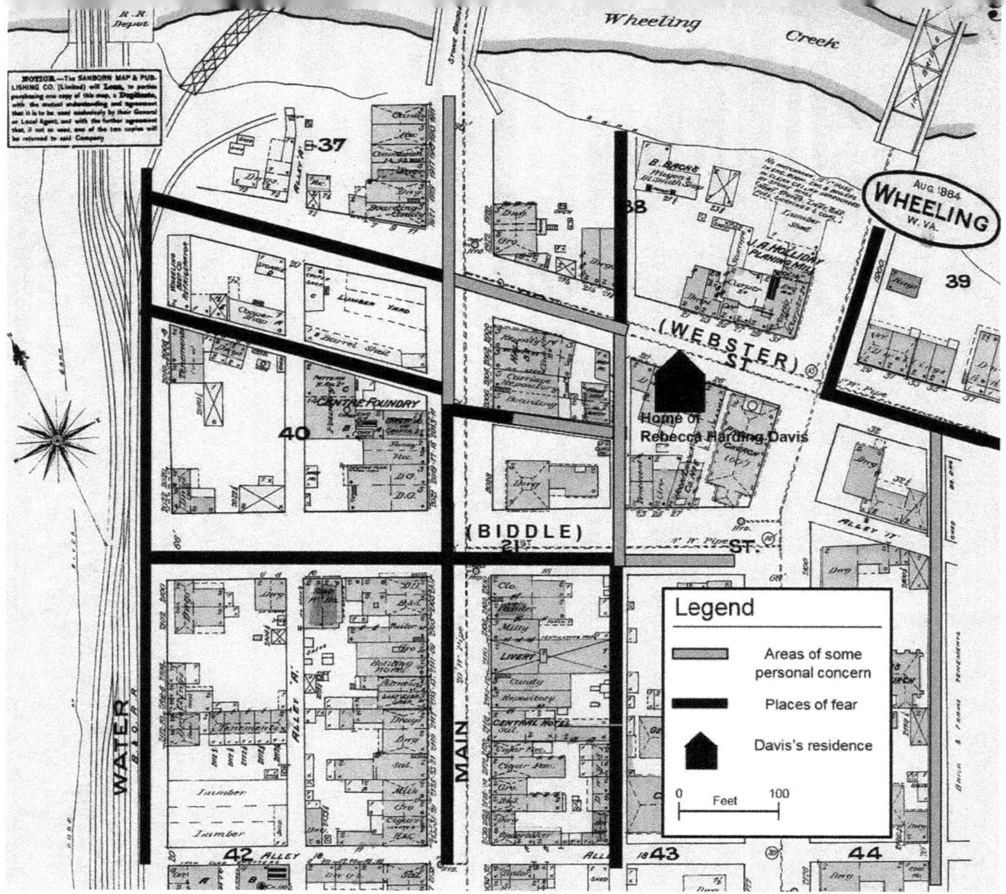

Figure 11.4 Davis's geography of fear in mid-nineteenth century Wheeling.

Source: © The Authors.

of past cultures through the spatial reading of literary texts is fraught with challenges and difficulties. Oftentimes the physical attributes of landscapes may have changed dramatically, or a text might describe places that have changed or been altered by subsequent physical, cultural, economic and social developments. However, the realities and accumulated experiences and symbolism of the inhabited landscape in historical writings are retained in plot, narrative and character interactions.

The pursuit of verisimilitude and hyperrealism should not be a goal for virtual immersive text analysis for this often leads to rather barren territory because the models will always be 'seen to stand in an inferior position to an original referent' (Gillings 20). Immersive geographies reside in the classic 'uncanny valley' where the perceived distinction between reality and representation ironically becomes more noticeable the closer one gets to representing reality (Mori et al.; Tinwell, Grimshaw and Williams). This is a particular issue in the field of robotics in seeking to make automatons resemble human likeness though its applicability to the representation of places is equally valid (Burleigh et al.; Zysk and Filkov). At some level the human mind is capable of filling-in the spaces of an obviously non-real representation and yet becomes uncomfortable and hypercritical the closer a building depiction or human portrayal comes to emulating the real. It is for this reason that human avatars are used with caution in this demonstration though the use of off-screen cues such as heckling or verbal abuse can greatly contribute to the inherent sense of a place. Although the virtual scenes stand as subjective renderings of a bygone place this should not detract from their value in literary geography. The seeming acceptance of traditional cartographic products as being objective and accurate representations of a mapped reality, whereas virtual representations and geovisualisation are to be seen as highly subjective, is disingenuous, as critical cartography has clearly demonstrated (Crampton; Crampton and Krygier).

This study in virtual literary geographies suggests how virtual immersive technologies provide a means for translating literary works into an innovative, multifaceted, digital environment. Using immersive geographies to map space and place creates new opportunities not only for extracting the spatial settings associated with literary works, but also for understanding the nuances of these locations and places. Drawing upon soundscapes and smellscapes opens up additional opportunities to pursue a sensuous and experiential GIS for literary geography. Whereas text mining runs the risk of 'casual ransacking', or extracting excerpts without the ability to interpret or reflect on the deeper meanings of literary works (Brosseau; Cooper and Gregory; Gregory and Cooper), this immersive approach demands reflection on the discursive dimensions of the text. Indeed, rather than exacerbate the quantitative-qualitative divide, immersive mapping draws upon both to gain a deeper understanding of place as embedded within literary works. The approaches outlined here enable the representations of place embodied in texts to extend beyond the physical and locational referents of space and allows us to peer into the symbolic, the emotional and the sensuous aspects of literary places. The extent to which insight is gained from such an approach to fictional writings remains open to discussion. Certainly, the immersive geographies of Davis's writings detailed here indicate that literary texts can articulate life-worlds and places by painting an emotional and evocative portrait of life in an industrialising town. It is suggested that such a representation will resonate more strongly with readers than a conventional mapping of space. In the absence of many other oral or observed writings, Davis's work and its immersive representation thus becomes a primary source and medium for exploring and understanding Wheeling in the mid-nineteenth century. The geosensory inputs of vision, sound and smell act to enhance the personal experience of Davis's Wheeling and contribute to that sense of place that she describes so evocatively in her writings. For many reasons, then, this chapter proposes that in addition to the identification and mapping of spatial associations embedded within digital texts, drawing on the thick description of texts through virtual immersive geographies may provide a complementary portrayal of place mediated by our own experiences of literature. Ultimately, our engagement with literature may be reinforced when the scholar also becomes an artist, and immersive studies such as this provide one small step toward that goal (Pocock 96).

Works Cited

Alves, Daniel, and Ana Isabel Queiroz. 'Studying Urban Space and Literary Representations using GIS Lisbon, Portugal, 1852–2009'. *Social Science History* 37.4 (2013): 457–81.

Baym, Nina, et al. *The Norton Anthology of American Literature.* 8th ed. 3 vols. New York: W.W. Norton & Co., 2011.

Bordessa, Ronald. 'The City in Canadian Literature-Realist and Symbolic Interpretations'. *Canadian Geographer-Geographe Canadien* 32.3 (1988): 272–6.

Brosseau, Marc. 'Geography's Literature'. *Progress in Human Geography* 18.3 (1994): 333–53.

Buchroithner, Manfred, (ed.) *True-3D in Cartography: Autostereoscopic and Solid Visualisation of Geodata.* New York: Springer, 2012.

Burleigh, Tyler J., Jordan R. Schoenherr and Guy L. Lacroix. 'Does the Uncanny Valley Exist? An Empirical Test of the Relationship between Eeriness and the Human Likeness of Digitally Created Faces'. *Computers in Human Behavior* 29.3 (2013): 759–71.

Caviedes, César N., (ed.) 'Geography and Literature'. *GeoJournal* 38.1 (1996): 1–2.

Cooper, David, and Ian N. Gregory. 'Mapping the English Lake District: A Literary GIS'. *Transactions of the Institute of British Geographers* 36.1 (2011): 89–108.

Crampton, J.W. 'Maps as Social Constructions: Power, Communication and Visualization'. *Progress in Human Geography* 25.2 (2001): 235–52.
Crampton, Jeremy W. and John Krygier. 'An Introduction to Critical Cartography'. *ACME: an International E-journal for Critical Geographies* 4.1 (2006). 11–33.
Cruz-Neira, Carolina, Daniel J. Sandin, Thomas A. DeFanti, Robert V. Kenyon and John C. Hart. 'The CAVE: Audio Visual Experience Automatic Virtual Environment'. *Communications of the ACM* 35.6 (1992): 64–72.
Darby, H.C. 'The Regional Geography of Thomas Hardy's Wessex'. *Geographical Review* 38.3 (1948): 426–43.
Davis, Rebecca Harding. *Bits of Gossip.* New York and Boston: Houghton, Mifflin & Co. 1904.
Dovey, Jon, and Helen W. Kennedy. *Games Cultures: Computer Games as New Media.* Maidenhead: Open University Press, 2005.
Fisher, Peter and David Unwin. *Virtual Reality in Geography.* London: Taylor and Francis, 2002.
Gillings, Mark. 'Virtual Archaeologies and the Hyper-real: Or, What Does it Mean to Describe Something as *Virtually*-real?' *Virtual Reality in Geography.* Eds, Peter Fisher and David Unwin. New York: Taylor & Francis, 2002. 17–34.
Gregory, Ian N., and David Cooper. 'Thomas Gray, Samuel Taylor Coleridge and Geographical Information Systems: A Literary GIS of Two Lake District Tours'. *International Journal of Humanities and Arts Computing* 3.1–2 (2009): 61–84.
Harris, Trevor M. and Vick Baker. 'Immersive Visualization System Promotes Sense of "Being There"'. *ArcNews* Winter 2006/2007.
Harris, Trevor M., and Paddington Hodza. 'Geocollaborative Soil Boundary Mapping in an Experiential GIS Environment'. *Cartography and Geographic Information Science* 38.1 (2011): 20–35.
Harris, Trevor M., Susan Bergeron and L. Jesse Rouse. 'Humanities GIS: Place, Spatial Storytelling, and Immersive Visualization in the Humanities'. *GeoHumanities: Art, History, Text at the Edge of Place.* Eds, Michael Dear, Jim Ketchum, Sarah Luria and Douglas Richardson. London: Routledge, 2011. 226–40.
Hart, John Fraser. 'The Highest Form of the Geographer's Art'. *Annals of the Association of American Geographers* 72.1 (1982): 1–29.
Hones, Sheila. 'Text as it Happens: Literary Geography'. *Geography Compass* 2.5 (2008): 1301–17.
Lando, Fabio. 'Fact and Fiction: Geography and Literature'. *GeoJournal* 38.1 (1996): 3–18.
Lewis, Michael and Jeffrey Jacobson. 'Game Engines'. *Communications of the ACM* 45.1 (2002): 27–31.
Lucchesi, Flavio. 'Geographic Memories in Travelogue Literature: The Australian Social Landscape in the Writings of Italian Travelers'. *GeoJournal* 38.1 (1996): 129–36.
Massey, Doreen. *For Space.* Sage, 2005. Print.
Moretti, Franco. *Atlas of the European Novel, 1800–1900.* London: Verso, 1999.
Mori, Masahiro, Karl F. MacDorman and Norri Kageki. 'The Uncanny Valley [from the Field]'. *Robotics & Automation Magazine, IEEE* 19.2 (2012): 98–100.
Noble, Allen G., and Ramesh Dhussa. 'Image and Substance: A Review of Literary Geography'. *Journal of Cultural Geography* 10.2 (1990): 49–65.
Olsen, Tillie. 'A Biographical Interpretation'. *Rebecca Harding Davis: Life in the Iron Mills and Other Stories.* New York: The Feminist Press, 1972. 67–174.

Pocock, Douglas D.C. 'Geography and Literature'. *Progress in Human Geography* 12.1 (1988): 87–102.

Porteous, J. Douglas. *Landscapes of the Mind: Worlds of Sense and Metaphor.* 1943. Toronto: University of Toronto Press, 1990.

Radford, Andrew. *Mapping the Wessex Novel: Landscape, History and the Parochial in British Literature, 1870–1940.* London: Continuum, 2010.

Saunders, Angharad. 'Literary Geography: Reforging the Connections'. *Progress in Human Geography* 34.4 (2010): 436–52.

Schweickart, Patrocinio P. and Elizabeth A. Flynn. *Reading Sites: Social Difference and Reader Response.* Modern Language Association of America, 2004. Print.

Sharp, Joanne P. 'Locating Imaginary Homelands: Literature, Geography, and Salman Rushdie'. *GeoJournal* 38.1 (1996): 119–27. Print.

Stainer, Jonathan. 'Localism, Signification, Imagination: De-Stabilizing Sectarian Identities in Two Fictionalized Accounts of "Troubles" Belfast'. *Social & Cultural Geography* 7.1 (2006): 103–26.

Thacker, Andrew. 'The Idea of a Critical Literary Geography'. *New Formations.* 57 (2005/6): 56–73.

Thorpe, James Ernest. *Relations of Literary Study: Essays on Interdisciplinary Contributions.* New York: Modern Language Association of America, 1967.

Tinwell, Angela, Mark Grimshaw and Andrew Williams. 'The Uncanny Wall'. *International Journal of Arts and Technology* 4.3 (2011): 326–41.

Yuan, May. 'Mapping Text'. *The Spatial Humanities: GIS and the Future of Humanities Scholarship.* Eds, David J. Bodenhamer, John Corrigan and Trevor M. Harris. Bloomington: Indiana University Press, 2010. 109–23.

Zysk, Wolfgang, and Rumen Filkov. 'Self-paced Learning with 3D Humanoid Virtual Tutors'. *European Scientific Journal* 4 (2013): 651–7.

12 Spatial Frames of Reference for Literature Through Geospatial Technologies

Gary Priestnall

Introduction

This chapter focuses on the use of 3D representations of landscape, where digital terrain data is used as a backdrop to place cultural heritage information including passages from works of literature, diaries and sketches, into a broader landscape context. Three case studies, which offer examples of different presentation techniques, one based on a computer monitor, one on a physical landscape model and one on a mobile device, are described. The studies engage with the same geographic area, upland Cumbria in North West England, where surface relief is a dominant feature of the landscape. Each case study, however, uses different technologies appropriate to the contexts of display and a set of stated aims. The chapter will summarise the affordances of each technique, present observations from their use and discuss some of the opportunities and design challenges of utilising geospatial technology and digital landscape models for communicating spatial context.

Through offering a backdrop to provide spatial context the techniques also enable alternative forms of georeferencing beyond the 'pin in the map' approach, which is perhaps the most common form of geospatial representation. Using single coordinate points or strings of points forming a route can mask the rich attribution of the object or experience which is being represented. The use of terrain-based frames of reference can communicate a landscape setting rather than a specific point location, and in so doing can promote richer reflections on the content, leaving more scope for alternative interpretations of literary references that are not tied to point-specific locations.

There are various technologies that enable digital landscape models to be used as a base for exploratory data analysis and geographic visualisation, from desktop geographic information systems (GIS) to fully immersive Virtual Reality. The focus for this chapter is on three technological approaches suitable for relatively easy deployment for informal learning and visitor experiences. They illustrate a range of techniques for visualising the relationship between landscape and literary texts using 3D models rather than 2D maps. The first utilises Google Earth, a virtual globe which uses combinations of aerial and on-the-ground perspectives and allows control over the way content can be displayed and referenced. The second employs a physical landscape model that is capable not only of giving aerial overviews, but also displaying imagery and maps, projected down onto it to augment the tactile relief representation. The third considers how we might augment a user's view of the actual landscape in the field by designing representations suitable for delivery through a mobile device.

Digital Approaches to Spatial Representation

The growing awareness of the benefits of using digital spatial representations and associated tools for analysis and presentation has been described by Goodchild and Janelle as integral to the 'spatial turn' that has taken place in the humanities and social sciences over the past twenty years (4–5). Similarly, Karl Offen has observed a 'cartographic turn' within literary studies, citing Franco Moretti's *Atlas of the European Novel, 1800–1900* as influential in promoting the use of maps as tools to reveal patterns and relationships (570). Furthermore, Santa Arias has described the benefits of cross-disciplinary explorations of spatial representations and processes and how the map is so much more than a 'backdrop' on which investigations are presented (40). The interest in computer-based mapping has formed an important part of the growing Digital Humanities field and the value of utilising tools for spatial analysis as well as digital maps has seen GIS be widely adopted. Notably, David Cooper and Ian Gregory have presented a literary GIS of the English Lake District which reconstructs journeys made by poets from place references and also derives maps of their emotional responses to the landscape. Technical challenges in this area lie not only in the geospatial methods for representing uncertainty in position (Guo, Liu and Wieczorek 1068–70), but also in the text processing methodologies which attempt to extract richer structures and meaning from texts and to associate these automatically with some kind of spatial footprint (Scheider and Purves 1–4). Defining the geographic footprint of literary references can be accomplished by exploiting existing gazetteers of town or street names, but doing this can be challenging when descriptions of natural features are involved. Curdin Derungs and Ross Purves, for instance, not only describe the merits of studying a large corpus of texts relating to Swiss alpine environments for revealing patterns within and between regions, but also acknowledge the need to extract greater knowledge about the context in which references are used within the text. For many of these studies the temporal sequences of events, including place name references, are important in forming the narratives seen as key to developing a richer notion of place (Tuan 691–2). David Cooper, for one, has described the complexity and richness of the relationship between literature and place by considering how the identity of the English Lake District is propagated through the texts of multiple authors (812–18).

The desire to explore spatial patterns within literary references and the large volumes of data that are often involved means that exploring more fuzzy or indeterminate spatial footprints for such references is a secondary concern, the focus being the clear presentation of such patterns over some kind of base map. Sebastian Caquard observes how the production of web-based maps has made this process more accessible although the style of base maps is often limited to standard feature sets. Base maps often serve to offer support navigation and to place facilities and services into their spatial context, and as such are often dominated by standard 2D cartographic symbols for roads, settlements and key natural features. Often the topographic characteristics of an area play an important part in the overall frame of reference by which people understand a landscape, and here the challenge is to portray this in an effective and understandable way whilst also displaying the other map content. The history of the development of techniques for portraying relief on maps, including contours, hatching and hill-shading, is reviewed by Peter Collier, David Forrest and Alastair Pearson. The flexibility offered by computer-based map layers and digital representations of the terrain surface itself has more recently facilitated the development of a wide variety of visualisation techniques that attempt to represent relief in more intuitive ways. Jonathan Raper, for instance, summarised the broader potential of multi-dimensional approaches to exploratory data analysis and presentation, emphasising the importance of being able to accommodate the requirements of temporal sequences of data (85–202).

Presenting relief in the form of perspective views, sometimes referred to as '3D maps' requires the development of a set of cartographic design guidelines especially given the opening up of the mapping process to non-cartographers (Häberling et al. 175–88). Much of this work, to date, has concentrated on the design of symbols, and related work on interactive 3D environments has considered design guidelines for landmarks within such visual scenes to aid navigation and orientation (Vinson 278–84). Research within experimental psychology and human-computer interaction continues to explore the effectiveness of landmarks in a range of contexts. For example Daniele Nardi, Nora Newcombe and Thomas Shipley consider the subtle interaction of slope and landmarks in helping people orientate themselves in a real physical space, and Caroline Snowdon and Christian Kray study the nature of landmarks that may be relevant to a user of mobile applications in natural environments. Acknowledging the varied nature of landmarks in many contexts of use, and also how they are represented symbolically, is becoming increasingly relevant to how we design geographic representations that aim to offer effective spatial frames of reference. Recently, geospatial technologies have been characterised by more open approaches to software development which encourage users to mix or 'mash up' digital maps with other forms of media, and to make alternative representations and visualisations possible, which can both stimulate and benefit from research in geographical information science (GISc) (Elwood 1–6).

The variety in the way spatial references are presented in literary texts suggests that there is much to gain from exploring alternative approaches to digital spatial representation in addition to those offered by desktop GIS. There are many types of technology that could be used for conveying spatial context to users through combinations of maps, 3D models and mixed media. These differ in their general capabilities, their level of interactivity and visual immersion and the contexts in which they are designed to be used, including desktop computers, screens for public display and personal mobile devices. Increased interactivity and mobility can bring new design challenges to maintain control over the mechanisms which normally help orientation and navigation such as symbol placement and the relationship between the user's view of the world and the perspective of the virtual representation. Daniel Montello describes some of these challenges in relation to 'you are here'-style maps which can be difficult to use if misaligned with the real environment they intend to represent. Where the environment is characterised by features with a significant vertical dimension various 3D representations can provide a more natural perspective for the viewer. There are, moreover, many technologies to ponder before arriving at something that may be considered fit for purpose (Shepherd).

Geospatial Technologies and Contexts of Use

There is a continuum of geospatial technologies from free and open software running on regular desktop computers through to costly specialist software and hardware, an example of the latter being fully immersive virtual environments. The nature of the intended audience and the likely context of use will also vary greatly. The focus in this chapter is on the application of technologies which involve modest implementation costs and where the intended audiences for the applications include the general public and student groups. Exploring the way geographic representations work in the context of everyday situations and non-expert audiences formed part of the proposed manifesto for map studies presented by Martin Dodge, Chris Perkins and Rob Kitchin. The three examples outlined in the following sections illustrate different forms of presentation that involve a desktop computer, a fixed situated display and a GPS-enabled mobile device. The case studies represent examples of how the same digital landscape data might be used in different ways to create aids to promoting spatial context designed around the various affordances of the different technologies.

The three case studies are:

- **Virtual Globes** – Google Earth was used to place a number of key locations from the writer and artist Edward Lear's travel diaries from 1836 into a broader landscape context, running as a virtual tour within a gallery space. It was also used to reconstruct some of the more arduous walking routes taken by Samuel Taylor Coleridge. These studies explored the use of a 3D digital landscape model and therefore involved some of the design considerations relevant to more immersive virtual environments.

- **Physical Landscape Models** – A physical relief model milled using digital terrain data where maps and images are projected down onto the model as part of an interactive display. This was designed to accompany an exhibition of documents relating to William Wordsworth, in particular several influential moments in Wordsworth's childhood which had strong connections with parts of the landscape.
- **Location-aware Mobile Computing** – Location-aware mobile phone-based guides designed to allow visitors to learn about various archival content whilst immersed in the landscape itself. Here the particular challenge was to support a direct engagement with features in the real landscape in which the user was situated.

Virtual Globes

Although other virtual globes were already in existence, the launch of Google Earth in 2005 had the most profound impact on the public's consciousness and helped to promote new explorations and sharing of spatial information. Within academia the exploitation of such technologies was recognised as one of the grand challenges facing GISc (Grossner, Goodchild and Clarke 145). Much of the potential of such technologies can be seen to lie in their ability to offer a platform for users to share geo-referenced information. In addition to this the relatively simple interface and highly interactive digital landscape model opened up possibilities for presenting information in a broader landscape context. Points, routes, areas, image drapes and 3D objects can all be georeferenced as 'places' on the landscape model and can also be organised into selectable groups in much the same way as layers in a GIS. In terms of promoting the spatial context of these places the inbuilt navigation tools are of interest: for example, the ability to invoke smooth flights between places by clicking on their entry on the legend. The *properties* of places include the style of both an icon and text label, a description field, but also the *view*. The view property stores the position, orientation and tilt of the camera view looking towards the object of interest. This can be easily stored as a *snapshot* of whatever the current viewing position and angle is, and so can be defined interactively to represent whatever broader landscape scene is deemed appropriate to convey spatial context. Such views need not focus on single objects of interest but can represent key vantage points onto any combination of other places, or just a view onto a landscape scene. Given that individual places can be linked together into a tour this means that any combination of places or vantage points can be sequenced together. Although much of the power of Google Earth lies in its interactive interface, there are circumstances where some control over what is presented may be required and therefore the creation of a *virtual tour* can be useful. Here the creator of the tour becomes the guide and, because the user will not be in control of orientating their view, we need to strive to support their sense of spatial orientation if they are to understand the spatial relationships being described.

An example is a virtual tour designed to run in the Wordsworth Trust Museum and Gallery in Grasmere, Cumbria, during the summer of 2009 (Priestnall and Cowton 148–51). The travel diaries and drawings of the writer and painter Edward Lear, who toured the Lake District in 1836, were the focus of an exhibition and there was a desire from the curator to place these physical exhibits into their landscape context. This also included a wish to raise awareness of how Lear's journey related to roads or places the visitors may well go on to see later that day, or that they knew from previous experience of travelling in the area. The travel diaries informed the placement of some of the drawings around the landscape model which were displayed within Google Earth as virtual billboards (Figure 12.1, left), as there was a requirement to demonstrate Lear's faithful representation of landform. The capability to drape images over the landscape model was used on one occasion where a vantage point used by Lear appeared to be in the middle of the Thirlmere reservoir which had been completed in 1894. By referencing a map from the 1860s over the landscape model, the location of the old road was revealed and the likely location where Lear stood to draw the landscape could be identified.

In order to attempt to convey how far apart the locations were, and how they related to features that may be familiar such as mountains, roads and the Wordsworth Trust itself, a decision was made to create an aerial flythrough visiting seven drawing locations. An important design element was the use of intermediary elevated viewpoints which aimed to re-align the viewer with the broader landscape and familiar landmarks before focussing in on the next drawing at ground level. Labelled pins marked key towns, mountains and lakes as well as major roads and viewpoints along the tour attempted to keep these visual anchors in view where possible in order to help provide some kind of frame of reference. Visitors commented (both in the visitor book and to museum staff) on how the display had raised awareness of Lear's representational style and the extent of his travels rather than on the novelty of using a virtual tour.

The flexibility to design viewpoints onto the landscape that either highlight specific known locations or more indeterminate areas or routes using the landscape model and landmarks as a reference frame, has potential to facilitate the exploration, discussion and presentation of literary references from multiple sources. By way of example, albeit focussing on quite obvious topographic references, the view in Figure 12.1 (right) illustrates a route and waypoints suggested from a combination of the notebooks of Samuel

Figure 12.1 Google Earth: Virtual Billboards (left) and exploring routes and relative positions (right).

Source: © The Author.

Taylor Coleridge and the journal of Dorothy Wordsworth (Coleridge 93–6; Wordsworth 15). They documented how Coleridge travelled on foot from his home in Keswick to visit the Wordsworths, illustrating his habit of taking arduous routes across the fells, and in this case partly at night. The use of key references from the text can be used to reconstruct the likely route, here shown in light a lighter shade. This annotated route can be easily distributed to other users of Google Earth via a KMZ file to allow them to explore it interactively, helping to convey the richness of activities and relationships described in the texts against the backdrop of a familiar frame of reference provided by the 3D model.

The types of views and tour segments described here can easily be shared and viewed within Google Earth itself, but they can also be incorporated into web pages by embedding the graphics window of Google Earth into a webpage containing links to the various viewpoints around the landscape model. In such a way a more interactive presentation could be created that connected original source text on a web page with the type of landscape views described earlier and so provide scope for viewers to interpret the relationship between text and landscape for themselves. In a teaching and learning context this could also be a useful platform to encourage students to think about the nature of landscape scenes which relate to vernacular descriptions of places and experiences found in literary texts.

Physical Relief Models

The use of Google Earth offers an example of using 3D landscape models to attempt to convey spatial context as mediated through some kind of screen or monitor. If the focus of a display was the wider landscape, and a situated display in a visitor centre or museum was required, then physical landscape models rather than screen-based representations could be considered. For over two hundred years physical landscape models have been used to put features of interest into a landscape context, often acting as 'you are here'-style maps. They offer a broad overview but also allow the viewer to examine detail from many different vantage points. The perception of depth offered by seeing the model with human stereo vision makes it easier for the viewer to appreciate even the most subtle of undulations in the landscape. The models are typically painted with some representation of land cover and are often limited to a certain number of features of interest being displayed over the model at any given time.

Combining the virtues of physical models with the flexibility of digital mapping has led to an exploration of projection-enhanced models in the form of the Projection Augmented Relief Model (PARM) technique as described in Priestnall et al. A few examples of projecting onto relief models can be found in the physical sciences, notably the *Illuminating Clay* project (Piper, Ratti and Ishii), where a small clay landscape model could be manipulated by hand, triggering changes to water flow which were re-projected down onto the model. The intention with the PARM approach was to explore the technique in a public context, initially focussing on communicating connections to landscape found in cultural heritage material.

The first installation based upon the PARM technique related to an exhibition of manuscripts by William Wordsworth that ran from Summer 2012 to Summer 2013 at the Wordsworth Trust Museum and Gallery. The focus of the installation became the 'Spots of Time', important events from Wordsworth's childhood which were later to influence his writing, especially his long, autobiographical poem *The Prelude*. The configuration of displays is shown in Figure 12.2 (left) and featured a physical relief model, a ceiling-mounted data projector, a monitor for displaying details of the manuscript, a small touch screen for users to initiate one of three sequences and an 'audio shower' to direct readings of the manuscript down towards the viewer. The physical relief model was created by converting digital terrain data to a form suitable for driving a Computer Numerical Control (CNC) milling machine, carving the relief from dense model board. The display was positioned about a quarter of the way around the visitor pathway through the museum spaces (Figure 12.2, right) after they had viewed general introductory material and also visited Dove Cottage, the Wordsworths' home from 1799 to 1808.

Three passages of poetry were chosen, each of which related in some way to part of the Lake District landscape. For each extract there was a physical display elsewhere in the gallery space featuring the original manuscript and associated items. As with the Edward Lear exhibition there was a desire to encourage visitors to examine the actual physical artefacts as well as be engaged by, and learn something from, the digitally enhanced exhibit. Specific learning objectives included an appreciation of Wordsworth's childhood memories and their connection to places in the landscape. Visitors were also invited to reflect upon Wordsworth's compositional process by looking at the way in which his manuscript material was repeatedly reworked across multiple versions of the same text.

The monitor display featured detailed extracts from the manuscript highlighting words as they were heard via the audio narration. At key moments the viewer's attention was drawn down to the model with a series of projected maps and images which aimed to highlight the part of the landscape to which the passage relates. The intention was to convey not only the general landscape position but also the mood of the passage and the wider context in which that place was being referenced in the manuscript. Plate 10 illustrates three pairs of images representative of the projected sequences for the three spots of time being presented.

The first sequence (Plate 10, left) conveys the joy of a young Wordsworth looking north towards Helvellyn over the valley of Grasmere, which seemed to him like paradise. This sequence used historical mapping combined with a torch-like illumination effect to convey the general area and direction. The second sequence (Plate 10, centre) provided a visualisation of an event

Figure 12.2 The 'Spots of Time' PARM display (left) and its position in the museum space (right).

Source: © The Author.

during a particularly harsh winter when, as a boy, Wordsworth joined the crowds skating on Esthwaite Water. Here the projected layer conveyed the prevailing weather conditions through a frozen effect before fading to focus on the lake itself. The third sequence (Plate 10, right) focused on the evening when the young Wordsworth took a boat for a row on Ullswater without permission and was being struck by guilt and the feeling of the mountains rising out of the darkness around him. The sequence began with a pulsating spot of light moving across the lake to draw attention to that location and to suggest the rhythmic action of rowing, followed by the progressive illumination of the mountains around the lake. Both effects were accompanied by an audio narration of the relevant passages from *The Prelude* (Book 1: 383–5, 404–12). This latter effect was created using a 3D modelling package employing the same digital terrain data as used to create the physical model. By inserting and animating virtual light sources above the terrain surface object a series of images were rendered from above and when projected onto the physical model gave the impression of an actual light source rising above that spot on the model.

Observations made, both directly and from video sequences, of people's interaction with the 'Spots of Time' installation suggested that it did promote discussion amongst groups related to features on the landscape, and pointing and tracing gestures that seemed more pronounced than when viewing a map on a wall or monitor, although demonstrating this in a more measurable way is a part of the ongoing research agenda with this technique. During the early demonstrations of the PARM display animated routes proved to be quite compelling for viewers and one area of ongoing development with the display is to convey more of the richness of the route by animating movement combined with progressively changing visibility maps (viewsheds) from the current point along the route. This conveys the changing vista when moving through the landscape and in some circumstances could offer a useful accompaniment to passages from literary texts.

Location-Aware Mobile Computing

Common to both the Edward Lear and 'Spots of Time' exhibits was an intention to raise awareness in visitors that many of the places highlighted could actually be visited during their travels later that day. The third case study focuses on the use of geospatial technology out in the field and once again considers how frames of reference derived from 3D landscape models could help people explore literature in the context of the actual landscapes that those texts describe.

Mobile devices equipped with GPS have for many years offered the ability to provide information about a user's surroundings including the concept of defining mediascapes where media elements can be automatically offered to users as they enter pre-defined trigger regions on the ground (Stenton et al.). This general concept of locative media has opened up possibilities for new forms of engagement with place references in literature

(Cooper and Priestnall 258–61), and the creative delivery of information *in situ*, even exploring literary narratives 'on the move', which 'allow readers and writers to see literature in relation to their lived environment' and could 'contribute to a vision of public spaces as inherently readable, playable and malleable by its inhabitants' (Løvlie 27). One of the challenges associated with locative media as noted by Løvlie, however, is the process of creating appropriate trigger zones, given that the 'author' can define their size, shape and position. This challenge is inextricably linked to the design of the media being used, in determining whether such media has situational relevance to a user at that time in that location.

One approach that has been explored is the use of 'you are here'-style terrain diagrams where key features are highlighted alongside landmarks which should be visible to the user, but also showing the user's location placed on more regional representations of the terrain (Plate 11, centre). The ordering of the sentences in this paragraph have been changed by the T&F production team in a way that disrupts the clear progression of ideas. Please revise as specified by the deletions and insertions that we have added. This approach was developed as part of an ongoing exploration of landscape references in the diaries and fieldnotes of Gordon Manley, the famous British climatologist, for the design of a locative media tour near the weather station at Great Dun Fell, a Pennine fell in the east of Cumbria. It forms an experimental accompaniment to the materials produced for the 'Troublesome Wind' self-guided walk created for the Royal Geographical Society / Institute of British Geographers (RGS/IBG) *Discovering Britain* series. A particular example (Plate 11, centre) draws the viewer's gaze towards a hill called Criffel, which is some distance away in southern Galloway, Scotland, and likely to be unfamiliar to most walkers on this route. The hill is referenced in a passage from Manley's diaries, which the user would have just heard. Landmarks likely to be more familiar and local to the user were used to help the orientation of a wire-frame graphic derived from the landscape model. Similar annotated wireframes have been used as digital transparencies placed over the camera view of a mobile phone (Plate 11, right), representing an attempt to make a scalable and accessible version of more technologically complex ways of augmenting landscape scenes (Priestnall et al. 1–4; Adams, Fitzgerald and Priestnall).

As orientation and tilt sensors converged with GPS and cameras on Smartphones, possibilities emerged for users to have a more direct engagement with the landscape scene. Augmented Reality (AR) browsers can display icons over the camera view indicating where information that may be of interest is located, but often these result in information overload and confusion when there appears to be nothing salient in the scene to which the user can relate. Once again digital landscape models can play a role, and with the Zapp application (Meek et al.), a line-of-sight determined by the phone's sensors is intersected with a digital terrain model held on the device to determine the distant coordinate the user is pointing at (Figure 12.3, left) which in turn can trigger information

Figure 12.3 The Zapp 'line-of-sight' mobile application.

Source: © The Author.

stored for that area. With this approach the only information the user can access relates to the area that is visible from their current location. This experimental app was designed to have a simple 'point and shoot' camera style (Figure 12.3, right) and also has interesting potential for the capture of data. Rather than record geographically-tagged information related to the user's current location, this approach can record information about more remote areas of interest and therefore the user's broader landscape context.

Discussion

Clearly the use of landscape models based almost entirely on ground elevation as presented here is more appropriate in the English Lake District than in many other environments. In areas which comprise low lying relief, or in urban environments, the characteristics of the landscape model, and the particular frames of reference used, will be different. Having said that, the three techniques presented here have been used in similar ways for studies on the University of Nottingham campus, building the landscape models from data at a finer scale and from other sources such as airborne laser-scanning. In such ways, literary references relating to urban environments could employ landmarks such as buildings, trees and other surface objects as frames of reference yet also explore the more indeterminate place references discussed in the chapter. Key to all three case studies has been the identification of techniques which promote a greater awareness of spatial context, either in the sense of the broad regional landscape or the immediate visual context of a user in the field. The case studies presented here demonstrate some experimental techniques for establishing frames of reference for the viewer so they may better appreciate the relative positions of landscape references, in some cases in relation to their current location. In Google Earth these techniques included the placement of aerial viewpoints containing familiar points of reference. For the physical model it was argued that the power of high definition stereo human vision allowed subtle depth perception cues to strengthen the construction of a 3D reference frame, in this case formed by the mountains and valleys but enhanced by the projection of other geographical content. Finally, digital terrains can be of use when helping the viewer in the field to make associations directly between literary material and the landscape features to which it might relate. The context of use in this case meant that there were design challenges in helping to direct the viewer's gaze in the correct direction and to improve the situational relevance of the information being presented.

Another theme emerging in these cases studies has been the potential for alternative forms of georeferencing beyond single point locations, and this seems particularly relevant to the more indeterminate or even contested nature of spatial footprints that might be associated with certain types of cultural heritage material including passages from literary works. In the case of the 'Spots of Time' display the projections onto the physical model aimed to provide some indication of the part of the landscape to which a passage of text related, in ways which deliberately avoided pinpointing particular locations. This was achieved using illumination effects, leaving something to the imagination of the viewer when it came to interpreting the landscape context. The physical landscape model was a complement to the written word which in this case came in the form of an audio narration. In a similar way the use of viewpoints in Google Earth not only provided a broader landscape context for points, routes or areas placed on the model, but also allowed more indeterminate landscape references to be made by using the view itself to suggest a spatial footprint. Virtual billboards were used in Google Earth to offer direct graphical augmentations of a landscape scene with historical images but they also have the potential to convey supplementary information about the view without necessarily pinpointing any specific locations. This technique was also used via locative media on a mobile device in the field, attempting to help the viewer make associations with the real landscape scene rather than the virtual one. As with the other techniques the intention was not to place literary references 'on the map' but to allow viewers to engage with the landscape in a way which left scope for their own interpretation and critical reflection.

The positioning capabilities of mobile devices allow users to investigate spatially referenced information out in the field and the examples presented above touched upon the challenges of making such information relevant to the landscape around the user. In addition to this, the ability to capture spatially referenced notes, sketches or voice recordings related to a user's reflections on a text whilst in the landscape offers an exciting opportunity to explore the richness of a text and the responses to that text. Any such use of technology in the field would clearly require careful design to ensure users were able to capture information easily without the device becoming the focus of attention. An avenue for further exploration here would be to use such techniques on fieldtrips as a way for students to synthesise multiple texts relating to a landscape with their own reflections on those texts or on the landscape itself. One design challenge here is to ensure that techniques promoted critical thinking in the students rather than simply deliver spatially referenced information in ways that suggested an unwarranted sense of truth or certainty.

Conclusion

This chapter has considered three case studies of geospatial technologies other than GIS, focussing on the communication of spatial context using digital landscape models to help provide a frame of reference. Although these are quite specific examples, it is hoped that they help to demonstrate the breadth of possibilities that geospatial technologies might offer when exploring literary works. There is a wide range of technologies each with a set of affordances suitable for different contexts of use, different audiences and different objectives. The three case studies presented here aimed to support informal learning in non-expert audiences but also illustrate the potential for promoting digital and spatial literacies. The breadth of techniques should be of interest for exploratory research and teaching across the arts, humanities and social sciences. Even within geography, a discipline where geospatial technology might be assumed to be widespread, there is scope to raise awareness. As Offen suggests 'historical geographers ought to explore more fully the possibilities that the technologies provide, especially when a multimedia approach might be more effective in conveying the range and quality of historical geographic research' (572). In addition to raising awareness of the capabilities of various geospatial technologies, there is further scope to develop simple tools, workflows and design guidelines that might promote easier adoption. Collaborations between users of geospatial technology and researchers in the arts and humanities will continue to be rewarding and productive for all concerned. We are also likely to see more use of such techniques becoming embedded within teaching and research training within arts and humanities programmes themselves. If there is an appreciation of the broad range of geospatial technologies available then appropriate choices can be made, whether in research, teaching or public display settings, which in turn should help to support exploration, interpretation and critical reflection.

Works Cited

Adams, Anne, Elizabeth Fitzgerald and Gary Priestnall. 'Of Catwalk Technologies and Boundary Creatures'. *ACM Transactions on Computer-Human Interaction* 20.3 (2013): art. 15. 34pp.

Arias, Santa. 'Rethinking space: An Outsider's View of the Spatial Turn'. *GeoJournal* 75.1 (2010): 29–41.

Caquard, Sebastien. 'Cartography I: Mapping Narrative Cartography'. *Progress in Human Geography* 37.1 (2011): 135–44.

Coleridge, Samuel T. *Coleridge Among the Lakes and Mountains: From his Notebooks, Letters and Poems 1794–1804.* Ed. Roger Hudson. London: The Folio Society, 1991.

Collier, Peter, David Forrest and Alastair Pearson. 'The Representation of Topographic Information on Maps: The Depiction of Relief'. *The Cartographic Journal* 40.1 (2003): 17–26.

Cooper, David. 'Poetics of Place and Space'. *Literature Compass* 5.4 (2008): 807–21.

Cooper, David, and Ian N. Gregory. 'Mapping the English Lake District: A Literary GIS'. *Transactions of the Institute of British Geographers* 36 (2011) 89–108.

Cooper, David, and Gary Priestnall. 'The Processual Intertextuality of Literary Cartographies: Critical and Digital Practices'. *The Cartographic Journal* 48.4 (2011): 250–62.

Derungs, Curdin and Ross S. Purves. 'From Text to Landscape: Locating, Identifying and Mapping the Use of Landscape Features in a Swiss Alpine Corpus'. *International Journal of Geographical Information Science* 28.6 (2013): 1272–93.

Dodge, Martin, Chris Perkins and Rob Kitchin. 'Mapping Modes, Methods and Moments: A Manifesto for Map Studies'. *Rethinking Maps: New Frontiers of Cartographic Theory.* Eds, Martin Dodge, Rob Kitchin and Chris Perkins. Abingdon: Routledge, 2009. 220–43.

Elwood, Sarah. 'Geographic Information Science: New Geovisualization Technologies – Emerging Questions and Linkages with GIScience Research'. *Progress in Human Geography* 33 (2008): 256–63.

Goodchild, Michael. F. and Donald G. Janelle. 'Toward Critical Spatial Thinking in the Social Sciences and Humanities'. *GeoJournal* 75.3 (2010): 3–13.

Grossner, Karl E., Michael F. Goodchild. and Keith C. Clarke. 'Defining a Digital Earth System'. *Transactions in GIS* 12.1 (2008): 145–60.

Guo Qinghua, Yu Liu and John Wieczorek. 'Georeferencing Locality Descriptions and Computing Associated Uncertainty Using a Probabilistic Approach'. *International Journal of Geographical Information Science* 22.10 (2008): 1067–90.

Häberling, Christian, Hansruedi Bär and Lorenz Hurni. 'Proposed Cartographic Design Principles for 3D Maps: A Contribution to an Extended Cartographic Theory'. *Cartographica* 43.3 (2008): 175–88.

Løvlie, Anders. S. 'Poetic Augmented Reality: Place-Bound Literature in Locative Media'. *13th International MindTrek Conference: Everyday Life in the Ubiquitous Era, Tampere* 30 September–2 October 2009: 19–28.

Meek, Sam, Gary Priestnall, Mike Sharples and James Goulding. 'Remote Mobile Capture of Points of Interest Using Line of Sight Modelling'. *Computers and Geosciences* 52 (2013): 334–44.

Montello, Daniel R. 'You are Where? The Function and Frustration of You-Are-Here (YAH) Maps'. *Spatial Cognition and Computation* 10 (2010): 94–104.

Moretti, Franco. *Atlas of the European Novel, 1800–1900.* London: Verso, 1998.

Nardi, Daniele, Nora S. Newcombe and Thomas F. Shipley. 'Reorienting with Terrain Slope and Landmarks'. *Memory and Cognition* 41 (2013): 214–28.

Offen, Karl. 'Historical Geography II: Digital Imaginations'. *Progress in Human Geography* 37.4 (2012): 564–77.

Piper, Ben, Carlo Ratti and Hiroshi Ishii. 'Illuminating Clay: A 3-D Tangible Interface for Landscape Analysis'. *Proceedings of the Conference on Human Factors in Computing Systems CHI '02* 2002: 335–62.

Priestnall, Gary, Elizabeth Brown, Mike Sharples and Gemma Polmear. 'A Student-led Comparison of Techniques for Augmenting the Field Experience'. *Proceedings of Mlearn* 2009

Priestnall, Gary, and Jeff Cowton. 'Putting Landscape Drawings in their place: Virtual Tours in an Exhibition Context'. *Geospatial Computing for the Arts, Humanities and Cultural Heritage Workshop at the Fifth IEEE e-Science Conference, Oxford, UK* 9–11 December 2009.

Priestnall, Gary, Jeremy Gardiner, Jake Durrant and James Goulding. 'Projection Augmented Relief Models (PARM): Tangible Displays for Geographic Information'. *Proceedings from Electronic Visualisation and the Arts (EVA) London* 2012.

Raper, Jonathan. *Multidimensional Geographic Information Science.* CRC Press, 2000.

Scheider, Simon, and Ross Purves. 'Semantic Place Localization from Narratives'. *ACM SIGSPATIAL COMP'13,* 5 November 2013, Orlando, FL, USA.

Shepherd, Ifan. D.H. 'Travails in the Third Dimension: A Critical Evaluation of Three-dimensional Geographical Visualization'. *Geographic Visualization: Concepts,*

Tools and Applications. Eds, Martin Dodge, Mary McDerby and Martin Turner. Chichester: John Wiley & Sons, 2008. 199–222.
Snowdon, Caroline, and Christian Kray. 'Exploring the use of landmarks for mobile navigation support in natural environments'. *Proceedings of the 11th International Conference on Human-Computer Interaction with Mobile Devices and Services (MobileHCI '09). ACM*, New York, NY, USA 2009.
Stenton, Stuart P., et al. 'Mediascapes: Context-Aware Multimedia Experiences'. *IEEE Multimedia* 14.3 (2007): 98–105.
Tuan, Yi-Fu. 'Language and the Making of Place: A Narrative-Descriptive Approach'. *Annals of the Association of American Geographers* 81.4 (1991): 684–96.
Vinson, Norman. 'Design Guidelines for Landmarks to Support Navigation in Virtual Environments'. *Proceedings of the Human Factors in Computing Systems Conference (CHI '99)*, Pittsburgh, USA 1999: 278–85.
Wordsworth, Dorothy. *The Grasmere Journals*. Ed. Pamela Woof. Oxford: Oxford University Press, 1993.
Wordsworth, William. *The Prelude: The Four Texts (1798, 1799, 1805, 1850)*. Ed. Jonathan Wordsworth. London: Penguin Classics, 1995.

13 Geovisuality

Literary Implications

Tania Rossetto

Introduction

Within literary studies, a distinction between the 'cartographic' and the 'visual' has frequently been acknowledged. Literary cartography, for instance, has been accused of dealing with abstract, distant *spaces*, whilst neglecting the analogic, close appreciation of *places* and *landscapes* (Cerreti). The writer has been alternately described as a painter (Cometa) and a cartographer (Turchi). The reader's visual imagination has been treated as something different from his or her spatial cognition. Research on space and language and on authors' and readers' imaginative geographies have distinguished 'survey knowledge', or map-like descriptions, from 'route knowledge', or tour-like descriptions (Mondada; Ryan, 'Cognitive Maps'), the former providing vertical projections and the latter providing embodied experience (Ryan, 'Space').

The boundary between the cartographic and the visual is challenged in our everyday experience when we use digital tools such as virtual globes or in-car satellite navigation systems, switch from map-views to street-views on Web-based mapping tools, engage with multimedia cartography in immersive ways or read maps through mobile devices whilst in close bodily contact with places and landscapes. Encompassing all these practices, 'Geographic visualisation', or 'Geovisualisation', has been defined as 'the application of any graphic designed to facilitate a *spatial* understanding of things, concepts, conditions, processes or events in the human world' (Dodge, McDerby and Turner 2). Conventional paper maps, as static devices with a normative planar view and consistent scale reduction, are only an archetypal form of geovisualisation, which now includes interactive, dynamic, disseminated map-like spatial images and digital media aimed at providing a *visual* understanding of space by employing different 'visual vocabularies' (3).

At a time when computing and mapping have become equally pervasive and ubiquitous, geovisual experiences have developed into familiar cultural and spatial practices. In light of this recent 'geovisual turn', the need to investigate how 'geovisuality' affects literary creativity as well as the analysis and reception of literary works emerges. This chapter aims to explore the possible literary implications of a specific focus on geovisuality from the point of view of a geographer interested in the relationship between cartography and literature, asking some questions and suggesting some paths for further research, without any ambition to systematise this area of interest. In other words, this chapter is a theoretical piece that aims to sketch out some trends in contemporary visual/spatial culture, comparing them with the realm of literature. Although the geovisuality-literature connection has been crucial to the recent development of the so-called 'spatial humanities', with the related advancement of refined qualitative geo-spatial technologies to meet humanist approaches, I contend that the geovisuality-literature connection can be explored in additional ways. These include the use of lay geovisual tools (such as virtual globes or cartographic apps) in both expert and non-expert experiences of reading literary works, the impact of a 'geovisual culture' on the imagination of the writer, the revival of pictorial literary maps, the representation of digital as well as pre-digital geovisual technologies and practices within the literary text, and the adoption of a geovisual frame of reference and lexicon in the interpretation of literature. Whether we consider the geo/cartographic realm or the literary one, however, embracing a geovisual perspective requires us to challenge some traditional binaries that revolve around the map/view opposition as well as the space/place opposition. The following sections elaborate on this challenge. The subsequent section concentrates on the geovisual implementation of geographic information systems (GIS)-based digital humanities, whilst the final section aims to open up additional paths towards a multifaceted appreciation of the geovisuality-literature connection.

Overcoming the Space/Place Opposition: Blurring the Map/View Boundary

The debate surrounding the meaning of space and place is longstanding, particularly but not only within geography. A widely acknowledged way to contrast these cardinal terms is to consider space as the domain of the quantitative and the abstract and place as the domain of the qualitative and the experiential. The analysis of space and place is what most commonly leads literary scholars to engage with geographical tools and approaches. This engagement has injected further complexity into the meanings ascribed to space and place to the extent that sometimes their meanings have been reversed, as is the case in Michel de Certeau's influential *The Practice of Everyday Life.* The opposition between space and place was emphasised during the 1970s and 1980s, when a humanist perspective, nurtured by phenomenological thinking, arose within geography. Neo-positivist spatial analysis was rejected in favour of a focus on places as the realms of everyday experience, subjectivity and meaning-making. The space-place binary was crucially connected with the emergence of the field of literary geography during those same years. As humanistic geography sought to bring human subjectivity and the sense of place back to the core of the discipline, it turned to literature as a site at which the objective world and subjective experience are blended (Brosseau 213).

Space and place are also implicitly associated with different scales (Agnew 319). Whilst arguing for an expurgation of the notion of scale from the geographic vocabulary in their famous article 'Human Geography Without Scale', Sallie Marston, John P. Jones and Keith Woodward included space and place within a list of binaries affiliated with the local-global distinction (421). The local-global scale binary is, in fact, associated with other binaries: difference-sameness, concrete-abstract, agency-structure, culture-economy, embodied-anonymous, here-there and responsible-detached. The authors came to propose a 'flat ontology', which conflates scales focusing on emergent, contingent spatial relationships as they occur in human practice. Moreover, they suggest that re-imagining the scale binary as conflated implies a re-imagination of the associated list of binaries.

Within the field of literary geography, place remains a prevailing concept, with the space-place divide serving as a pervasive frame. Tracing the emerging intellectual terrain of the so-called 'geohumanities', Dear considers place as an analytical 'primitive' and states that 'researchers commonly distinguish between *space* as an abstraction and *place* as a social construct, that is, what humans create out of space' (313). Within literary studies, it has been recently reaffirmed that despite the fact that place seems to have come under attack by the delocalising forces of globalisation, place, in its experiential dimension, remains crucial to the 'strategic role that imaginative literature plays' within the postmodern world (Prieto 15, 13). Bertrand Westphal's *Geocriticism*, indeed, is a place-centred form of literary criticism. '*Platial* experiences in literature', then, have been emphasised by approaches that, by exploring the body and physicality within literature, search for immersive sensations rather than detached representations of space within texts (Moslund 30). In what follows, I argue that the kind of re-imagination of binaries proposed by Marston, Jones and Woodward could be applied to an additional opposition – namely, the map/view opposition – and that these conflations have implications for literature, since they connect spheres, such as the cartographic and the visual, which are traditionally conceived as being separate within literary works and literary reception.

Literature between the Cartographic and the Visual

Generally speaking, we map spaces and view places and landscapes. Indeed, the role of landscape in literary criticism, as well as in geography, has its own long tradition given the evident connection of landscape with the visual and aesthetics. Nonetheless, even within geographical research, the concepts of place and landscape frequently overlap, particularly when used within a practical, embodied approach. Without challenging the place/landscape distinction (as in Setten), the binary opposition to which I am referring puts space at one side and place and landscape on the other.

Traditionally, Western mapping practices have been viewed as fundamentally irreconcilable with the expression of lived places and emotional landscapes. Expressive forms, such as writing, painting or photography, have been considered more suitable ways of grasping human experience: the map is detachment, whereas place and its narration are involvement; the map is an act of measuring, whereas narrative is an act of rendering the qualities of places; maps communicate spaces of homogeneity, whereas literary texts express the emotional variations of landscapes (Pearce). One of the first reactions of geographers when the idea of mapping literary works was advanced in the well-known work of Franco Moretti was to criticise this act as an attempt to apply a Cartesian grid to literary texts, thus neglecting the qualitative dimension of both textual and real places. Claudio Cerreti, for instance, saw Moretti as influenced by 'spatial analysis' rather than geography, with his denotative geometry of the 'extensio' of spatial relations being far from a geography of the connoted 'intensio' of places. Similarly, it has frequently been remarked that although GIS analyse *space* through digital mapping, the humanities are predominantly focused on issues of *place* and the sense of place (Harris, Bergeron and Rouse 228; emphasis added).

In parallel to or in association with the space-place opposition within literary analysis, there is also a focus on antithetic 'perspectives' and 'points of views' from above and from the ground. It has been noted that literary geography integrates these 'opposed perspectives' by reflecting upon the *different* senses of place and space we obtain from aerial and street level views via 'wordmaps' and 'verbal snapshots', respectively (Luria 69). De Certeau's famous opposition between the map-like view of Manhattan from the top of the World Trade Center and the immersive perception of the city from the street level has become a major point of reference in this sense. For de Certeau, within modernity, the 'map' has suppressed the 'itinerary', which stands for the human perspective and narrative. Very recently, for instance, Eric Prieto recalled de Certeau's map/itinerary opposition to propose investigating 'different narrative modes and perspectives' to further develop the geocritical approach (24).

As has been noted (Ng-Chan), with the advent of virtual globes such as Google Earth, de Certeau's observation may be more relevant than ever. These new devices, in fact, seem to allow a convergence between the 'map' and the 'itinerary' through montages, combinations and assemblages of dual points of view.[1] Analysing the documentary film *Patience (After Sebald)* (2011), as well as the connection between narrative cartography and

1 An incisive and similar re-visiting of de Certeau is advanced by Lammes within computer games research.

cartographic cinema, Taien Ng-Chan sees the possibility of relating the itinerary back to the map actualised within the cine-cartographic mediation of Sebald's literary work *The Rings of Saturn* by director Grant Gee, which includes a Google Maps mash-up from Barbara Hui's *Litmap* project. Interestingly, Ng-Chang also relates the map (space)/itinerary (place, landscape) binary and de Certeau's famous passage with the opposition between the 'grid map' (functionalistic spatial devices that inhibit imagination) and 'the story map' (forms of spatial expression that embody personal experience and provide room for emotions), which Robert Macfarlane discusses in his 2007 work *The Wild Places.* Sébastien Caquard ('Cartography I') compares this opposition between the grid and the story map with geospatial imagery, concluding that the new cartographic environment in which we live is challenging this opposition: 'the distinction between the story map and the grid map is blurred by the fact that more and more often, they coexist within a common platform such as Google Maps' (140). Interestingly, to denote the difference between a professional 'cartographic perspective' and an 'artistic/humanistic perspective' on cartography, in a recent book, Sébastien Caquard, Laurene Vaughan and William Cartwright do not use the map/view opposition but introduce the idea of 'mapping from above' and 'mapping from the ground' (Caquard et al. 3). As Les Roberts has noted, 'at a practical as well as a cultural level, a growing convergence between *visual* culture, mapping and cartography has blurred the epistemological boundaries that police understandings of what we might consider to be a "map" as distinct from, say, an "image"' (4). In these various interventions within the realm of geohumanities, we can appreciate how the cartographic and the visual are becoming increasingly intertwined.

New geo-spatial technologies and new everyday mapping practices are the basis of this 'blur'. The 'geovisual turn' is one of the most prominent of these novelties. Geographic visualisation includes a range of innovative methods for displaying spatial data within interactive environments, which provide possibilities for private exploration and manipulation; multiple views of the same data; and the mixing of maps with other graphics, texts, iconic displays and multimedia materials in a way that facilitates visual thinking. Within geovisual environments, therefore, the map is able to convey the figurative, to marry geometry with geography and to become an emotional, imaginative, embodied entity, as the concept of 'affective geovisualization', articulated by Jim Craine and Stuart C. Aitken, suggests. Sarah Elwood, for instance, has highlighted the relationship between visual and geoweb imagery, along with the consequent fluid combination of cartographic and photographic visualisation. LaDona Knigge and Meghan Cope suggested the conflation of the 'ground' and the 'God's eye view' in the concept of 'grounded visualization'. This tendency towards mixing the two perspectives is particularly emphasised within the theory and practice of qualitative GIS, which originates moments of 'hybrid geography' in its attempt to collide different epistemologies whilst overcoming the qualitative/quantitative binary (Leszczynski). An example is the imagined grid developed by Jin-Kyu Jung and Sarah Elwood which consists of a special GIS layer where data, such as photographs, sketch maps, texts, audio clips or video, can be stored and geo-located, manipulated by any spatial operations and displayed on a map. As the same work by Jung and Elwood demonstrates, the hyperlinking of verbal texts to spatial objects in a GIS is another common way to integrate the qualitative into the quantitative. This is not, however, the only way in which the colliding tensions operating within qualitative GIS may involve literature. What, then, are the consequences of this geovisual trend within the field of literary studies? What kind of tensions between the two different spheres to which the concept of geovisualisation alludes might emerge from a specific focus on 'geovisualisation and literature'?

Digital Humanities Beyond the Mapping/Phenomenology Divide

Intertwining notions of landscapes and maps, Gary Lock has noted that GIS is a 'technology that can go beyond the traditional map into new realms of representation' (89). Although recognising the tension between the objective approach of traditional 2D GIS and the interpretative approach of humanities disciplines in more recent technological developments, Lock sees the potential to create a continuum between 'observation' and 'inhabitation', between mapping on one hand and phenomenological and non-representational theories on the other. Because 'a traditional two-dimensional map, whether hand-drawn or digital, is not a good representation of inhabited place', incorporating this extra sense of place is one of the challenges for GIS (Lock 91). Another task is to integrate the cognitive and cultural filters activated during spatial perception, which leads to a topological, rather than topographical, approach. For Lock, the analytical conception of scale must be integrated with a phenomenological one, as well as with multiscalar techniques. Following this perspective, going 'toward *place*-centred representations' (95; emphasis added) means going toward the merging of GIS and virtual reality (VR). The re-emergence of phenomenology as a way to overcome merely representational understandings of landscape in cultural geography and other humanist disciplines, Lock contends, challenges the quantitative basis of digital representation and requires further development and creativity in geospatial and geovisualisation technologies.

Similarly, it has been suggested that new geospatial technologies, such as immersive and experiential visualisation, in conjunction with GIS, could more profitably contribute to the humanities by capturing the perceptual, experiential, emotional, symbolic nature of place or landscape, which the two-dimensional Cartesian plane of the map misses (Harris, Bergeron and Rouse). The phenomenological approach to the interpretation of place and landscape, following Trevor M. Harris, Susan Bergeron and L. Jesse

Rouse, requires not only embedding sources such as texts, photographs, drawing, audio or video within qualitative GIS, but also shifting to GIS-based geovisualisation and immersive technologies, which provide dynamic, interactive, multidimensional and sensuous representations. The concept of 'deep mapping', for instance, has been employed (Bodenhamer, Harris and Corrigan) as a means to evoke experimental ways of digitally translating 'from textual to visual communication' that are less GIS-dependent: beyond the absolute space based on Euclidean coordinate systems, deep maps are prefigured as complex, creative spatial tools which 'are meant to be visual, time-based, and structurally open' (Bodenhamer, Harris and Corrigan 174).

New forms of cartographic visualisation are thus developed in order to capture the fragmented spatio-temporal dimension of narrative (Caquard and Fiset). In this sense, for instance, David Cooper and Ian N. Gregory explicitly called for a turn towards a 'qualitative literary GIS'. In my opinion, however, the application of these GIS technologies to upgrade traditional handmade literary cartography and to provide more sensitive, hybrid means via which to render the spatial and visual dimensions of literary texts is only one of the possible approaches that could emerge from the application of the 'geovisual turn' within the field of literary studies.

Geovisuality and Literature: Tentative Considerations

Although one emergent tendency within digital literary cartography is to develop highly specialised qualitative GIS/geovisual techniques to promote a 're-cartographisation' of traditional literary mapping (see Rossetto, 'Theorizing Maps'), the engagement of expert and non-expert readers with popular web-based 'neogeographical' tools also requires consideration. Within literary studies, indeed, there is a growing interest in neogeography, which has been inaugurated by the introduction of Google Maps and refers to a set of web-based technologies defined in opposition to classical GIS (Young and Levin 152). Indeed, despite criticisms of Google Earth (Mittman), some literary scholars have begun to embrace this platform as a means of teaching and analysing literature. Convinced that technologies enhance the reading experience, Jerome Burg coined the term 'Lit Trips' to describe the use of Google Earth and other integrated web-based resources to teach literature to high school students. He also created the Google Lit Trips website (Berson and Berson), which provides pre-made Lit Trips and guides teachers in designing their own Lit Trips. Erin Sells, in contrast, assigned groups of students to create interactive online maps for various characters in Virginia Woolf's *Mrs Dalloway* using Google Earth, subsequently assembling those individual maps into a comprehensive one. In her view, in comparison to the map embedded in the edition of Woolf's novel they were using, students' online maps were able to create 'a far denser and richer visual representation of the novel' (Sells 29).

The use of web-based spatial applications to map literature in such ways has been termed 'bookmapping technologies' (Cavanaugh and Burg). An alternative methodology is showcased by Rick Mott who has created a website with various digital tools (maps and photographs, as well as video and audio clips, animation and 360-degree panoramas) to offer a range of 'contextual material' in order to understand the role of landscape within Leslie Silko's 1977 novel *Ceremony*. Mott's digital resource illustrates how the practice of making literary maps, and placing the quotations from texts within them in such a way that the text becomes a layer of the map, is not the only way to experience or study literature through spatial multimedia. Very often readers do not reduce the text to a digital map: they do not practice a Morettian 'distant reading'. Rather, they use digital geovisual devices dynamically in conjunction with the primary text at hand. The text remains 'intact', and a multifaceted geovisual practice is activated by the processes of reading and interpreting.

The following passage from a newspaper article on literature and everyday digital practices effectively illustrates these processes:

> So imagine, instead, reading *The Grapes of Wrath* on an iPad, then pausing when the Joad family arrives at a Hooverville, bringing up a Google Earth satellite image of that exact location and tapping the screen to zoom-in for the 3-D street-view. Tap on yet another link and you could perhaps listen to some music from this era – think, *Brother, can you Spare a Dime?* or a Duke Ellington number. (Farquharson)

As this passage suggests, although we tend to consider literary mapping only within isolated personal computing systems, we should more carefully consider ubiquitous computing and mobile devices. Lay reading practices, the use of portable literary audio-guides and more general contemporary literary tourism practices should be considered in the light of a geovisual appreciation of literature. These increasingly common practices, it seems to me, could be investigated in connection with studies of the 'microprocesses of reading' literary texts from both cultural and cognitive points of view (Elfenbein 484). If studies of the experience and embodiment of literature find interesting intersections with investigations of visuospatial thinking, as the focus on space within cognitive narratology demonstrates, it could also be interesting to investigate the embodiment of maps and corporeal experience through portable geovisual devices (Rossetto 'Embodying the Map').

Another way that the relationship between the geovisual turn and literature could be investigated, I argue, is via the impact of 'geovisuality', or a 'geovisual culture', on the writing, reading and interpretation of literature. The famous distinction introduced by Hal Foster (Mirzoeff) between vision (sight as a physical process) and visuality (sight as a social phenomenon) could be applied to geovisual technologies, thus promoting research on 'geovisuality' rather than the strictly 'geovisual'. Martin Dodge, Chris Perkins and Martin Kitchin point out that maps and mapping have often been neglected in visual studies (224). There is a need to ensure that geovisual technologies are not marginalised in this way, but are instead allowed to figure prominently in critical analyses of contemporary visual culture. If the cognitive impact of these new technologies is considerable (Montello), then how do they help to shape the visual-spatial imaginations of the writer and, by extension, how are they represented within the creative texts? Moreover, how does geovisuality become part of the visual culture of both the expert and non-expert reader?

A few years ago, introducing a special issue on photography and literature for the journal *Poetics Today*, Silke Horstkotte and Nancy Pedri provided a review of emerging 'text-and-image studies in the humanities' that was particularly useful in contextualising specific research on geovisuality and literature. Interestingly, the authors mentioned 'literature and maps' amongst the topics of study concerning 'word-and-image relations'. It could be illuminating to compare the studies reviewed in this article regarding – how writing calls forth images, how literary texts are visually read, how visual perceptions intersect with language, how verbal descriptions convey visual perceptions and how visual culture is narrativised – along with the notion of geovisuality/geovisual culture. A range of open questions emerges. How do cultural artefacts that collapse the distinction between the cartographical and the visual enter the study of visual culture and literature? Can literature be read 'geovisually'? How are geovisual media included (via transcription, transposition, interference or ekphrasis) within the verbal medium? Can we identify forms of 'geovisual writing'? Is there a mutual 'inter-art integration' between geovisual media and literature? Do geovisuality and literature produce forms of 'mixed media art'?

Today, there are creative ways in which mixed forms of cartographic artwork incorporate literature, images and maps: Margaret Pearce integrated narrative directly within map design; Amy Wells proposed a 'collage' map made up of different visual materials, which reflects the private geography of an author (183); and Laura Canali,[2] a professional cartographer working for the Italian journal of Geopolitics, *Limes*, designed literary maps that implement cartographic language in very creative, artistic ways. On can, however, also refer to the emerging field of infographics, which is rich in applications creatively inspired by a cartographic flavour and devoted to literature.[3] This production, in truth, has an antecedent in the past and recent production of literary 'pictorial maps' for touristic or educational purposes, which traditionally mixed cartographic and figurative images (see Hopkins and Buscher).

2 See 'Cartografie dell'immaginario'; 'Andrea Zanzotto, geopoeta della Grande Guerra'; Canali and Miglio.

3 See, for instance, the work by the Density Design team on Charles Dickens's London (Density Design).

An additional way to explore the connection between geovisuality and literature is to consider geovisual practices as they 'emerge' from literary texts. The traditional research subject of 'maps in literature' (Muehrcke and Muehrcke), therefore, could be reanimated via research on 'geovisual practices in literature'. As I suggest elsewhere (Rossetto 'Theorizing Maps', but see also Prieto; Caquard and Cartwright), in literary texts, we encounter map-readers in action and emergent mapping practices, amongst which geovisual tools increasingly appear. Below, I provide a literary account of a geovisual practice in which the interaction between in-car satellite navigation, the mental map of the writer and the visual appreciation of the surrounding urban landscape converge in a vivid way. This piece, extracted from the Italian writer Gianrico Carofiglio's urban novel *Né qui né altrove. Una notte a Bari*, embeds crucial aspects of the impact of satnav tools on spatial awareness and graphicacy (Axon, Speake and Crawford). The passage verbalises and provides a surrogate experience of an embodied performance of both place-landscape/space and their geovisualisation, continuously shifting between immersion and detachment.

> Something was going on that evening. After seeing the palms, I started to pay attention to the satnav and how it worked. On the monitor, I saw the 3D colour map of the area I was driving through. I watched the street we were going through, then the parallel and the perpendicular ones. Via Piccinni was on the left; Strada del Palazzo dell'Intendenza was on the right, beyond the limits of the old city. And then, there were the perpendicular via Sparano da Bari, via Andrea da Bari and via Roberto da Bari . . . I had passed by those streets . . . thousands of times. However, that evening I felt as if I was discovering those streets for the first time thanks to the bright monitor. The names of the streets materialised on the coloured screen and they became real entities for the first time. They were no longer blurred and evanescent. The names of the roads, the sense of the places, my very position in those places acquired a new, concrete meaning . . . . The satnav told me that, parallel to Piazza Massari, there was via Boemondo. That was the first time I had noticed the street. I mean, I perfectly knew that, at a certain point, there was a street dedicated to Boemondo D'Altavilla, Prince of Antiochia. I knew that via Boemondo was near to Piazza Massari, Corso Vittorio Emanuele, Castello Svevo and the Prefect's Office. Despite this, only then did I become aware of the position of streets and squares; only then did I make sense of those places by noticing the links among points in space . . . . The streets and the city seemed much more real than in those days when I could go past them, perceive their smell, hear their sounds, touch their walls. (Carofiglio 58–60, 115)[4]

4 The translation from the Italian is by Annalisa Andrigo, to whom I am grateful. These passages were previously highlighted by Papotti within an essay on the emerging 'urban narrative' genre in Italy.

Adopting an interpretative mode analogous to the place-centred approach of the French 'geocriticism' (see Prieto 'Geocriticism', 21), here, one can adopt a sort of 'cartocriticism', a cartographically-centred mode of analysis that is less concentrated on poetics than on cartographic (or specifically geovisual, as in this case) practice. A recent Italian collection of essays on 'maps in literature' (Guglielmi and Iacoli) provides some interesting paths towards a cartocritical analyses.[5] Whereas some of the essays consider the literary treatment of cartographic themes to investigate the author, in some of them the literary analyses provide insights into the investigation of cartography in itself. I have applied this approach more explicitly to the mapping practices appearing in Cormac McCarthy's *The Road* (see Rossetto 'Theorizing Maps').

A further way to engage with geovisuality and literature emerges from the fact that, as seen above, the term 'geovisual' evokes the collapse of a range of binaries. Since Google Earth was launched in 2005, unprecedented 'slippy' maps have become a part of our everyday lives. As has been noted, because 'users can shift seamlessly between the God-like perspective of the whole globe and the viewing of details normally available only through physical presence', 'Google Earth is a major change from the traditional strong distinction between maps and globes and the actual landscapes' (Jensen 129, 131). Following Veronica Della Dora, the simple act of zooming in or switching to Street View merges two traditions of spatial representation that differ in scale and mode: 'quantitative vs. qualitative, maths vs. art, space vs. place, specialised training vs. amateur skills, . . . the grid vs. pictorial vignettes' (6). Amongst these binaries, the opposition between authority and emancipation stands out. Significantly, an awareness of the critical implications of this quotidian 'geovisual turn' came early in postcolonial criticism: a field of literary studies that has traditionally been particularly sensitive to the geopolitics of cartography. Contributing to a postcolonial literature companion with a specific focus on the 'geovisual turn', David Howard noted that postcolonial theorists have a particular interest in maps because 'cartography has consistently provided the graphic arm of colonial enterprise' (148). Saliently, for Howard, digital geovisualisation and collaborative mapping processes, which divorce cartography from colonial authority, have led to a 'renewed politics of maps', in which postcolonial thinkers, practitioners and writers are increasingly involved (151). To provide an example, in Belfast-based writer Ciaran Carson's work, Neal Alexander has identified a convergence of 'map as imposition' and 'map as resistance', as well as 'map as controlling and totalizing' and 'map as open and rhizomatic' (Deleuze and Guattari's famous term). Alexander associates the aerial view with the first attitude and the street view with the second one. The rhizomatic, dynamic depiction of the city by Carson, Alexander

5 For an English review, see Rossetto, 'Italian Literary Mappings'.

states, envisages 'the germ of an utopian mapping, one that is predicated not upon elevated distance but on an approach to the city at the level of its streets' (527). Here, a critical practice forged by a geovisual imagery is clearly at work.

The interest in the comparison between maps and views, as well as aerial perspectives and grounded visions, should be linked with the fact that literary critics are embedded within a geovisual culture. Such work has frequently been carried out. Consider, for instance, Grahame Smith's analysis of the dual, binary literary-cinematic treatment of London and Paris provided by Charles Dickens in *A Tale of Two Cities*, with London visually experienced and written at the street level (as a labyrinth) and Paris perceived and verbally depicted from above (as a bird's eye view). Benjamin A. Brabon, to provide just one more example, examined Ann Radcliffe's romances, suggesting that the author combines panoramic surveys of space and landscape vistas, thus acting as a literary cartographer and also as a literary painter, an 'amalgamation' that Brabon names using the term 'Gothic cartography' (840). However, the map/view opposition seems to be increasingly scrutinised through literary repertoires within geoliterary research. Giancarlo Alfano, whilst introducing his recent book entitled *Paesaggi, Mappe, Tracciati (Landscapes, Maps, Layouts)* (after Moretti's well-known *Graphs, Maps, Trees*), analyses the shift from classical literary accounts that adopt views from above and tidy, static descriptions of landscapes contemplated from fixed vantage points (that is, a cartographic, or 'perspective', paradigm) to contemporary accounts adopting mobile gazes, internal and unstable perspectives and immersive and fragmented visions (that is, a phenomenological, 'optic' or even 'haptic' paradigm). Alfano, moreover, refers to an 'archaeology of the mobile gaze', tracing this tradition back to particular genres, such as travel literature, or highlighting cases in which this shift from the cartographic to the phenomenological happens within a single literary work.

Some literary analyses, in fact, retrace an 'archaeology of the geovisual' in the past in a way that can be compared to the work carried out by Caquard in the field of cinemaps ('Foreshadowing Contemporary Mapping'). Geovisualisation, in fact, has also been viewed as the complement to the more general '*art* approach that cartography had used previously', at a time when painting and map-making were not yet clearly separate activities (Cartwright, Miller and Pettit 27). Thus, current geovisual tools, as well as many other forms of new digital media, can also be considered to be the 're-mediation' (Bolter and Grusin) of past forms of geovisualisation. Literature, after all, with its multi- and trans-scalar verbal navigations, is also a very fluid geovisual tool. It would be interesting to research how literature in the pre-digital era adopted formal 'geovisual' strategies. Within the *Mapping the Lakes* research project, carried out at Lancaster University through the development of qualitative literary GIS and interactive literary maps based on Google Earth, the research team quotes a description extracted from

William Wordsworth's *Guide to the Lakes*, which literalises a form of visual and spatial experience now felt by Google Earth users:

> To begin, then, with the main outlines of the country; I know not how to give the reader a distinct image of these more readily, than by requesting him to place himself with me, in imagination, upon some given point; let it be the top of either of the mountains, Great Gavel, or Scawfell; or, rather, let us suppose our station to be a cloud hanging midway between those two mountains . . . ; we shall then see stretched at our feet a number of valleys, not fewer than eight, diverging from the point, on which we are supposed to stand, like spokes from the nave of a wheel. (*Mapping the Lakes*)

Google Earth, which 'offers a movement beyond the cartographical limitations of GIS through its visual representation of the contoured landscape', re-mediates the literary text, allowing the user 'to follow the Wordsworthian spatial model' (*Mapping the Lakes*).

With regard to pre-digital geovisual practices, Alison Byerly has focused on the textual evocations of two phenomena crucial to the Victorian era's visual culture (and fiction): namely, the panorama viewed by the stationary spectator and the mobile scenic experience made possible by hot-air balloon travel. Byerly argues that narrative strategies informed by these visual perspectives provided forms of virtual travel. Whereas panoramas provided an immersive, experiential perception of landscape, ballooning provided a topographical, detached perception of it. This second practice, Byerly shows, was continually related to the panoramic one within textual descriptions. Balloon ride accounts, which frequently refer to the map-like appearance of the country when seen from above, prefigured visual practices that are very similar to the navigation of virtual globes, with the combination of oblique and vertical views and the shifting of scales of observation. An interesting verbalisation of the balloon experience that directly connects it to satellite technology at a time when this technology had not yet entered our everyday visual/mapping practices is offered by Italian writer Andrea Zanzotto. In his essay 'Venice, Maybe' (1978), which was included in a famous photo-book on Venice, after imagining an ideal 'immersive' approach to the watery city on board a floating slab of earth moved by the tide, he literalises a hypothetical departure from it as follows:

> One should perhaps say goodbye to Venice, its estuaries and surroundings, not from the height of an aeroplane, with its frantic roar and superabundance of crude energy, but from that of a benign gas balloon taken in tow by a satellite, whirling and tipping in search of larger horizons. The mysterious scarab of the city, wounded and wounding, burns far below in the middle of the lagoon, nourished by it and yet seemingly immolated within itself . . . . We see winding watercourses, blunt

> seawalls, improbable sea-vegetation; and in the pallor, say, of the grass contained inside an orthogonal design of wooden piles we may able to guess at an extinct topography. As we whirl higher and higher we are able to observe better that strangely tense and yet delicate embrace between land and sea, and its hundred casual positions, the elbowing of sweet waters by salt ones, and also the soft and apparently salubrious tints of chemical deposits. (Zanzotto)

Here, it seems that the 'as if' character of intermedial references is at work, as described by Irina Rajewski, since the literary text embeds a sense of geovisual qualities, imitating by textual means an imaginary mix of old (the balloon) and new (the satellite) geovisual practices (54–5).

Conclusion

This chapter has traced a series of lines of reflection and intervention regarding the geovisuality-literature connection. One of the major challenges that a geovisual paradigm presents to the humanities is overcoming the 'view from above/from the ground' dichotomy, which has often been associated with many binary oppositions, such as technical/humanist, quantitative/qualitative, maths/art, grid/pictorial and space/place-landscape. Although many studies of visual culture and literature that follow a historical perspective tend to separate maps from visual media, more specific attention to the 'geovisual' dimension could be used in exploring hybridised forms within past scopic regimes and visual techniques, acknowledging literary accounts of geovisual practices in the pre-digital era. Literature also reflects contemporary geovisual ways of experiencing and seeing space and provides narrations of these new mapping/visualising practices. Current geovisual practices should thus be analysed in terms of their verbalisation through literary texts. This is of great interest for map scholars as well because literary texts elicit practices that are often mute, emphasise neglected aspects of mapping devices, give shape to cartographic emotion and verbalise map embodiment and geovisual affective experiences. Those same devices, experiences and emotions are now increasingly involved in everyday reading practices, as well as in the teaching of literature. A geovisual culture can be seen at work when literary critics use the geovisual lexicon in their analyses. Literary scholars should pay attention to this phenomenon not only by developing highly sophisticated geovisual tools in collaboration with cartographers (or map artists and graphic designers) but also by acknowledging the pervasive presence of lay geovisual practices that inform both reading and writing experiences. The aforementioned aspects are some of the various possible directions into which we can explore and investigate the connection between geovisuality and literature. The tentative review that was provided within this chapter, which clearly requires further practical research, mainly intends to point out the need for a specific focus on the geovisual as an increasingly common dimension of our everyday visual and spatial practices. The relationship between geovisuality and literature appears to be a plausible object of interest for studies in intermediality, but it could also be at the centre of a productive space of confrontation between map scholars and literary scholars: a space where the exchange of different competencies, creative suggestions or even destabilising visions should be reciprocal.

Works Cited

Agnew, John A. 'Space and Place'. *Handbook of Geographical Knowledge.* Eds, John A. Agnew and David N. Livingstone. London: Sage, 2011. 316–30.

Alexander, Neal. 'Mapping Junkspace: Ciaran Carson's Urban Cartographies'. *Textual Practice* 21.3 (2007): 505–32.

Alfano, Giancarlo. *Paesaggi Mappe Tracciati. Cinque studi di letteratura e geografia.* Napoli: Liguori, 2010.

Andrea Zanzotto, Geopoeta della Grande Guerra' *Limes: Rivista Italiana di Geopolitica.* 20 May 2014. Web. 20 March 2015. <http://www.limesonline.com/rubrica/andrea-zanzotto-geopoeta-della-grande-guerra?refresh_ce>.

Axon, Stephen, Janet Speake and Kevin Crawford. '"At Next Junction, Turn Left": Attitudes Towards Sat Nav Use'. *Area* 44.2 (2012): 170–77.

Berson, Ilene R. and Michael J. Berson. 'Integrating Literature and the Social Studies with Google Lit Trips'. *Social Education* 75.2 (2011): 111–13.

Bodenhamer, David J., Trevor M. Harris and John Corrigan. 'Deep Mapping and the Spatial Humanities'. *International Journal of Humanities and Arts Computing* 7.1–2 (2013): 170–75.

Bolter, Jay D. and Richard Grusin. *Remediation. Understanding New Media.* Cambridge: MIT, 2000.

Brabon, Benjamin A. 'Surveying Ann Radcliffe's Gothic Landscapes'. *Literature Compass* 3.4 (2006): 840–45.

Brosseau, Marc. 'Literature'. *International Encyclopedia of Human Geography.* Eds, Rob Kitchin and Nigel Thrift. Amsterdam: Elsevier, 2009. *ad vocem.*

Byerly, Alison. '"A Prodigious Map Beneath His Feet": Virtual Travel and the Panoramic Perspective'. *Nineteenth-Century Contexts* 29.2–3 (2007): 151–68.

Canali, Laura, and Camilla Miglio. 'Cartographies de l'Imaginaire'. *Géocritique: État des lieux/ Geocriticism: A Survey.* Eds, Clément Lévy and Bertrand Westphal. Limoges: Pulim, 2014. 200–14.

Caquard, Sébastian. 'Foreshadowing Contemporary Digital Cartography: A Historical Review of Cinematic Maps in Films'. *The Cartographic Journal* 46.1 (2009): 46–55.

———. 'Cartography I: Mapping Narrative Cartography'. *Progress in Human Geography* 37.1 (2011): 135–44.

Caquard, Sébastien, and Jean-Pierre Fiset. 'How Can We Map Stories? A Cybercartographic Application for Narrative Cartography'. *Journal of Maps* 10.1 (2014): 18–25.

Caquard, Sébastien, Laurene Vaughan and William Cartwright. 'Mapping from Above/Mapping from the Ground: Mapping Environmental Issues in the City'.

Mapping Environmental Issues in the City. Arts and Cartography Cross Perspectives. Eds, Caquard, Sébastian, Laurene Vaughan and William Cartwright. Berlin: Springer, 2011. 1–8.

Caquard, Sébastien and William Cartwright. 'Narrative Cartography. From Mapping Stories to the Narrative of Maps and Mapping'. *The Cartographic Journal. Cartography and Narratives* Special Issue 51.2 (2014): 101–6.

Carofiglio, Gianrico. *Né qui né altrove. Una notte a Bari.* Rome-Bari: Laterza, 2008.

'Cartografie dell'immaginario'. *Limes: Rivista Italiana di Geopolitica.* Dec. 2015. Web. 20 March 2015. <http://www.limesonline.com/>.

Cartwright, William, Suzette Miller and Christopher Pettit. 'Geographical Visualization: Past, Present and Future Development'. *Journal of Spatial Science* 49.1 (2004): 25–36.

Cavanaugh, Terence, and Jerome Burg. *Bookmapping. Lit Trips and Beyond.* Washington, DC: International Society for Technology in Education, 2011.

Cerreti, Claudio. 'In margine a un libro di Franco Moretti. Lo spazio geografico e la letteratura'. *Bollettino della Società Geografica Italiana* 3.1 (1998): 141–8.

Cometa, Michele. *Parole che dipingono. Letteratura e cultura visuale tra Settecento e Novecento.* Roma: Meltemi, 2004.

Cooper, David, and Ian N. Gregory. 'Mapping the English Lake District: A literary GIS'. *Transactions of the Institute of British Geographers* 36.1 (2010): 89–108.

Craine, Jim, and Stuart C. Aitken. 'The Emotional Life of Maps and Other Visual Geographies'. *Rethinking Maps. New Frontiers in Cartographic Theory.* Eds, Martin Dodge, Rob Kitchin and Chris Perkins. London: Routledge, 2009. 149–66.

Dear, Michael. 'Afterword. Historical Moments in the Rise of the Geohumanities'. *GeoHumanities. Art, History, Text at the Edge of Place.* Eds, Michael Dear et al. London: Routledge, 2011. 309–14.

De Certeau, Michel. *The Practice of Everyday Life.* Trans. Steven F. Rendall. Berkeley: University of California Press, 1984.

Deleuze, Gilles and Félix Guattari. *A Thousand Plateaus.* Capitalism and Schizophrenia. Vol. 2. Trans. Brian Massumi. Minneapolis: University of Minnesota Press, 1987.

Della Dora, Veronica. 'A World of "Slippy Maps": Google Earth, Global Visions, and Topographies of Memory'. *Transatlantica* 2 (2012): 1–20.

Density Design. 'CHARLES DICKENS – La Lettura #10'. Bēhance. 2015. Web. 20 March 2015. <https://www.behance.net/gallery/2984617/CHARLES-DICKENS-La-Lettura-10-Corriere-della-Sera>.

Dodge, Martin, Chris Perkins and Rob Kitchin. 'Mapping Modes, Methods and Moments: A Manifesto for Map Studies'. *Rethinking Maps. New Frontiers in Cartographic Theory.* Eds, Martin Dodge, Rob Kitchin and Chris Perkins. London: Routledge, 2009. 220–43.

Dodge, Martin, Mary McDerby and Martin Turner, (eds) *Geographic Visualization. Concepts, Tools and Applications.* Chichester: Wiley, 2008.

Elfenbein, Andrew. 'Cognitive Science and the History of Reading'. *PMLA* 121.2 (2006): 484–502.

Elwood, Sarah. 'Geographic Information Science: Visualization, Visual Methods, and the Geoweb'. *Progress in Human Geography* 35.3 (2010): 401–8.

Farquharson, Vanessa. 'What the Joads saw: Does Google Lit Trips Enrich Literature?' *National Post* 28 January 2011. n. pag. 20 March 2015.

Guglielmi, Marina and Giulio Iacoli, (eds) *Piani sul mondo. Le mappe nell'immaginazione letteraria.* Macerata: Quodlibet, 2013.
Harris, Trevor M., Susan Bergeron and L. Jesse Rouse. 'Humanities GIS. Place, Spatial Storytelling, and Immersive Visualization in the Humanities'. *GeoHumanities. Art, History, Text at the Edge of Place.* Eds, Michael Dear et al. London: Routledge, 2011. 226–40.
Hopkins, Martha E. and Michael Buscher. *Language of the Land. The Library of Congress Book of Literary Maps.* Washington: Library of Congress, 1999.
Horstkotte, Silke and Nancy Pedri. 'Introduction: Photographic Interventions'. *Poetics Today* 29.1 (2008): 1–29.
Howard, David. 'Cartographies and Visualization'. *A Concise Companion to Postcolonial Literature.* Eds, Shirley Chew and David Richards. Oxford: Blackwell, 2010. 141–60.
Knigge, LaDona, and Meghan Cope. 'Grounded Visualization: Integrating the Analysis of Qualitative and Quantitative Data Through Grounded Theory and Visualization'. *Environment and Planning A* 38.11 (2006): 2010–37.
Jensen, Jakob L. 'Augmentation of Space: Four Dimensions of Spatial Experience of Google Earth'. *Space and Culture* 13.1 (2010): 121–33.
Jung, Jin-Kyu and Sarah Elwood. 'Extending the Qualitative Capabilities of GIS: Computer-Aided Qualitative GIS'. *Transactions in GIS* 14.1 (2010): 63–87.
Lammes, Sybille. 'Terra Incognita: Computer Games, Cartography and Spatial Stories'. Eds, Marianne van den Boomen et al. *Digital Material. Tracing New Media in Everyday Life and Technology.* Amsterdam: Amsterdam University, 2009. 223–38.
Leszczynski, Agnieszka. 'Quantitative Limits to Qualitative Engagements: GIS, Its Critics, and the Philosophical Divide'. *The Professional Geographer* 61.3 (2009): 350–65.
Lock, Gary. 'Representations of Space and Place in the Humanities'. *The Spatial Humanities. GIS and the Future of Humanities Scholarship.* Eds, David J. Bodenhamer, John Corrigan and Trevor M. Harris. Bloomington: Indiana University, 2010. 89–108.
Luria, Sarah. 'Spatial Literacies. Geotexts'. *Art, History, Text at the Edge of Place.* Eds, Michael Dear et al. London: Routledge, 2011. 67–70.
Macfarlane, Robert. *The Wild Places.* London: Granta, 2007.
Mapping the Lakes: A Literary GIS. N.d. Web. 20 March 2015. <http://www.lancaster.ac.uk/mappingthelakes/>.
Marston, Sallie A., John Paul Jones III and Keith Woodward. 'Human Geography Without Scale'. *Transactions of the Institute of British Geographers* 30.4 (2005): 416–32.
Mirzoeff, Nicholas. 'On Visuality'. *Journal of Visual Culture* 5.1 (2006): 53–79.
Mittman, Asa. 'Inverting the Panopticon: Google Earth, Wonder and Earthly Delights'. *Literature Compass.* 9.12 (2012): 938–54.
Mondada, Lorenza. *Verbalisation de l'espace et fabrication du savoir.* Lausanne, Université de Lausanne, 1994.
Montello, Daniel R. 'Cognitive Research and GIScience: Recent Achievements and Future Prospects'. *Geography Compass* 3.5 (2009): 1824–40.
Mott, Rick. 'Digitizing Leslie Silko's Laguna Landscape'. *Image [&] Narrative* 14 (2006). n. pag. 20 March 2015.
Moslund, Sten P. 'The Presencing of Place in Literature. Toward an Embodied Topopoetic Mode of Reading'. *Geocritical Explorations. Space, Place, and Mapping in Literary and Cultural Studies.* Ed. Robert T. Tally Jr. New York: Palgrave Macmillan, 2011. 29–43.

Muehrcke, Phillip C. and Juliana O. Muehrcke. 'Maps in Literature'. *The Geographical Review* 63.3 (1974): 317–38.
Ng-Chan, Taien. 'Mapping Out *Patience*: Cartography, Cinema, and W.G. Sebald'. *Proceedings of the International Cartographic Conference* 2013: n. pag. 20 March 2015.
Papotti, Davide. 'Racconti di città: strategie di interpretazione urbana nella collana "Contromano"'. *La geografia del racconto. Sguardi interdisciplinari sul paesaggio urbano nella narrativa italiana contemporanea*. Eds, Davide Papotti and Franco Tomasi. Bruxelles: Peter Lang, 2014. 35–57.
Pearce, Margaret. 'Framing the Days: Place and Narrative in Cartography'. *Cartography and Geographic Information Science* 35.1 (2008): 17–32.
Prieto, Eric. 'Geocriticism, Geopoetics, Geophilosophy, and Beyond'. *Geocritical Explorations. Space, Place, and Mapping in Literary and Cultural Studies*. Eds, Robert T. Tally Jr. New York: Palgrave Macmillan, 2011. 13–27.
Rjewski, Irina O. 'Intermediality, Intertextuality and Remediation: A Literary Perspective on Intermediality'. *Intermédialités* 6 (2005): 43–64.
Roberts, Les, (ed.) *Mapping Cultures: Place, Practice, Performance*. Basingstoke: Palgrave Macmillan, 2012.
Rossetto, Tania. 'Embodying the Map. Tourism Practices in Berlin'. *Tourist Studies* 12.1 (2012): 28–51.
———. 'Theorizing Maps with Literature'. *Progress in Human Geography* 38.4 (2014): 513–30.
———. 'Italian Literary Mappings'. *Society and Space*. 2014. Web. 20 March 2015. <http://societyandspace.com/2014/10/22/rossetto/>.
Ryan, Marie-Laure. 'Cognitive Maps and the Construction of Narrative Space'. *Narrative Theory and the Cognitive Sciences*. Ed. David Herman. Chicago: University of Chicago, 2003. 214–42.
———. 'Space'. *The Living Handbook of Narratology*. Eds, Hühn, Peter et al. Hamburg: Hamburg University. 20 March 2015.
Sells, Erin D. 'Mapping *Mrs. Dalloway*: Teaching Woolf with Google Earth™'. *Virginia Woolf Miscellany* 73 (2008): 28–30.
Setten, Gunhild. 'Fusion or Exclusion? Reflections on Conceptual Practices of Landscape and Place in Human Geography'. *Norwegian Journal of Geography* 60.1 (2006): 32–45.
Smith, Grahame. *Dickens and the Dream of Cinema*. Manchester: Manchester University Press, 2003.
Turchi, Peter. *Maps of the Imagination: The Writer as Cartographer*. San Antonio, TX: Trinity University, 2004.
Wells, Amy. 'La cartographie comme outil d'analyse littéraire: des cartes métaphoriques aux cartes SIG'. Eds, Maleval, Véronique, Marion Picker and Florent Gabaude. *Géographie poétique et cartographie littéraire*. Limoges: Pulim, 2012. 169–86.
Westphal, Bertrand. *Geocriticism: Real and Fictional Spaces*. Trans. Robert Tally Jr. New York: Palgrave Macmillan, 2011.
Young, Sarah J., and John Levin. 'Mapping Machines: Transformations of the Petersburg Text'. *Primerjalna književnost* 36.2 (2013): 151–62.
Zanzotto, Andrea. 'Venice, Maybe'. *Living Venice*. Photographs by Roiter, Fulvio. English version ed. by Maurice Rowdon. Udine: Magnus, 1978. n. pag.

14 'Setting the globe to spin'

Digital Mapping and Contemporary Literary Culture

David Cooper

Introduction

In an essay published in *The Guardian* in April 2014, Rachel Hewitt offered a lament for the declining popularity of the Ordnance Survey paper map: an elegy which was predicated on both an unapologetic nostalgia for remembered places and an understandable celebration of the tactility and texturality of the material map. There is much to admire in Hewitt's polemical and poetical defence of the 'emotional, physical and imaginative' power of the paper map ('Turn Around When Possible' 3). At the same time, however, there is a danger that the fetishising of paper maps can lead to a reactionary dismissal of the new imaginative geographies that have been opened up by geospatial technologies. There is a need, then, to counter such cartographic conservatism by exploring how the practices of contemporary creative writers have been shaped by the emergence of a suite of digital technologies. How, for instance, have contemporary poets represented geospatial technologies such as Google Earth in conventional literary texts? Is it possible to identify some examples of how writers have actually produced new digital maps as part of their own creative practices? By extension, in what ways have writers sought to harness the potential of digital technologies to think about the embodied materiality of what it means to be-in-the-world? This chapter addresses these questions by examining some key examples of how contemporary British writers – including Paul Farley, Charles Cumming and, most importantly, Rachel Lichtenstein – have either represented or used digital maps and mapping practices in poetic, fictional and non-fictional place writings. Moreover, this chapter is undergirded by the belief that the conceptual thinking and digital practices of critical literary map-makers will be enriched through an engagement with the self-reflexive and processual work of contemporary creative practitioners.

The Poetics of Geospatial Technologies: Paul Farley's 'Google Earth' and Neil Rollinson's 'GPS'

As Jerry Brotton points out, Plato, in the *Phaedo*, imagined seeing the world from above as 'a gleaming, perfect sphere'; yet it was not until the early

1970s that an image of the whole earth – 'the geographer's ultimate object of study' – was first made available via a photograph taken from NASA's Apollo 17 spacecraft (Brotton 405). Now, in the second decade of the twenty-first century, however, anyone with a sufficiently high-speed internet connection is able to undertake a dizzying journey over the whole world via Google Earth: a virtual experience whose extraordinary popularity is attributable, according to Vittoria Di Palma, to 'its propensity to tap into deeply rooted and widely held fantasies of unfettered vision' (Di Palma 240). Paul Farley's 2012 poem, 'Google Earth', opens with a breathless acknowledgment of the ways 'the world's most popular geospatial application' (Brotton 405) has radically recalibrated the popular geographical imagination:

> Now I'm a hand setting the globe to spin,
> finding a country, starting to zoom in
> now I'm an eye. Now I'm a meteorite. (Farley 16)

Immediately, then, Farley's poem is alert to the experiential and imaginative possibilities opened up by a technology that valorises 'the instantaneity of seamless, hassle-free travel without leaving the safety of a home computer' (Presner et al. 85).

Farley's 'Google Earth', though, moves beyond an uncomplicated celebration of this totalising global vision to highlight a series of interlinked practices that are integral to an interrogation of the ways desktop geospatial technologies are shaping the contemporary geographical imagination. First, there is the issue of control as, through its powerful and seamless zoom functionality, Google Earth empowers users to adjust the representational scale and, with disorientating speed, to focus in on the particularities of landscape (Di Palma 263). Farley's Emersonian all-seeing 'I'/'eye' begins by 'finding a country' and then, at the beginning of the poem's second tercet, zooms in to sweep over a patchwork of distinct topographies: 'The scars of business corridors, the white | clay works, national parkland, estuaries' (Farley 16). In this instance, then, the manipulable geographical perspectives afforded by Google Earth allow Farley to validate the vision of the English landscape he and Michael Symmons Roberts articulate in their topographical prose book, *Edgelands* (2011): a collaborative text in which seemingly difficult-to-reconcile terrains are shown to exist cheek-by-jowl within the over-populated, contested space of the writers' native country.

Farley's speaker does not only utilise the zoom function to offer an objective landscape character assessment, however, but also uses the geospatial technology to reflect on the self-in-landscape: a second key practice. According to Brotton, a dualism characterises the egocentrism of online mapping. In one sense, Google Earth elevates users 'like gods, inviting them to take flight and look down upon the earth from a divine viewpoint' (Brotton 9). At the same time, the user's instinctive desire is almost invariably 'to first locate him- or herself on the digital map, by typing in their home address before anywhere else, and zooming in to see that location' (Brotton 9).

Although Farley's speaker does not zoom in on his 'home address', the egocentrism signalled by the use of the first-person in the opening line continues throughout the poem as the speaker's gaze moves from the generic ('black holes/ of reservoirs, flight paths of major roads') to the particular as he virtually travels 'down lost bus routes, birth streets, | the school roof still in bad need of repair' (Farley 16). Google Earth, therefore, enables the speaker – in 'a timeless act of personal reassurance' (Brotton 9) – to layer, imaginatively, his remembered self onto and into digital cartographic space. As a result, Farley's speaker moves from using the titular technology for the digital envisioning of the wider contemporary landscape to the egocentric and nostalgic revisiting of (virtual) place.

'Google Earth' touches upon a third theme that is integral to thinking about desktop digital literary mapping practices: the relationship between geospatial technologies and the embodied situatedness of everyday life. As Di Palma points out: 'The view from above . . . displaces us as viewers: it offers an extended panorama, and diminishes the importance of our quotidian concerns by decentering our habitual perspective' (Di Palma 262). Over the course of the poem, Farley's speaker places especial emphasis on the personal disembodiment of the desktop traveller as he is able to imagine himself variously as a 'meteorite', 'a drop of rain' and 'a balloon by Odilon Redon' (Farley 16): images which correspond with Brotton's suggestion that Google Earth is capable of offering the user 'a transcendent moment of contemplation, beyond time and space, seeing everywhere from nowhere' (Brotton 9). The poem is destabilised by a tension, however, as the description of how 'there are never clouds | because the west wind of the Internet | blows silently' (Farley 16) suggests a diminishing of the phenomenological experience of what it means to be-in-the-world. Whereas Percy Bysshe Shelley famously apostrophised the 'west wind' as 'Wild Spirit, which art moving everywhere – | Destroyer and Preserver' (Shelley 577), the virtual world presented through Google Earth is uncannily suspended in both time and space. The user envisions a dynamic sense of movement as he or she freely flies over represented space; but the geographical environment through which the user creates a path is both cloudless and eerily devoid of activity.

Farley's 'Google Earth' exemplifies how contemporary poets are still using traditional literary forms and formats to reflect upon the new spatialities opened up by digital technologies. In this specific instance, Farley's poem is organised into eight tercets followed by a stand-alone climactic line and there is a frequent – if variable – use of end-rhyme. In addition, the short lyric was written for the printed page and appeared in the poet's fourth major collection, *The Dark Film*, published by Picador. An analogous example of such literary formalism is provided by another English writer, Neil Rollinson, in 'GPS'; although, saliently, Rollinson removes the poetry of geospatial technologies from the sequestered space of the home and, instead, takes it out into-the-field. In 'GPS', Rollinson similarly employs tercets (although he offers ten rather than eight) when recounting getting lost on a walk up the Lake District fell of High Raise just to the north of William Wordsworth's Grasmere: 'You walk | by your nerves, the map redundant as you sink | into darkness and panic. Thank God | for your GPS' (Rollinson 8). Here, the egocentrism afforded by digital mapping technologies is ironically celebrated as it enables Rollinson's speaker to situate himself in the Cumbrian uplands and to plot his 'route step by step' as, eventually, he descends 'into a gentler landscape' (Rollinson 8). Whereas Farley meditates upon the God-like geographical visions opened up to the desktop traveller, therefore, Rollinson's speaker jokingly thanks God for the fact that handheld Global Positioning System technology allowed him to find his way home – 'buzzing, walking on air | all the way to the pub' (Rollinson 8) – and, by extension, the opportunity to narrate his spatial story. 'GPS', then, is concerned with how contemporary technologies can shape embodied landscape practices and (self-dramatising) spatial performances; yet, in formal terms, the poem remains resolutely traditional.

Geo-Locating the Places of Fiction: Charles Cumming's 'The 21 Steps'

In sketching out some key tropes in contemporary literary practices, it is necessary to move from the consideration of poems *about* the digital representation of geographical space to offer some reflections on what might be described – in an adaptation of a term used by Sarah Luria – as *digital* geo-texts (Luria 67–70). In other words, there is a need to engage with a selection of projects in which actual geospatial visualisations are produced and, on occasion, proffered as an intrinsic part of the reading experience.

Perhaps unsurprisingly, initial experiments in the making of creative digital literary maps were underpinned by an essentially positivist celebration of the potential for geo-locating the places of fiction. 'The 21 Steps' (2008), written by Charles Cumming, formed part of 'We Tell Stories': an online project in which six authors were each commissioned by Penguin Books to write new short fiction inspired by an earlier text. Cumming – the author of a series of successful thrillers – took John Buchan's 1915 adventure novel, *The Thirty-Nine Steps*, as the imaginative springboard for his own born-digital short story. Of greater relevance, in the current context, is that Google Earth, Google Maps and Google Street View were all employed to provide the user with 'slippy' (della Dora 2012) geovisualisations of the movements of the narrator, Rick Blackwell, within Cumming's intertextual remaking of the classic man-on-the-run narrative.

On entering 'The 21 Steps' pages of the 'We Tell Stories' website, the user is immediately offered a satellite view of an as-yet-unnamed cityscape and is provided with the information that: '*The 21 Steps* is told by following the story as it unfolds across a map of the world' (Cumming). Crucially, then, the user's starting-point is the cartographical, rather than textual,

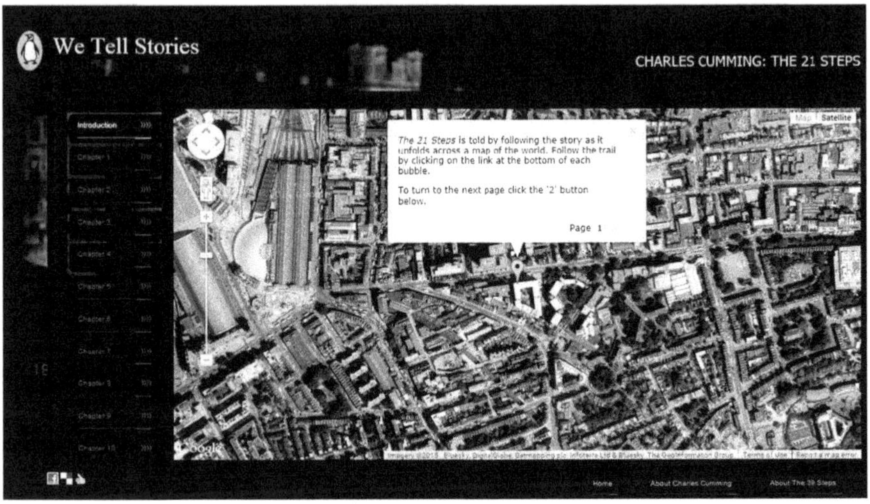

Figure 14.1 Screenshot of the Introduction to Charles Cumming's 'The 21 Steps'.
Source: © Penguin Books Ltd.

representation of urban space; he or she begins with the digital mapping of the built environment rather than the written word. This creative preoccupation with location is underscored when the user navigates his or her way through further prefatory information to reach the opening lines of the first of Cumming's twenty-one chapters. Conventional Google markers pin-point a site in the bottom right-hand corner of an unidentified large building viewed from above and, with a knowingly enigmatic turn-of-phrase, the accompanying text bubble states: 'I was the wrong man in the wrong place at the wrong time' (Cumming). Clicking on the second page of the chapter brings up more specific geographical information and roots the narrative in the networked space of central London: 'I'd only come to St Pancras to see what all the fuss was about'. The narrator then goes on to reveal how, in the absence of having 'anything better to do', he had taken a ten-minute walk from his flat in order to sample the Champagne Bar at the recently opened Eurostar terminal in St Pancras Station (Cumming). It is immediately clear, then, that 'The 21 Steps', through its synthesis of Google geo-visualisations and the written word, is a geotext predicated on an imaginative attentiveness to place.

As the story unfolds, the user is taken on a short trip along Euston Road to the British Library before being led on slightly longer journeys in and around London; and it is on these more extensive journeys over and above the labyrinthine city that the creative potentiality of Google Earth technology begins to emerge. As Di Palma points out, Paul Virilio argues that 'speed – the defining characteristic of our contemporary virtual and digital age – has made formerly key notions like public and private, and the distinctions between them, entirely defunct' (Di Palma 263–4). As a result, the cinematic speed with which the user of 'The 21 Steps' is transported across and down into the cityscape seems to be entirely appropriate for a text that self-consciously and playfully seeks to update a classic novel of surveillance and espionage for the digital age. It is an idea that is reinforced by Cumming's own reflections on the project: 'Plot was everything. Suspense was everything. It was all about pace and movement' (Black 2008). As with Buchan's Richard Hannay, Cumming's narrator soon leaves London to head north to Scotland; but whereas the protagonist of *The Thirty-Nine Steps* takes the train to the rural terrain of Galloway, Rick Blackwell, flies up to the urban space of Edinburgh from Heathrow Airport. On this journey north, the user's view of the landscape automatically shifts to Google Maps; but when Blackwell enters Edinburgh, the user is once again afforded satellite imagery and visualisations of the detailed particularities of this built environment.

In what ways, then, does Cumming's project signal a new form of creative practice? In her article, 'Cartographies of Digital Fiction: Amateurs Mapping a New Literary Realism', Annika Richterich examines the 'mashed-up' nature of 'geomedia fictions': creative projects which offer a movement towards a 'literary neogeography' by making maps an essential component of the digital artefact (Richterich 238–9). In charting the emergence of this new digital genre of 'geomedia fictions', Richterich focuses on two key projects which invite the reader to move through both textual and cartographical representations of geographical space: the novel, *Senghor on the Rocks,* which first appeared in digitised and cartographised form in 2008; and, saliently, 'The 21 Steps'. Richterich, in exploring Cumming's move towards a creative 'literary neogeography', both places critical emphasis on the way the 'map itself acts as graphical user interface' and highlights how the project opens up a new generic classification of 'fiction in interactive maps' (Richterich 242): a formulation which reinforces how the digital architecture of 'The 21 Steps' privileges the cartographic over the textual.

Yet, as well as gesturing towards the possibilities of a new form of creative digital literary mapping, Cumming's project simultaneously raises a series of issues and questions. 'The 21 Steps' circumnavigates the limitations of the conventional plane-view mapping of landscape by allowing the user, through Google Street View, to move through the (digital) built environments of London and Edinburgh. Yet, although Google Street View enables the user to place him or herself onto a virtual Euston Road, the interior space of the British Library – in which dialogue takes place and action unfolds – remains frustratingly out of bounds. This sense of readerly frustration is in imaginative accordance with the cardinal themes of Cumming's digital rewriting of Buchan's espionage novel; and the way Blackwell is thwarted in his attempts to enter Gallery 66 of the National Gallery – the usually public space is 'Closed for Refurbishment' – highlights how, for both reader and narrator, particular interiors remain tantalisingly out of sight. In other digital literary mapping projects, however, it might be difficult to offer a similar theorisation of the relationship between textual content and spatial visualisation and this inability to enter the intimate spaces of architectural interiors might simply highlight the intrinsic limitations of such geo-visualisation practices. Additional problems of spatial presentation are identified by Richterich who argues that the creative digital literary mappings showcased in 'The 21 Steps' project 'primarily utilise maps in a pictorial sense and therefore tend to generate rather basic mappings' (Richterich 243). For Richterich, then, the richly complex, non-linear geographies almost invariably embedded in literary texts demand the exploratory development of new cartographic 'aesthetics' (Richterich 243) that transcend the limitations of an exclusive reliance on Google data and technologies.

Geo-Visualisation as a Promotional Tool: Zadie Smith's *NW*

In an illuminating interview, Cumming stresses that 'The 21 Steps' was a collaborative project – undertaken with the programming agency, Six to Start, as well as Penguin – in which the narrative was shaped by the demands for the creative production of geographical data which could be readily visualised in digital form (Black). As Richterich succinctly puts it: 'The proposed design, thus, dominated the writing process from the start' (Richterich 243).

Richterich endeavours to further unpick the knotty entanglements of process and product by asserting that the project's 'focus on visual and media performance can be traced back to economically grounded PR-deliberations and the company's search for new markets, target groups and future business models' (Richterich 243). Richterich's framing of 'The 21 Steps' as a pilot project for reaching new audiences through digital technology is given credence by the way in which major publishing houses have been increasingly drawn to the use of geo-visualisation tools in the online promotion of new work by leading contemporary novelists.

A prominent example is offered by Penguin's online presentation of *NW* (2012): Zadie Smith's Joycean novel about the practice of everyday life in the London borough of Willesden. The homepage of the section of Penguin's website dedicated to Smith's novel features a masthead in which the title of the text and the name of the author are placed in a central position and are draped over a barely discernible Google Map of London. Crucially, Google markers are presented in the four corners of this masthead with each marker referring to a key location in Smith's novel: Willesden Lane; 274 Kilburn High Road; Camden Lock and 37 Ridley Avenue. The user, upon clicking on one of the markers, calls up an audio recording of Smith reading a geo-specific extract from her novel. At the same time, the text which Smith reads pops up – in a range of font sizes – upon a series of still images of the named location.

To draw upon the example of Willesden Lane, which is located in the northwest of the map presented on the homepage, the imaginative effect of this mode of presentation is two-fold. First, the polysensory synthesis of text, audio recording and photographic images underscores the profound boredom of everyday life for the young characters who find themselves growing up in what they perceive to be this unremarkable London suburb. The relative peripherality of this place, which is signalled by its location on the homepage Google Map, is reinforced as the user is presented with a fusion of words and images characterised by a monotonous greyness. Alongside this, though, the user is immediately offered a multi-media vision of Willesden Lane as a globally networked place: a location in which a young Sikh contemplates combating the perceived tedium of everydayness by taking money from the till in the family shop; a site in which another character, Michel, laments the fact that the pastries made off Willesden Lane will never be as good as those experienced back home in France. Smith's imaginative vision of Willesden Lane, therefore, corresponds with Doreen Massey's exploration of Kilburn High Road as a 'pretty ordinary place' which is characterised by the heterogeneity of its inhabitants and the porosity of its perceived boundaries: an understanding which underpins Massey's influential articulation of a 'progressive sense of place' (Massey 152–4). Here, however, my principal interest does not lie in a detailed analysis of the literary geographies embedded within Smith's text but, instead, rests on the way Smith's publishers use geo-spatial technologies as a visualisation and – crucially – promotional tool. That is to say, both the difficult-to-read map and the streetscape photography serve purely illustrative functions and the digital literary mapping practice is primarily underpinned by commercial imperatives rather than creative impulses. The fact that Penguin turns to these technologies highlights an awareness of their prominence in the contemporary spatial imagination; but, in summary, the online project is not supported by any consideration of what geo-visualisations are and what they can (and cannot) do.

A Processual Mapping of Place: Rachel Lichtenstein's *Diamond Street*

In contrast, it is possible to identify contemporary practitioners whose work is characterised by a commitment to thinking self-reflexively about the ways geospatial technologies might shape the creative process and, by extension, the imaginative understanding of the texturalities of place. The writer, artist and curator, Rachel Lichtenstein, is perhaps most widely known for *Rodinsky's Room* (1999): a book – co-written with Iain Sinclair – predicated on the archival documentation of a room on Princelet Street, Spitalfields, East London which had once been occupied by the orthodox Jewish scholar, David Rodinsky, but which had been left abandoned for twenty years. *Rodinsky's Room* announced what have subsequently emerged as signature themes and tropes in Lichtenstein's work across a range of media: a preoccupation with the interpenetration of the real and imagined geographies of East London; a psychogeographic identification and exploration of uncanny traces; and a distinctive imbrication of topographical, biographical and – crucially – autobiographical modes of expression. In *Diamond Street,* Lichtenstein's geographical gaze remains fixed on London but, this time, shifts westwards to Hatton Garden: a street in Holborn which has been the centre of the city's diamond and jewellery trade for two centuries. Lichtenstein enacts a deep mapping of this circumscribed place through a combination of pedestrian practice, archival research, and personal recollection and reflection. Arguably the most striking strand of Lichtenstein's methodology, however, is her incorporation of extensive interviews with people who live and work in Hatton Garden. The result is the creation of a genuinely polyvocal textual account of what the book's subtitle describes as 'the hidden world' of Diamond Street.

Of even greater relevance, given the thematic focus of this chapter, are the roles played by cartographies and the cartographic imagination in this textual enacting of deep topography. Lichtenstein's 'map-mindedness', to apply an expression used by Hewitt, is foregrounded in the opening sentences of the first chapter which invoke a traditional paper map (*Map of a Nation* 203):

> Hatton Garden is the fold in the map, a place on the edge of different borderlands that sits on the city fringe, somewhere between Clerkenwell, Holborn and Farringdon. In the *London A-Z* the street is defined as a single, long, straight road, just within the borough of Camden. The southern end intersects with Holborn Circus, a busy roundabout where five other roads meet, and the northern end flows directly into Clerkenwell Road. (Lichtenstein 1)

Lichtenstein's imaginative excavation of place, then, begins with an attempt to locate Hatton Garden on an authoritative, planar cartographic representation of the city. Right from the start of the text, though, the reader is alerted to the fact that Lichtenstein is acutely sensitive to the 'slipperiness' of paper maps (Bushell 2012). That is to say, Lichtenstein eschews a positivist understanding of maps as unambiguously presenting and generating geographical knowledge in favour of an articulation of cartographical representations of space as partial and incomplete. This awareness of the limitations of maps is reinforced in the following paragraph as the objective *London A-Z* 'definition' of Hatton Garden as a linear street is blown apart by the alternative autobiographical understandings of place offered by Lichtenstein's interviewees: 'I found, however, that after speaking with people who work there today, the geography of the street became more fluid'. In the practice of everyday life, then, ' "the Garden" describes a wider area, including Hatton Garden itself, and a number of surrounding and interconnecting roads' (Lichtenstein 1). As a result, the geographies of quotidian embodied experience serve to problematise, and to recalibrate, the official geographies that have been mapped out in authoritative cartographic representations of the city.

Yet, instead of completely rejecting paper maps for their lack of geographical verisimilitude, Lichtenstein, throughout *Diamond Street*, pores over a succession of historic cartographies. Her archival research begins, for instance, with a visit to the manuscripts reading room at the Guildhall Library where she moves from *The City of London from Prehistoric Times* – 'an atlas of reconstructed maps put together by academics at Oxford University in the late 1980s' – to 'the oldest printed map of the city' which is 'known as the Agas map or Civitas Londinium' and which 'was probably surveyed between 1570 and 1605' (Lichtenstein 2–3). In Chapter 11, the author advances her cartographic reconstruction of place by consulting a range of maps from the end of the sixteenth century: 'A Plan of the City and Liberties of London after the dreadful conflagration in the year 1666' which is on display in the Museum of London; Ogilby and Morgan's map of the City of London from 1676; and 'a facsimile of a detailed survey of the completed Hatton Garden estate, made in 1694' by a property developer (Lichtenstein 103–4).

Later on, Lichtenstein is similarly entranced when, on a visit to the Natural History Museum, she encounters an 'illustrative drawing, dated 1638, plotting the sites of the natural springs along the route of the Fleet': perhaps the most famous of London's lost rivers (Lichtenstein 257). In each instance, Lichtenstein's geographical imagination is stimulated by the practice of tracing Hatton Garden's past through the palimpsestic, richly intertextual history of its maps (Cooper and Priestnall). At the same time, the author's creative reading of such maps remains consistently underpinned by an awareness of these cartographies as socially constructed texts that were produced in specific contexts and for particular reasons. Instead of dismissing such cartographies for their geographical inaccuracy, Lichtenstein highlights how an understanding of their errors, omissions and infelicities is, in fact, crucial to an understanding of the hidden histories of Hatton Garden. Map-making is a form of storytelling; and stories, of all kinds, are integral to the development of Lichtenstein's sense of the making and re-making of this particular place.

Lichtenstein's conception of cartography, therefore, corresponds with a 'processual' understanding of maps 'as mappings that ceaselessly unfold through contingent, citational, habitual, negotiated, reflexive and playful practices, embedded within relational contexts' (Kitchin et al. 480). So far, however, this discussion has focused exclusively on Lichtenstein's engagement with paper maps; but, over the course of *Diamond Street*, the author becomes increasingly alert to the potentialities and problems associated with contemporary digital mapping practices. Her imaginative interest in Geographic Information Systems (GIS) is first piqued in a meeting with Jon

Chandler at the Museum of London's Archaeological Department. Lichtenstein, as an excavator of place, is instantaneously attracted to Chandler's description of how GIS technology can transcend the representational fixity of paper maps by enabling the layering of historic and contemporary information onto the one cartographic space. She is similarly drawn to Chandler's demonstration of the ways in which GIS can allow for the georeferencing of historic maps and can enable the user to oscillate between different representational scales at the click of a button.

Chandler invites Lichtenstein to play with the technology herself, and the writer's geographical imagination is ignited by the overwhelming amount of historic spatial data she accesses on the computer screen. Most memorably, she stretches 'a historic hand-drawn map' over a contemporary map of Holborn and discovers that the 'boundary walls of this [sixteenth-century] estate encircled the same area known as "the Garden" today': 'It was as if an invisible perimeter from the past still existed there, constructed from generations of whispered memories about the place' (Lichtenstein 25). In contrast to the more extreme uncritical positivism associated with some quantitative GIS research, however, Lichtenstein's response to the imaginative geographies opened up by the layering of digital maps is tempered by Chandler's cautionary assertion that the 'technology has its limits: the maps are only as good as the information gathered and can never tell the full story' (Lichtenstein 25).

Lichtenstein returns to digital mapping technologies in Chapter 24 – 'Virtual Drift' – of *Diamond Street* in which she engages with the experimental practices of the New York-based digital media artist, Mary Flanagan. Lichtenstein is particularly drawn to Flanagan's '[borders]' project which 'consists of video documentation of a series of virtual walks she has taken through reconstructed sites from the past in popular, shared, online, multi-user worlds'; walks which are 'purely experiential' and in which the artist roams 'without purpose through imagined landscapes, where she becomes drawn into the seductive beauty of the digital image, often finding herself lost' (Lichtenstein 292). This fascination with Flanagan's digital remaking of the classic psychogeographic *dérive* prompts Lichtenstein to invite 'the *flâneur* of the Internet to take a virtual walk with me [using Google Street View] around the perimeter of the area known as the Garden today' (Lichtenstein 292–3). Lichtenstein's digital tour of the area is informed by her familiar sensitivity to the inherent limitations of mappings as producers of geographical knowledge. This sensitivity is explicitly articulated when Flanagan demonstrates how she is unable to 'enter the dark interior of one of the shops' on Hatton Garden (Lichtenstein 298) and Lichtenstein reflects on the intimate spaces which even Google Street View is unable to visualise: 'The interior spaces, the narrow alleyways, the small backstreets and pedestrianised areas are inaccessible because the vehicles [used by Google to map the urban landscape] are too wide to enter' (Lichtenstein 300).

The notion that it is the writer of deep topography who is able to enrich the placial narrative by penetrating these concealed spaces is implicitly reinforced by the methodological approaches showcased in the two chapters that immediately follow 'Virtual Drift'. In Chapter 25 ('Full Circle'), Lichtenstein resumes her interlocutory role as she listens to the personal testimonies of different workers associated with Holts Lapidary Showroom: a project which necessarily involves her moving beyond the geological window display to enter the shop; and a project in which she is led 'downstairs for a tour of Holt's workshops' (Lichtenstein 307). This motif of underground space – a world which is physically inaccessible to Google's vehicles – recurs in the following chapter ('Subterranea') in which the author's imaginative gaze shifts downwards to the labyrinthine network of 'underground tunnels', 'subsurface passageways', 'disused nineteenth-century railway tunnels', vaults and 'unused underground lines' (Lichtenstein 316) that lie beneath the streets of this part of central London. Through her own situated practices as a deep topographer, therefore, Lichtenstein seems to be questioning the authenticity of the type of disembodied, digitised spatial voyeurism performed by Flanagan.

Yet, crucially, Lichtenstein remains simultaneously open to the creative possibilities facilitated by Google Earth: a technology which, in spite of 'its complex privacy issues . . . is also truly magical . . . as it gives a sense of an almost godlike power' (Lichtenstein 293). Lichtenstein is initially attracted to the way Google Earth recalibrates her spatial understanding of her home city by providing aerial images of sites ordinarily viewed from street level. More particularly, Google's satellite imagery facilitates a psychogeographic reconfiguration of the symbolic geometry of central London. For example, Lichtenstein's quotidian understanding of 'the circular shape of Holborn Circus' is destabilised as, when viewed from above, the roundabout 'looked something like the star shape of the Seven Dials junction in Covent Garden': a defamiliarising vision which, in turn, prompts the writer to reflect on 'recent researches in the maps reading room of the British Library, where I had studied Sir Christopher Wren's original hand-drawn maps for rebuilding the city after the Great Fire' (Lichtenstein 294). Google Earth, then, serves to enrich the author's engagement with, and understanding of, the palimpsestic histories woven within the material fabric of this particular urban place.

This process of defamiliarisation is advanced further as Flanagan shifts to Google Street View and her avatar simply drifts through this digital landscape. Lichtenstein is transfixed by Flanagan's capacity to construct a relatively nuanced reading of place from the comfort of her chair as, for instance, she swiftly offers the critical assessment that this is a schizophrenic space: 'I see modernization here . . . The street seems to be divided into the old and the new . . . It looks like a historic place, but split in half' (Lichtenstein 297). Although Flanagan's place-specific observations are fascinating to both author and reader, of greater interest – in the context of the

Figure 14.2 Aerial view of Holborn Circus through Google Earth.

Source: © 2015 Google.

current chapter – is the methodological playfulness through which she uses digital technologies to open up new perspectives on the built environment. Throughout the Street View tour, Flanagan's sense of place is generated by the 'magical' ability to move her avatar in and around, up and down, the digitised Hatton Garden in acts of virtual urban exploration: she floats skywards to investigate 'the architecture above shop level'; she climbs 'to the top of the buildings, nosing in windows, spotting a few signs for offices and workshops'; and she jumps 'like a superhero on to the top of [a] viaduct' (Lichtenstein 298–304). Moreover, Flanagan alerts Lichtenstein to the fact that even the default position of Google Street View offers an alternative vision of the city to the ordinary perspective of the pedestrian since it 'is slightly elevated from the normal eyeline of street level, as the images are taken from cameras strapped to the top of a van, meaning that the details of the architecture above the shops comes more sharply into focus' (Lichtenstein 298). For Flanagan, then, Google Street View offers a portal into a landscape with which she is personally unfamiliar; but, for Lichtenstein, the digital technology is a tool for opening up new ways of seeing the undiscovered country of the nearby.

Lichtenstein, inspired by these guided excursions into virtual terrains, has expanded her own practices through the production of 'The Diamond Street App'. Developed in collaboration with a team of digital media partners, the app allows users to access a multi-media timeline which begins with a detail from the 1570 Agas map of London and which ends with a black-and-white photograph of Hatton Garden taken in 2012. As well as providing a different site-specific image, each marker on the timeline invites the user to click on a diamond icon to pull up a short passage written by Lichtenstein or the textual transcription of an interview with a local resident. By clicking a second diamond, the user is able to access audio recordings of readings and conversations focusing on such sites as St Etheldreda's Chapel, Bleeding Heart Yard and Hockley-in-the-Hole. Elsewhere on the app, the user is able to develop his or her sense of the surface and subterranean geography of this part of London by accessing four short films narrated by Lichtenstein. The overall effect is the production of a digital resource that moves towards the spatial humanist's understanding of a 'deep map' as 'a finely detailed, multimedia depiction of a place and the people, animals, and objects that exist within it' (Bodenhamer et al. 3).

It is striking that, when accessed from a physical distance, the user's entry point for 'The Diamond Street' app is temporal rather than spatial. That is to say, history – in the form of the timeline – is the core organising principle. The digital experience is radically transformed, though, if the user engages the app's GPS mode whilst walking within the boundaries of the original Hatton Garden estate. As Lichtenstein explains: 'sounds, images, film and text will emerge in relation to specific locations, giving readers a deeper and more immersive experience of both the book and the places described within' (Lichtenstein, 'Diamond Street App'). In this mode, therefore, the organising principle is geographical situatedness; and, crucially, the user is encouraged to retrace his or her steps if the site-specific information begins to fade as he or she moves away from a named location. In the first instance, Lichtenstein frames the app as a digital supplement

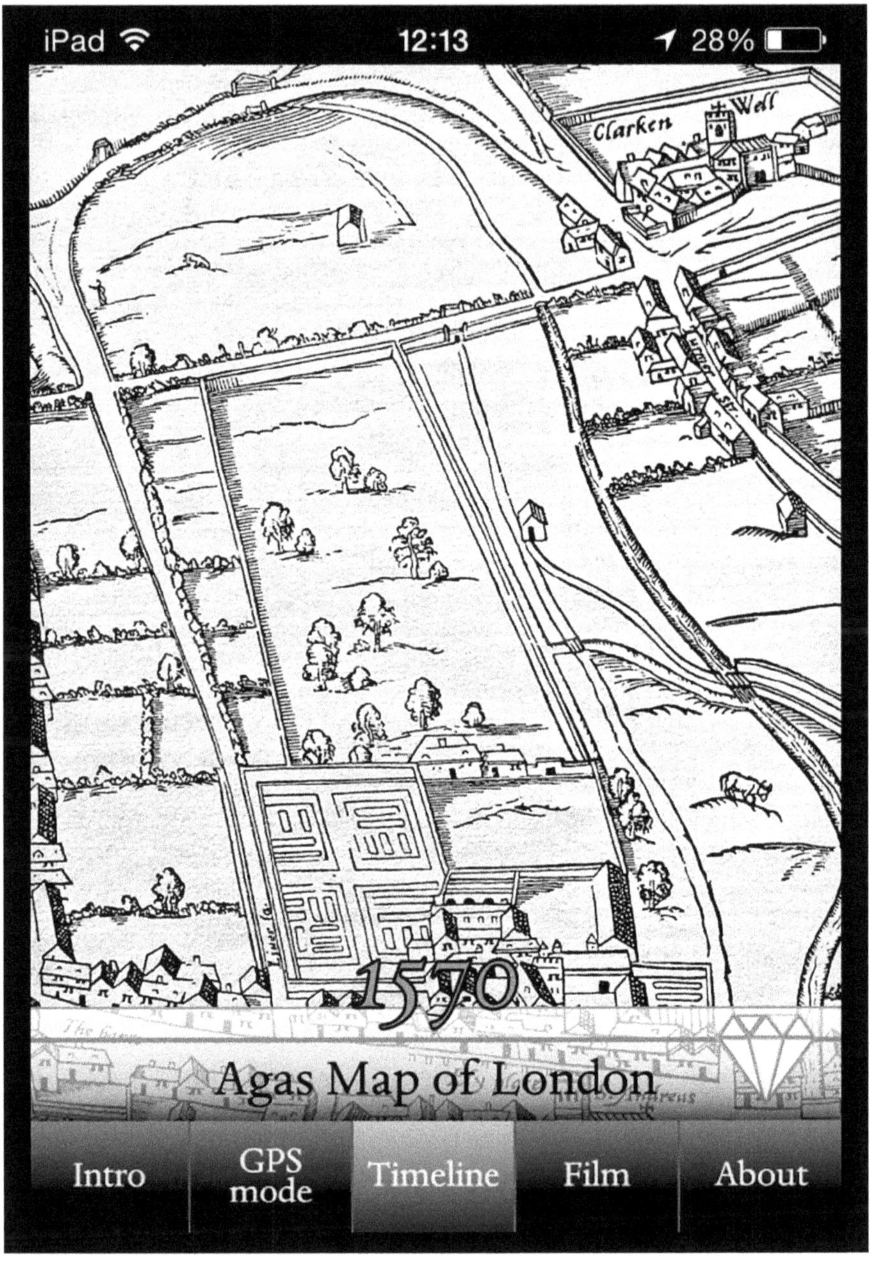

Figure 14.3 Detail from 1570 Agas map in Rachel Lichtenstein's *The Diamond Street App*.

Source: © Rachel Lichtenstein.

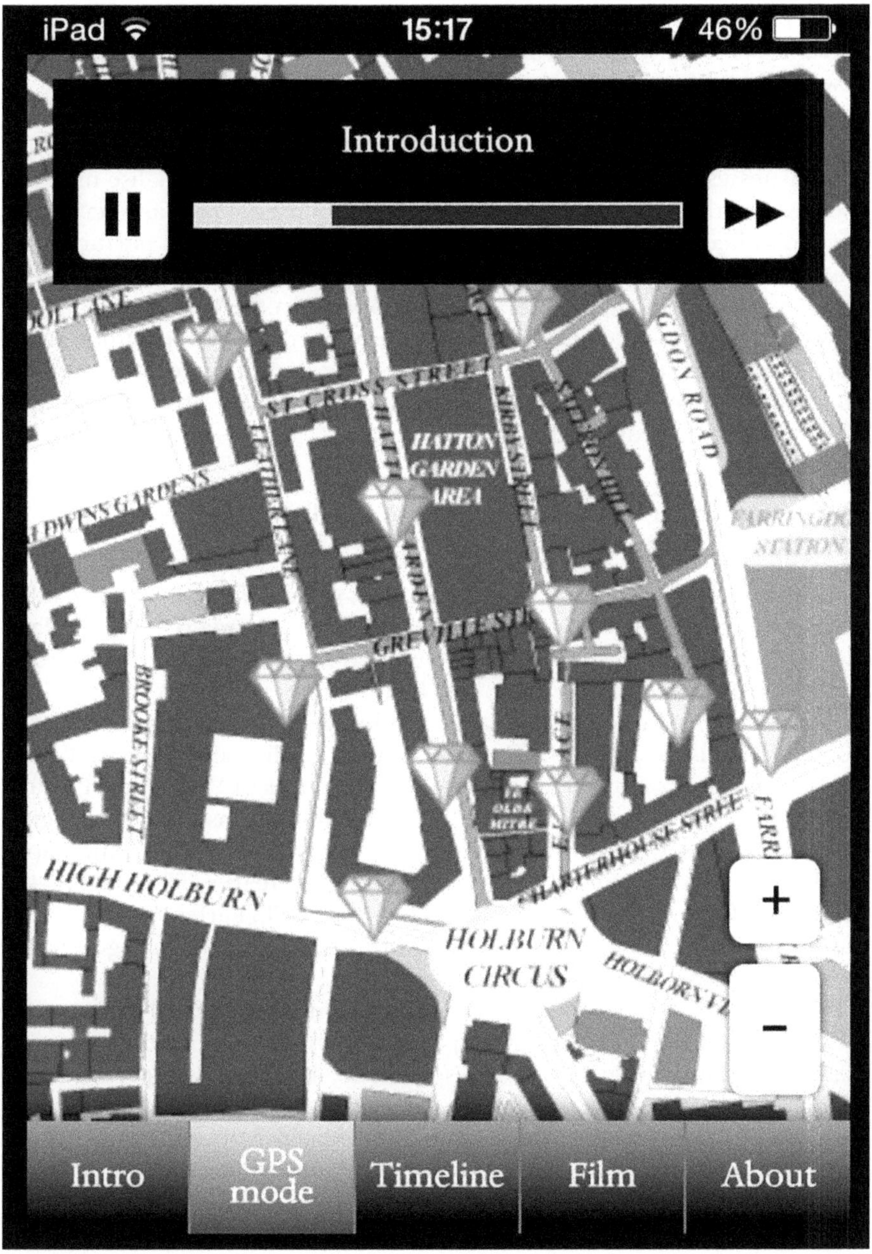

Figure 14.4 Screenshot from GPS mode of Rachel Lichtenstein's *The Diamond Street App.*

Source: © Rachel Lichtenstein.

to the book. As she goes on to acknowledge, however, the app can also be understood to be a new form of difficult-to-define creative practice: 'Part new media experience, part walking tour, this location-based app fuses text, documentary film, image and theatre with real-time interaction' (Lichtenstein, 'Diamond Street App'). Accessed in GPS mode, then, Lichtenstein's app harnesses geospatial technologies to enable users to visualise material and imaginative connections between the palimpsestic past and, crucially, the embodied present. Under the influence of Flanagan's performative practices, therefore, Lichtenstein's restless cartographic imagination here turns to in-the-field digital mapping and opens up the prospect that such geospatial technologies might contribute towards a 're-enchantment of place': a possibility which runs counter to the dominant techno-sceptical tropes and discourses of much contemporary writing about landscape, space and place (Evans and Robson).

Literary Mapping in the Digital Age: A Playfully Anxious Creative-Critical Practice

Towards the end of 'Google Earth', the focus of Paul Farley's speaker shifts from the nostalgic revisiting of childhood places to sweep over 'the lake isles of Lough Gill, | Adlestrop's dismantled barrow, a hill | on the road north of Poughkeepsie' (Farley 17). Here, Farley illustrates how the use of digital maps can be incorporated within everyday reading practices as Google Earth is used to zoom down into Irish, English and American rural landscapes haunted by the ghostly poetic presences of W.B. Yeats, Edward Thomas and Anthony Hecht. That is to say, he demonstrates how digital maps can be used to provide contemporary visualisations of landscapes invoked in earlier poems of place: in this context, Yeats's 'The Lake Isle of Innisfree', Thomas's 'Adlestrop' and Hecht's 'A Hill'. Farley's poem, therefore, is a creative text concerned, at least in part, with the way geospatial technologies can open up new forms of critical practice.

Conversely, critical literary map-making might be conceived as a creative act. In a theoretical discussion of paper maps, Denis Cosgrove proposes that: 'As a graphic register of correspondence between two spaces, whose explicit outcome is a space of representation, mapping is a deceptively simple activity' (Cosgrove 1). From a positivist perspective, then, maps are predicated on the practices of objectively surveying and measuring the material landscape; and their functional fidelity is based on an intellectual and practical faith – from both the map-maker and the map-user – in the power of geometry. Taking his lead from J.B. Harley's influential deconstructive readings of maps, however, Cosgrove problematises this cartographic positivism to posit a pluralistic understanding of map-making which allows for the influence of the 'spiritual, political [and] moral . . . the remembered, the imagined, the contemplated' (Cosgrove 2). As a result, Cosgrove asserts that: 'Acts of mapping are creative, sometimes anxious, moments in coming to knowledge of the world, and the map is both the spatial embodiment of knowledge and a stimulus to further cognitive engagements' (Cosgrove 2). For Cosgrove, then, mapping – both reading and making – is always a creative act.

By extension, therefore, critical literary mapping – the geo-visualising of a literary text or texts – involves at least three layers of creativity. First, reading itself can be framed as a dynamically creative act through which the critic reflects on the spatial particularities of the chosen text or texts in order to determine *what* is to be mapped. Then, he or she needs to decide *how* these attributes are to be presented in cartographic or geovisual form. Third, the critical literary map is subsequently recreated as it is brought-into-being with each singular use. Such reflections, therefore, manifestly take critical digital literary map-making further away from the positivist assumptions that underpinned the first tentative moves towards the development of literary GIS (Cooper and Gregory).

I want to end this chapter, then, by suggesting that there are salient synergies between the exploratory literary mappings of some creative practitioners – as exemplified by the cartographic adventures of Rachel Lichtenstein – and the self-consciously playful practices of some contemporary digital humanists. That is to say, some digital humanists are drawing upon, and developing, creative strategies to advance critical thinking about literary geographies; and, at the same time, some creative writers are using geospatial technologies to think critically about geographical experience and its literary articulation. As part of this convergence of critical and creative practices, there has been a shared move towards what Edward S. Casey labels 'the main unit of landscape' (Casey 350): place. Many spatial humanists have subscribed to the kind of macro-mapping exemplified by the work of Franco Moretti in which the cardinal concept is 'space': 'an abstraction from place', to apply Casey's definition, within which 'locations and positions . . . are sites' (Casey 353). An alternative model, though, is offered by the emerging practice of deep mapping: 'a new *creative* space' (emphasis added), in the words of David Bodenhamer, which draws upon the traditions of phenomenological geography to explore 'the connectedness and ties between human emotion and the physical fabric of landscape' (Bodenhamer 21). In addition, and connected with this move towards the placial, there has been an increasing interest – from both critical and creative practitioners – in the in-the-field mapping possibilities afforded by mobile technologies. In his attempt to disambiguate the twin terms, space and place, Casey emphasises the significance of embodiment: so, 'space . . . is experienced in disembodied detachment' (Casey 353); whereas 'place . . . is the scene of situatedness [and is] experienced by the entire body' (Casey 350). Mobile devices, therefore, facilitate a rapprochement between digital cultures and being-in-the-world by enabling users to access representational material whilst located in, and moving through, place. Finally, and perhaps most importantly, digital literary map-making – in all its forms – is increasingly informed by a processual understanding of maps and mapping practices: a conceptual framing which celebrates the imaginative potential of maps whilst simultaneously acknowledging that all geovisualisations are partial and provisional. For many creative practitioners and digital humanists, then, literary mapping in the digital age can be understood to be a playfully anxious act.

Works Cited

Black, Doug. 'Charles Cumming: The 21 Steps'. *Cool Hunting*. 3 April 2008. Web. 20 March 2015. <http://www.coolhunting.com/culture/charles-cumming>.

Bodenhamer, David J. 'Narrating Space and Place'. *Deep Maps and Spatial Narratives*. Eds, David J. Bodenhamer, John Corrigan and Trevor M. Harris. Bloomington: Indiana University Press, 2015. 7–27.

Bodenhamer, David J., John Corrigan and Trevor M. Harris. 'Introduction: Deep Maps and the Spatial Humanities'. *Deep Maps and Spatial Narratives*. Eds, David J. Bodenhamer, John Corrigan and Trevor M. Harris. Bloomington: Indiana University Press, 2015. 1–6.

Brotton, Jerry. *A History of the World in Twelve Maps*. London: Allen Lane, 2012.

Bushell, Sally. 'The Slipperiness of Literary Maps: Critical Cartography and Literary Cartography'. *Cartographica*. 47.3 (2012): 149–60.

Casey, Edward S. *Representing Place: Landscape Painting and Maps*. Minneapolis: University of Minnesota Press, 2002.

Cooper, David and Ian N. Gregory. 'Mapping the English Lake District: A Literary GIS'. *Transactions of the Institute of British Geographers*. 36.1 (2011): 89–108.

Cooper, David and Gary Priestnall. 'The Processual Intertextuality of Literary Cartographies: Critical and Digital Practices'. *The Cartographic Journal*. 48.4 (2011): 250–62.

Cosgrove, Denis. 'Introduction: Mapping Meaning'. *Mappings*. Ed. Denis Cosgrove. London: Reaktion, 1999. 1–23.

Cumming, Charles. 'The 21 Steps'. *We Tell Stories*. N.d. Web. 20 March 2015. <http://www.wetellstories.co.uk/stories/week1/>.

Di Palma, Vittoria. 'Zoom: Google Earth and Global Intimacy'. *Intimate Metropolis: Urban Subjects in the Modern City*. Eds, Vittoria Di Palma, Diana Periton and Marina Lathouri. London: Routledge, 2009. 239–70.

Dora, Veronica della. 'A World of "Slippy" Maps: Google Earth, Global Visions, and Topographies of Memory'. *Transatlantica*. 2 (2012): n. pag. Web. 20 March 2015. <https://transatlantica.revues.org/6156>.

Farley, Paul. 'Google Earth'. *The Dark Film*. London: Picador, 2012. 16–17.

Farley, Paul and Michael Symmons Roberts. *Edgelands: Journeys into England's True Wilderness*. London: Jonathan Cape, 2011.

Hewitt, Rachel. *Map of a Nation: A Biography of the Ordnance Survey*. London: Granta, 2010.

———. 'Turn Around When Possible'. *The Guardian: Review*. 19 April 2014: 2–3.

Kitchin, Rob, Justin Gleeson and Martin Dodge. 'Unfolding Mapping Practices: A New Epistemology for Cartography'. *Transactions of the Institute for British Geographers*. 38.3 (2013): 480–96.

Lichtenstein, Rachel. *Diamond Street: The Hidden World of Hatton Garden*. London: Hamish Hamilton, 2012.

———. *The Diamond Street App*. 2015. Web. 20 March 2015. <http://www.diamondstreetapp.com/the-diamond-street-app/>.

Luria, Sarah, 'Geotexts'. *GeoHumanities: Art, History, Text at the Edge of Place*. Eds, Michael Dear, Jim Ketchum, Sarah Luria and Douglas Richardson. London: Routledge, 2011. 67–70.

Massey, Doreen. *Space, Place, and Gender*. Minneapolis: University of Minnesota Press, 1994.

Presner, Todd, David Shepard and Yoh Kawano. *HyperCities: Thick Mapping in the Digital Humanities*. Cambridge, MA: Harvard University Press, 2014.

Richterich, Anna. 'Cartographies of Digital Fiction: Amateurs Mapping a New Literary Realism'. *The Cartographic Journal*. 48.4 (2011): 237–49.

Robson, Di and Gareth Evans (eds). *Towards Re-Enchantment: Place and Its Meanings*. London: Artevents, 2010.

Rollinson, Neil. 'GPS'. *Amphibians*. Grasmere: The Wordsworth Trust, 2007. 8–9.

Shelley, Percy Bysshe. 'Ode to the West Wind'. *Shelley: Poetical Works*. Ed. Thomas Hutchinson, rev. G.M. Matthews. Oxford: Oxford University Press, 1970. 577–9.

Smith, Zadie. 'NW'. *Penguin Books USA*. 2012. Web. 20 March 2015. <http://www.penguin.com/static/pages/features/zadie_smith/>.

Index

Note: Tables, figures, maps, and illustrations are indicated by italic page numbers. Color plates are indicated by "*1p*" for example meaning page 1 of the plates section.

For Product Safety Concerns and Information please contact our EU representative GPSR@taylorandfrancis.com
Taylor & Francis Verlag GmbH, Kaufingerstraße 24, 80331 München, Germany

www.ingramcontent.com/pod-product-compliance
Ingram Content Group UK Ltd.
Pitfield, Milton Keynes, MK11 3LW, UK
UKHW020019200926
475702UK00014B/1968
9780815366782